Excel

SUCCESS ONE® HSC

PHYSICS

Past HSC questions
2006–2018 by Module
Past 2019–2024 HSC papers

PASCAL
PRESS

ISBN 978 1 74125 769 4

Pascal Press
PO Box 250
Glebe NSW 2037
(02) 9198 1748
www.pascalpress.com.au

Publisher: Vivienne Joannou
Commissioning and project editor: Mark Dixon
Edited by Karen Enkelaar and Rosemary Peers
Answered checked by Neville Warren, Colin McAuliffe and Adam Sloan
Cover design by Michael Sherman
Typeset by lj Design (Julianne Billington) and Grizzly Graphics (Leanne Richters)
Cover photo by *The Sydney Morning Herald*
Printed by Vivar Printing/Green Giant Press

Foreword

Congratulations on selecting Physics as part of your HSC program of study. This course will prepare you well for further tertiary study or a career in academia. It will also give you a lifelong understanding of the key principles and thinking skills of Physics.

Excel *Success One HSC Physics* is a valuable learning tool that has been developed to help you with your HSC preparation. This edition has been kept up to date with the inclusion of the 2024 HSC Examination paper with sample answers. Your teacher will have many suggestions for how to use this resource to help support your preparation for the HSC Examination.

This edition also contains selected questions from past HSC papers which are relevant to the new Year 12 Physics syllabus. We have also included extra questions written specifically for the new content with sample answers provided.

On behalf of the Science Teachers' Association of New South Wales we hope that your year will be enjoyable and productive, and the results a just reward for your efforts.

STANSW thanks those members who continue to be involved in the production of this book.

Margaret Shepherd
STANSW President
January 2025

The Mark Maximizer Guide was written by Diane Alford.

HSC EXAMINATION PAPER QUESTIONS

SAMPLE ANSWERS

The sample answers to HSC questions contained in this publication are examples of answers which the authors believe would score full marks. However, they are not endorsed by NESA.

Contributors to the sample answers include Mark Butler, Terri Patterson, Peter Roberson, George Pinniger and Stephen Fogwill.

COMMENTS

The Science Teachers' Association of NSW (STANSW) and Pascal Press welcome constructive comments on the questions and answers for future editions of this book.

FORMAT

This guide's format has been constructed to support students' revision for the HSC Examinations based on the Physics Stage 6 syllabus 2017 (updated January 2018):

- The **first section** contains objective-response and short-answer questions from past HSC Examination Papers (2006–2018) as well as sample questions. This section is arranged according to the syllabus Module topics.
- The **second section** consists of the 2019–2024 HSC Examination papers.

Contents

Mark Maximizer Guide

This Mark Maximizer Guide will arm you with strategies and tips so you can maximize your marks by making every minute of your HSC Examination count. And it has some great tips for your exam preparation too. Read it. Use it.

The structure of the HSC Examination paper

All Science exams consist of two sections. The following table summarises the structure, which changed in 2019. All questions are compulsory.

Section	*Mark*	*Question type*	*Answers*	*Recommended time*
I	20 (1 mark per question)	Objective response	Fill in the responses on the answer sheet.	35 minutes
II	80 (varying mark values with at least two items worth 7–9 marks)	Short answer: Questions may have parts. There will be at least 20–25 items. A variety of stimulus material such as text, diagrams, pictures, graphs, photographs and illustrations that are essential for answering the question may be included.	Answer questions in the spaces provided, showing all working and calculations. Space provided is a guideline for expected answer length but extra space is provided at the back of the booklet. If you use extra space make sure you clearly indicate the question number or part. Some answers may require a diagram or graph while in other answers you may choose to support it with diagrams or other graphical material.	2 hours 25 minutes

Time allocation

The examination begins with 5 minutes reading time. You will not be able to write in this time.

Rather than starting to read the objective-response questions at the beginning, you could:

- spend this time reading some of the short-answer questions in Section II,

or

- select some of the easy questions to attempt earlier in the exam after completing the objective-response questions, to build your confidence for the more difficult questions later.

Time management

A planned approach to time management in the exam is important. There are several strategies you could use:

- Prepare an exam timetable

  The table above shows the suggested times for each section as recommended by NESA. If you choose to proceed through the exam in an orderly sequence of sections or parts, you can work out an exam timetable before the exam. You will then know when you should be changing sections and check that you are on schedule. Try to be a bit ahead of schedule so that you have time to check answers, make corrections or return to more difficult questions.

Calculate time allocated per mark

If you do not choose to work through the exam in a fixed sequence, you may wish to calculate time allocations for questions based on the 100 marks and 180 minutes in which to record answers. Each mark should take 1 minute 48 seconds to earn. The following table gives calculations of approximately how long you should spend on questions of a particular mark value.

Question value (marks)	*Suggested time*
1	1 min 45 s
2	3 min 30 s
3	5 min 20 s
4	7 min 00 s
5	9 min 00 s
6	10 min 45 s
7	12 min 45 s
8	14 min 30 s
9	16 min 00 s
10	18 min 00 s

- Don't panic!

  Worrying about time too much will only add to your stress and waste time. There is no need to time yourself for each question, and particularly not for the low-value questions. However, if you panic or lose track of time or waste too much time on a question of relatively small mark value, you may be severely disadvantaged.

Analysing the questions

Questions in the HSC are designed to enable students to show their achievement of a range of outcomes over a range of levels. They are not trying to trick you, and the question will state clearly all that is required of you. If marks are to be awarded for giving an answer in the correct units, then the question will clearly state that you need to give the answer in the correct units.

- Highlight key terms

  Highlight or underline the key terms or things you have to do (the verbs) and the key ideas or concepts that make up the question — a useful strategy.

- Learn the lingo

  The glossary of terms that are often used in questions will help you analyse what is required of the question (see the Glossary of key words on page ix).

- Remember that questions are planned to vary in their level of difficulty

  Don't panic or waste excessive time on a question that you find particularly difficult. The following table may give you some ideas on how to find the easier questions that you can attempt first. Then, when you have finished everything that you can do, you can move on to spend time with the very difficult questions.

Question type	*Key words that indicate level of difficulty*
Difficult or complex questions often demanding high levels of skills to use the facts	assess, construct, critically analyse, critically evaluate, evaluate, justify, propose, recommend
Questions of medium levels of difficulty requiring some thinking and using the facts	account for, analyse, apply, assess, calculate, classify, compare, contrast, deduce, demonstrate, discuss, examine, explain, estimate, extract, extrapolate, how …, interpret, investigate, predict or recommend, sketch, why ...
Simple or direct questions of low levels of difficulty based on remembering and communicating the facts	account, clarify, define, describe, identify, outline, recall, recount or summarise, state, what ..., which ...

Also, there may be variation in difficulty within the topics you have studied. A relatively easy remembering-type question may become a little more difficult if it is about a particularly difficult idea.

- Look at question structure and mark value

The structure of the question may also contribute to differences in levels of difficulty. The mark value for the question may give you some clues. For example if the question is only two lines long but is worth six marks, that is a clue that it may require extra care to ensure all aspects of the question have been covered.

Many students find the questions that are divided into subsections easier to attempt. They feel that each small portion of the question can be 'bitten off' at a time. Often the answer to one subsection helps to provide some ideas to help with the other subsections.

The more difficult integrated questions may seem a bit more difficult to swallow, and when you do, they may give you indigestion. Practising these questions using examples from this book will really help you develop your skills and help avoid the need for antacid tablets!

- Answer all parts

A common error is that students often only answer one part of a two-part question. This is very easy to do if the parts are not clearly numbered but are written as sentences following each other. It is worth making a note on the paper of any multiple-part questions and make sure you return to the question to finish all sections.

Glossary of key words

For each subject you should prepare summaries and a glossary of the key definitions. You cannot answer questions if you are unfamiliar with the content of the subject.

You also need to be familiar with the language of exam questions. This involves understanding the verbs that are the key words that tell you what you have to do to earn the marks. Below is a table of the key verbs that may be contained in questions and the definitions of those words.

Key word	*Definition*
Account	Account for: state reasons for, report on Give an account of: narrate a series of events or transactions
Analyse	Identify components and the relationship between them; draw out and relate implications
Apply	Use, utilise, employ in a particular situation
Appreciate	Make a judgement about the value of
Assess	Make a judgement of value, quality, outcomes, results or size
Calculate	Ascertain/determine from given facts, figures or information
Clarify	Make clear or plain
Classify	Arrange or include in classes/categories
Compare	Show how things are similar or different
Construct	Make; build; put together items or arguments
Contrast	Show how things are different or opposite
Critically (analyse/ evaluate)	Add a degree or level of accuracy depth, knowledge and understanding, logic, questioning, reflection and quality to (analyse/evaluate)
Deduce	Draw conclusions
Define	State meaning and identify essential qualities
Demonstrate	Show by example
Describe	Provide characteristics and features
Discuss	Identify issues and provide points for and/or against
Distinguish	Recognise or note/indicate as being distinct or different from; to note differences between
Evaluate	Make a judgement based on criteria; determine the value of
Examine	Inquire into
Explain	Relate cause and effect; make the relationships between things evident; provide why and/or how
Extract	Choose relevant and/or appropriate details
Extrapolate	Infer from what is known
Identify	Recognise and name
Interpret	Draw meaning from
Investigate	Plan, inquire into and draw conclusions about
Justify	Support an argument or conclusion
Outline	Sketch in general terms; indicate the main features of
Predict	Suggest what may happen based on available information

Key word	*Definition*
Propose	Put forward (for example a point of view, idea, argument, suggestion) for consideration or action
Recall	Present remembered ideas, facts or experiences
Recommend	Provide reasons in favour
Recount	Retell a series of events
Summarise	Express, concisely, the relevant details
Synthesise	Put together various elements to make a whole

Answering the question

Objective-response questions

Many people find objective-response (multiple-choice) questions relatively easy. Recording the answer is certainly easy, but they can be deceptive in their level of difficulty. Some make up for the time saved recording the answer by requiring time to think about the answer. Some of the distracters (i.e. the wrong answer options) may be based on common mistakes or misconceptions and may lead you away from the correct answer.

Careful reading and interpretation of questions are critical. Sometimes you may be left with two similar answers and you have to dig into your deeper understanding to work out the implications of the subtle difference. The more practice you get on objective-response questions, the more skilful you will become. In this book we have provided explanations for making the correct choices; you do not have to do this in an exam.

Short-answer questions

- Underline the verb and analyse the question

  After you have analysed the question, link it to the specific module and the concepts of that module. If the question is in parts, look at the parts and see if they give you clues about the other sections of the question. If the question gives you the freedom to select the format of your answer, choose a format that you feel most comfortable with. For example, you may prefer to express some ideas in a table, flowchart or diagram, or in writing. This may be especially important in those questions with a high mark value that require you to integrate ideas.

  Use the mark value and space provided to give you a rough guide to the number of points in the answer and the length of the answer. If it is worth six marks, you may need to include at least six key points or one point with a set of arguments, supportive statements, reasons or counter-arguments.

- Read through your answers

  It is important to read through your answers to short-answer questions to check that you have clearly communicated your understanding and addressed the demands of the question. Check that any assumptions you have made have been stated and that your thinking is clear. If you are running out of time it is important to attempt questions in a very concise manner; even if you remain uncertain as to what the question is about, think back to the module and try to make some links to key ideas.

- Attempt all questions

  It is essential you attempt all questions. You do not lose marks for guessing or recording wrong answers. Remember, however, if you ramble on with a lengthy, confused or contradictory answer you may make a satisfactory answer meaningless and therefore worth no marks. Clear and concise answers are always preferable. If you are running out of time, writing key points, tables or diagrams may help communicate an answer or earn partial marks.

  Avoid persevering with one question to make it perfect if it means you leave out several others. The mark value of questions can serve as a guide for prioritising your attempts at questions when running out of time. You will do yourself a great disservice if you do not attempt questions worth six or more marks.

- Show your working

Common mistakes

Examiners provide reports after each HSC Exam that indicate areas of strengths and weaknesses in student responses. From these common mistakes, it is clear that students generally can benefit by:

- more precise and scientific use of terminology, especially that specified in the syllabus
- more precision with graphs and labelling of diagrams
- avoiding answers that are too long or detailed or that use ambiguous terms
- correct interpretation of questions, e.g. listing instead of describing, or giving a description not an explanation
- answering in specific terms not generalities
- ensuring all parts of multi-part questions are attempted
- being able to clearly describe procedures, apparatus, etc. for mandatory practical activities as a firsthand experience.

Golden rules

Preparing for the exam

- Make certain you have packed all the equipment you need such as the correct calculator, ruler and pencil for diagrams.
- Obviously the exam will reward those students who have a thorough knowledge of the whole course. Do all assignments and assessment tasks to your best ability and use them in your revision.

- Get as much practice at completing and getting feedback on exam-type questions as you possibly can. This book is perfect for this!
- Make certain you complete, know and revise the mandatory practicals from the syllabus so you can describe the procedure or evaluate its methods.

In the exam

- Don't give up! If you find it difficult, so will most other students. Attempt all questions with concise, clear answers.
- Use the Periodic Table and data sheets provided to assist with calculations. Clearly set out working and show substitutions to relevant formulae as appropriate and show units.
- Don't round off during calculations but use your calculator memory and present your final answer to the correct number of significant figures.
- Check final answers to calculations make sense.
- Have a plan that ensures you manage time well. If the option is your particular strength, you may start with it, but do not spend more than 45 minutes on it. Alternatively, start with the objective-response because the correct answer must be there, and it can help you get over your nerves.
- Know yourself and work to your strengths. If you want to experiment with different strategies for tackling the exam, do that in the trial exam and reflect on how the plan worked.
- At the end re-read the questions and the answers you have written. Check that you have really communicated what you intended to write.
- Don't panic! It can lead you to make some silly decisions. There will be some difficult questions in the exam but there will also be some easy ones.

Finally, GOOD LUCK!

CHAPTER 1

Module 5
Advanced Mechanics

Part A Objective-response questions

1 This diagram shows the path of a cannonball, fired from a cannon.

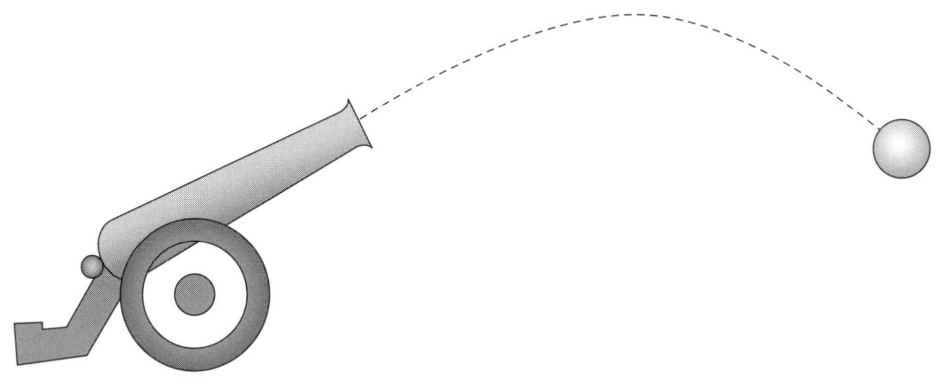

Which set of vectors represents the horizontal and vertical components of the cannonball's velocity along the path?

	Horizontal	*Vertical*
(A)	⟶ → → ⟶	↓ ↓ ↓ ↓ (all long)
(B)	⟶ → → ⟶	↑ (long) ↑ (short) ↓ (short) ↓ (long)
(C)	⟶ ⟶ ⟶ ⟶	↑ (long) ↑ (short) ↓ (short) ↓ (long)
(D)	⟶ ⟶ ⟶ ⟶	↓ ↓ ↓ ↓ (all long)

(q6, 2013 HSC)

2 A ball is launched horizontally from a cliff with an initial velocity of u ms^{-1}. After two seconds, the ball's velocity is in the direction 45° from the horizontal.

What is the magnitude of the velocity in ms^{-1} at two seconds?

(A) u

(B) $1.5u$

(C) 19.6

(D) 27.7

(q20, 2014 HSC)

3 A marble rolls off a 1.0 m high horizontal table with an initial velocity of 4.0 ms^{-1}.

How long will it take the marble to hit the floor?

(A) 0.20 s (B) 0.25 s (C) 0.45 s (D) 3.20 s

(q15, 2011 HSC)

4 A satellite is in a high orbit around the Earth. A particle of dust is in the same orbit.

Which row of the table correctly compares their potential energy and orbital speed?

	Potential energy	*Orbital speed*
(A)	Different	Same
(B)	Different	Different
(C)	Same	Same
(D)	Same	Different

(q6, 2014 HSC)

5 A projectile is launched from a cliff top. The dots show the position of the projectile at equal time intervals.

Assuming negligible air resistance, which diagram best shows the path of the projectile?

(A)

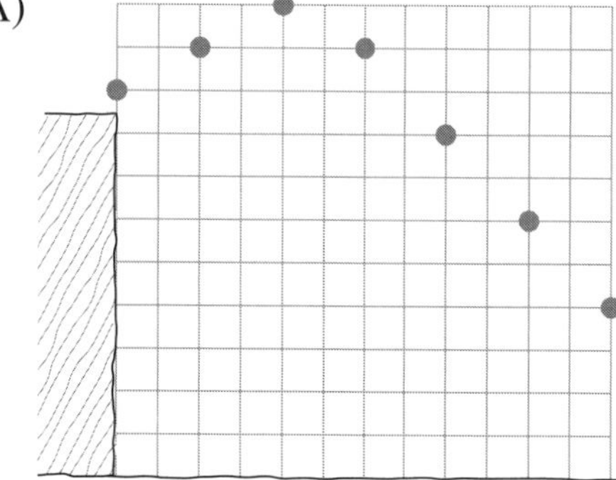

(B)

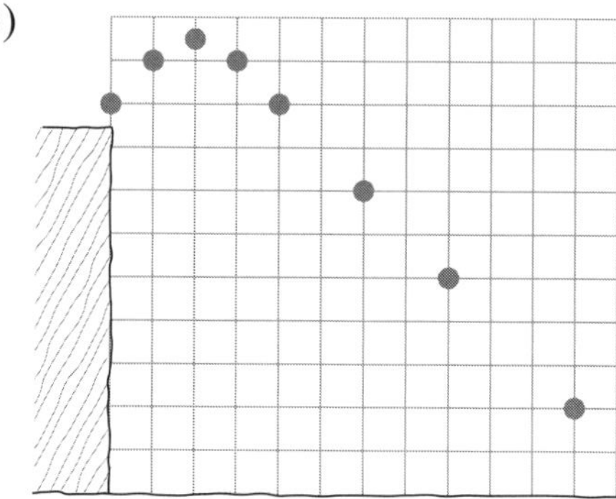

(C)

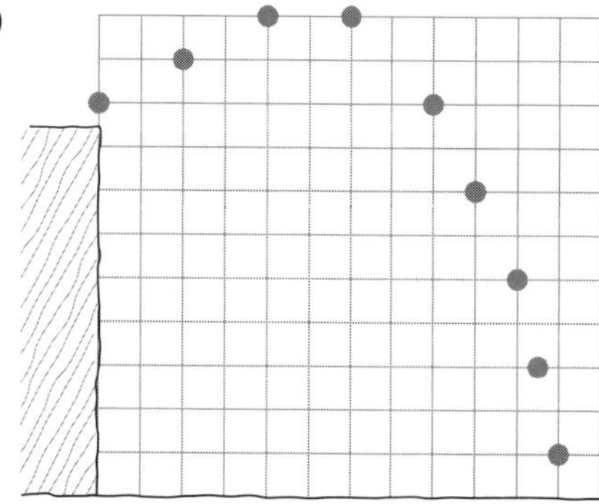

(D)

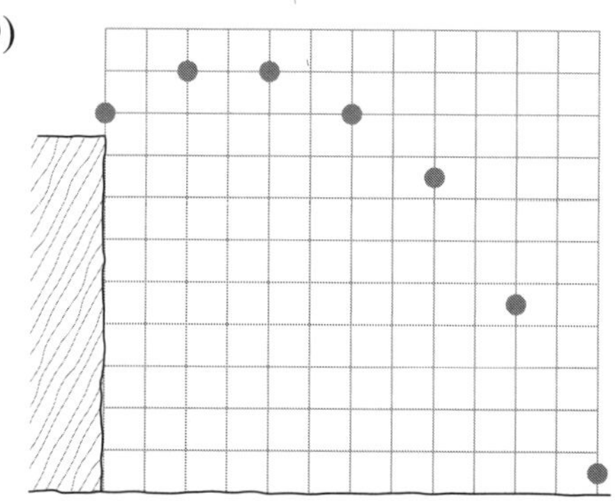

(q4, 2015 HSC)

6 An astronaut working outside a spacecraft in orbit around Earth is not attached to it.

Why does the astronaut NOT drift away from the spacecraft?

(A) The force of gravity acting on the astronaut and spacecraft is negligible.

(B) The spacecraft and the astronaut are in orbit around the Sun with the Earth.

(C) The forces due to gravity acting on both the astronaut and the spacecraft are the same.

(D) The accelerations of the astronaut and the spacecraft are inversely proportional to their respective masses.

(q19, 2015 HSC)

7 A projectile was launched from the ground. It had a range of 70 metres and was in the air for 3.5 seconds.

At what angle to the horizontal was it launched?

(A) 30° (B) 40°

(C) 50° (D) 60°

(q20, 2015 HSC)

8 A satellite orbits Earth with period T. An identical satellite orbits the planet Xerus which has a mass four times that of Earth. Both satellites have the same orbital radius r.

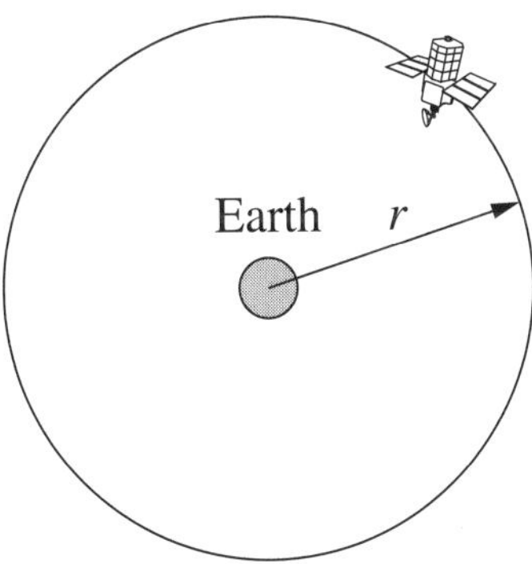

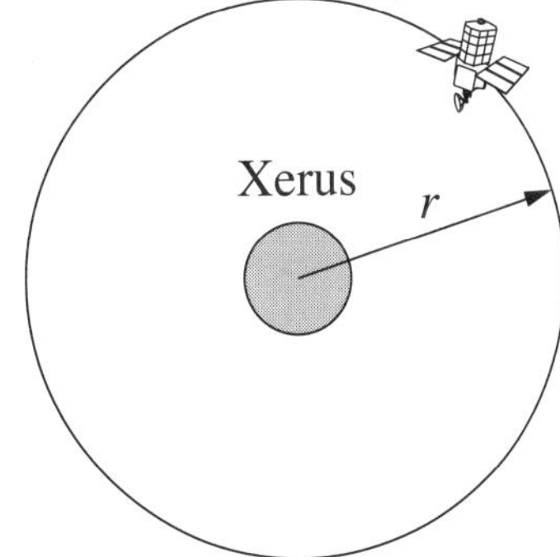

What is the period of the satellite orbiting Xerus?

(A) $\frac{T}{4}$

(B) $\frac{T}{2}$

(C) $2T$

(D) $4T$

(q14, 2016 HSC)

9 A projectile was launched horizontally inside a lift in a building. The diagram shows the path of the projectile when the lift was stationary.

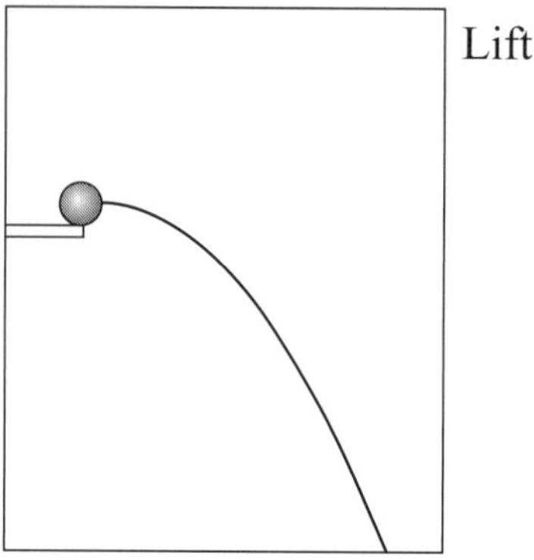

The projectile was launched again with the same velocity. At this time, the lift was slowing down as it approached the top floor of the building.

Which diagram correctly shows the new path of the projectile (dotted line) relative to the path created in the stationary lift (solid line)?

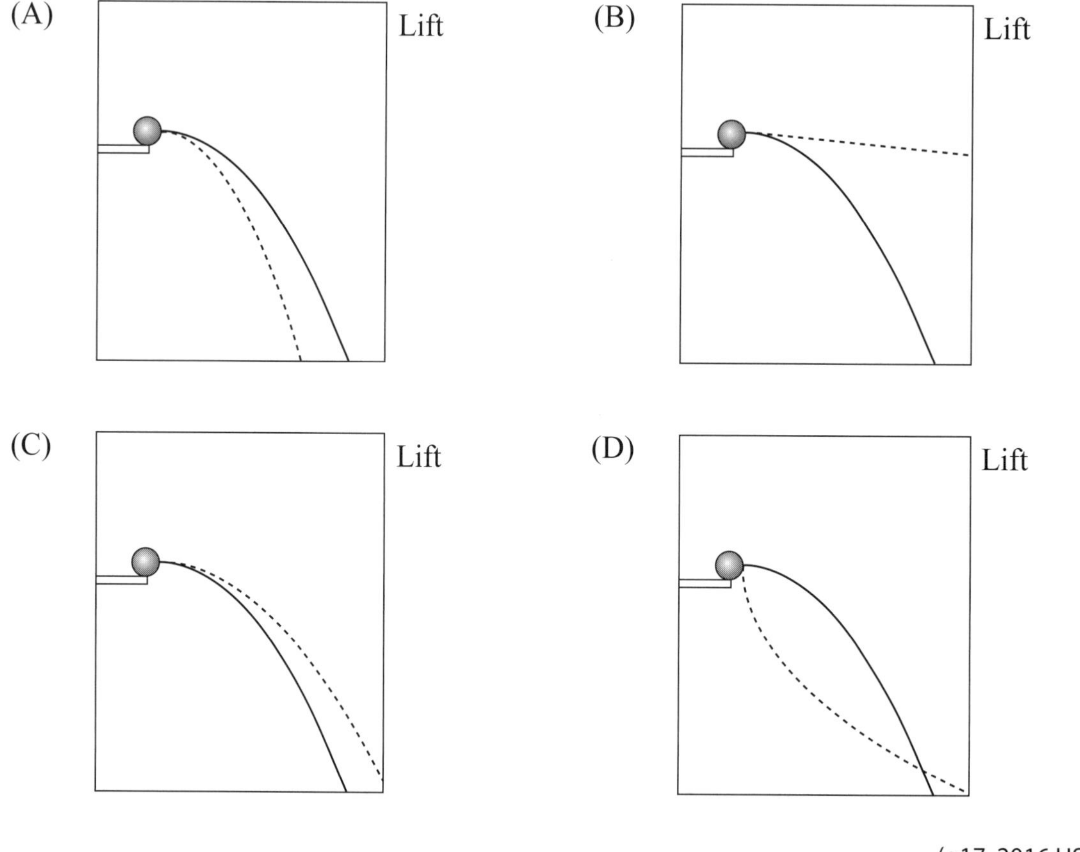

(q17, 2016 HSC)

10 A motorcycle travels around a vertical circular path of radius 3.6 m at a constant speed. The combined mass of the rider and motorcycle is 200 kg.

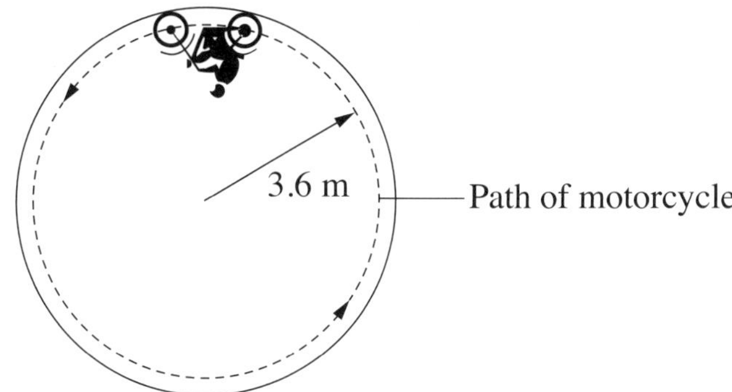

What is the minimum speed, in ms^{-1}, at which the motorcycle must travel to maintain the circular path?

(A) 0.42

(B) 1.9

(C) 5.9

(D) 35

(q18, 2016 HSC)

11 An astronaut with a mass of 75 kg lands on Planet *X* where her weight is 630 N.

What is the acceleration due to gravity (in ms^{-2}) on Planet *X*?

(A) 0.12

(B) 8.4

(C) 9.8

(D) 735

(q4, 2017 HSC)

12 Consider two satellites in Earth orbit. Satellite *A* has a mass *m*, an orbital radius of *r* and a total orbital energy of *E*. Satellite *B* has a mass of 4*m* and an orbital radius of $\frac{r}{2}$.
What is the orbital energy of satellite *B*?

(A) 8*E*

(B) 4*E*

(C) 2*E*

(D) 16*E*

(Sample question)

13 Friction between the tyres and the road is required for a car to safely negotiate a corner. What coefficient of friction μ is required for a car of mass m to safely negotiate a horizontal (unbanked) corner of radius r?

(A) $\mu > \dfrac{v^2}{rg}$

(B) $\mu < \dfrac{v^2}{rg}$

(C) $\mu > \dfrac{mv^2}{r}$

(D) $\mu < \dfrac{mv^2}{r}$

(Sample question)

14 A satellite orbits Earth with an elliptical orbit that passes through positions X and Y.

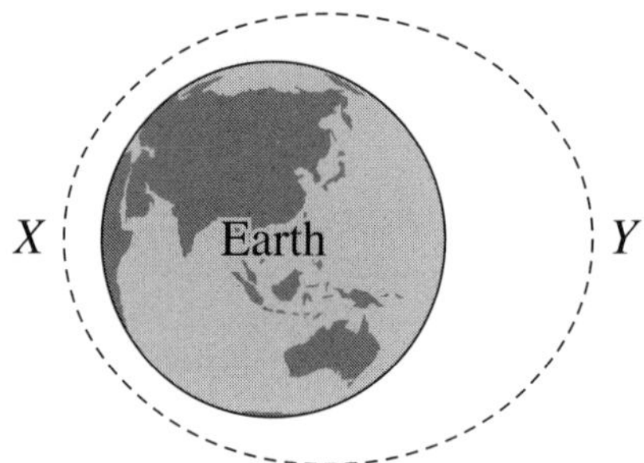

Which row of the table correctly identifies the position at which the satellite has greater kinetic energy and the position at which it has greater potential energy?

	Greater kinetic energy	*Greater potential energy*
(A)	X	X
(B)	X	Y
(C)	Y	X
(D)	Y	Y

(q12, 2017 HSC)

15 A car travelling at a constant speed follows the path shown.

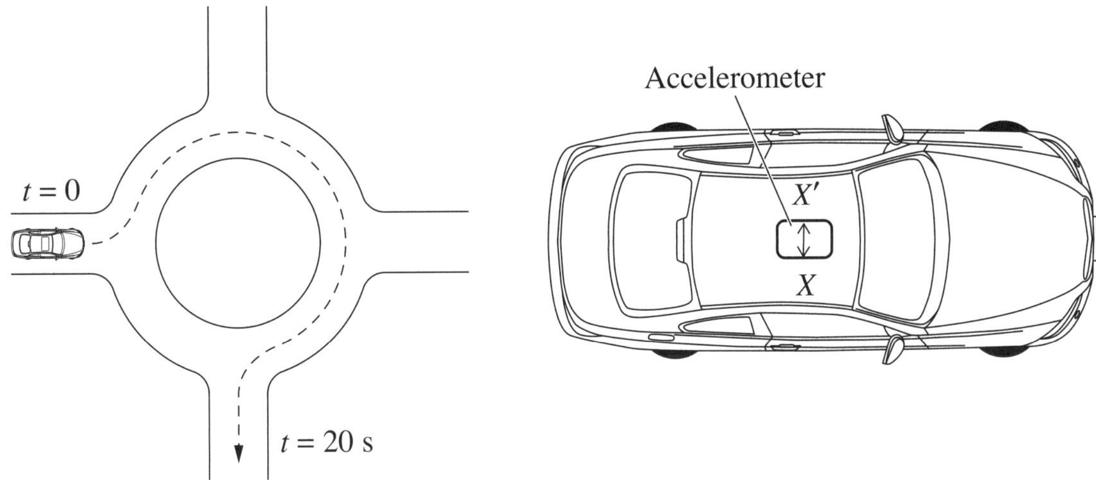

An accelerometer that measures acceleration along the X–X' direction is fixed in the car.

Which graph shows the measurements recorded by the accelerometer over the 20-second interval?

(A)
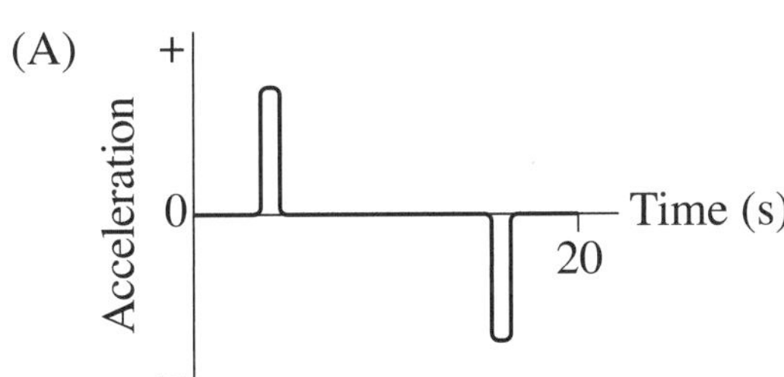

(B)
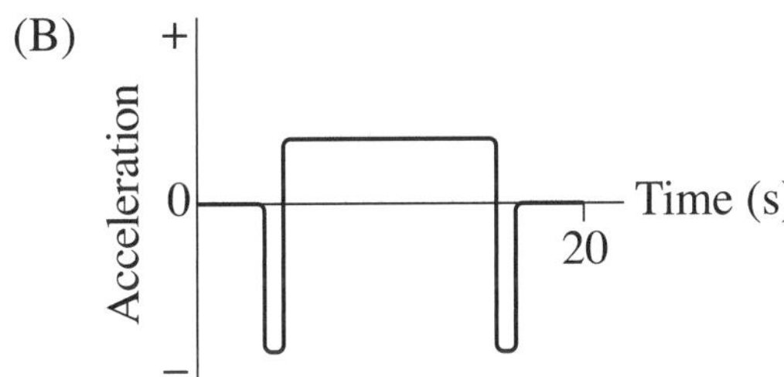

(C)
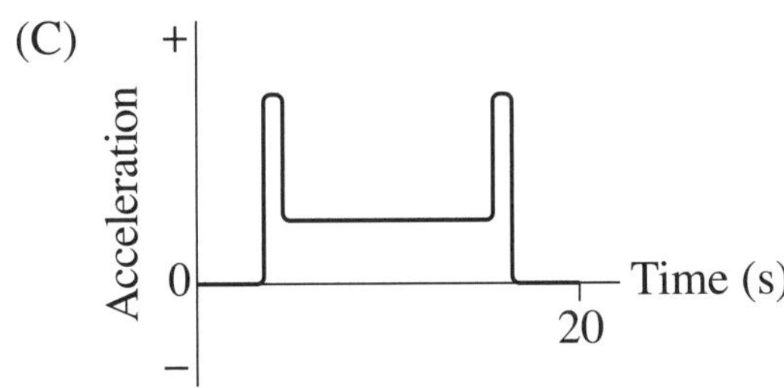

(D)
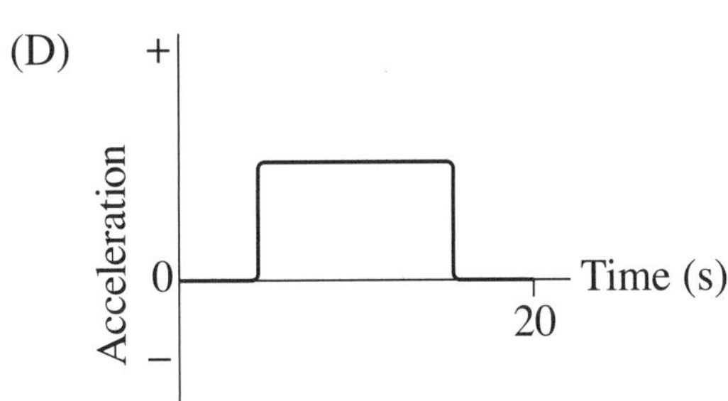

(q15, 2017 HSC)

16 A satellite orbits Earth as shown.

Which diagram correctly shows the direction of the satellite's acceleration?

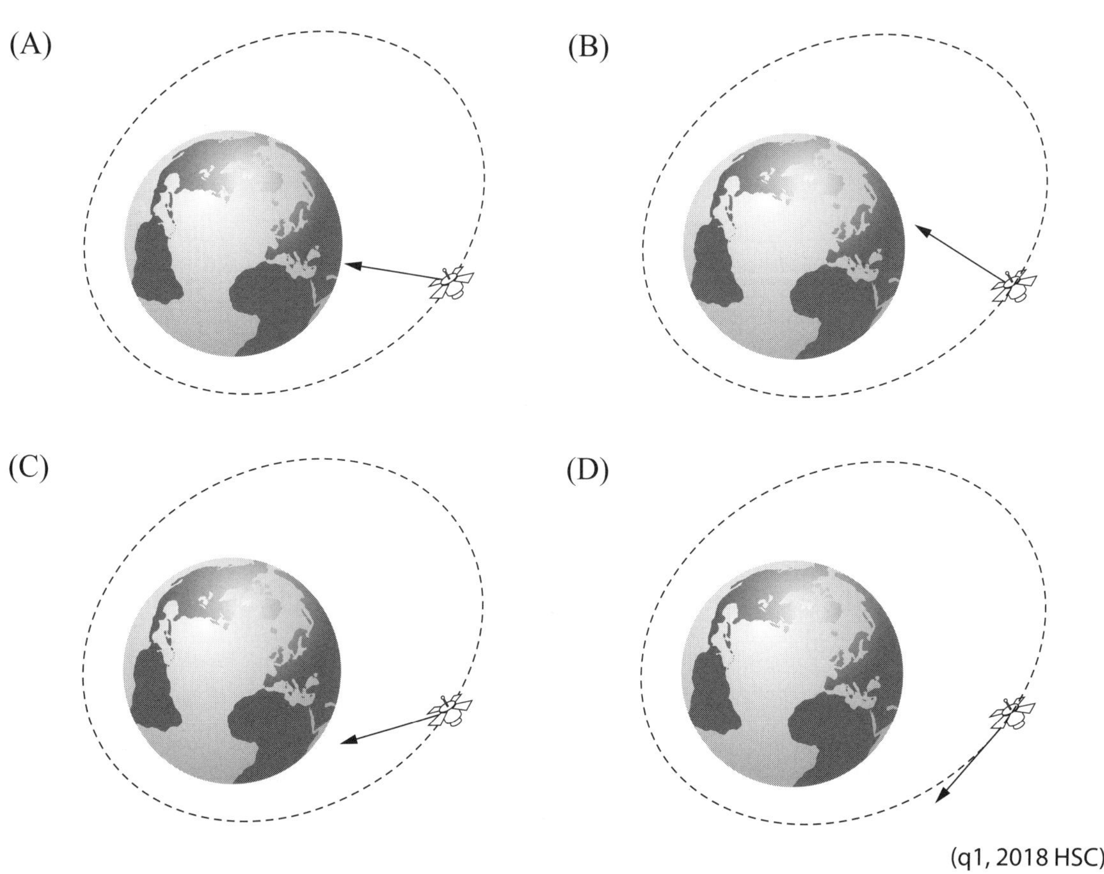

(q1, 2018 HSC)

17 The graph shows the altitude of the International Space Station (SS).

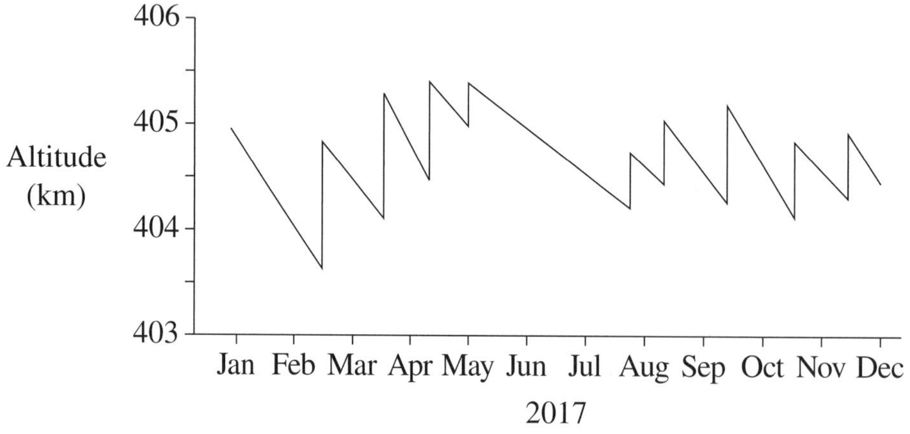

The altitude can only be boosted by supply craft visiting the ISS.

Why does the altitude decrease in the times between height boosts?

(A) Momentum of the ISS is being transferred to air molecules.

(B) The mooon's gravity changes the net force on the ISS as it orbits Earth.

(C) The decrease in altitude makes it possible for a supply craft to reach the ISS.

(D) The total mass of the ISS changes with the deliveries from each supply craft.

(q3, 2018 HSC)

18 A planet X has twice the mass and twice the radius of Earth.

What is the magnitude of the gravitational acceleration close to the surface of planet X?

(A) $\frac{1}{2}g$

(B) $1\,g$

(C) $2\,g$

(D) $4\,g$

(q7, 2018 HSC)

19 During the launch of a space vehicle from Earth, an astronaut feels an increased downward g force.

In which of the following situations would a person also feel an increased downward g force?

(A)

Roller-coaster speeding up

(B)

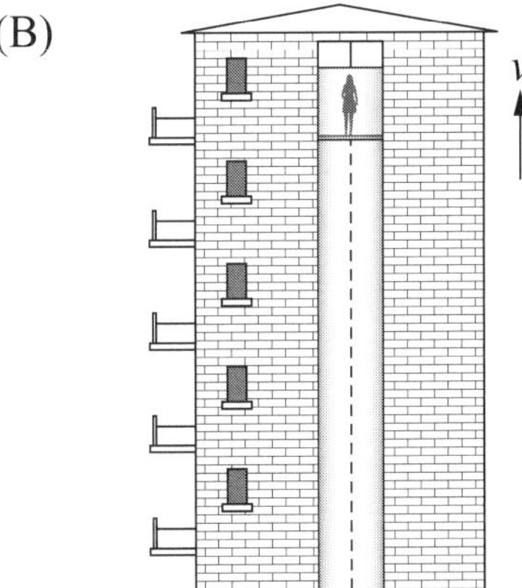

Lift slowing down

(C)

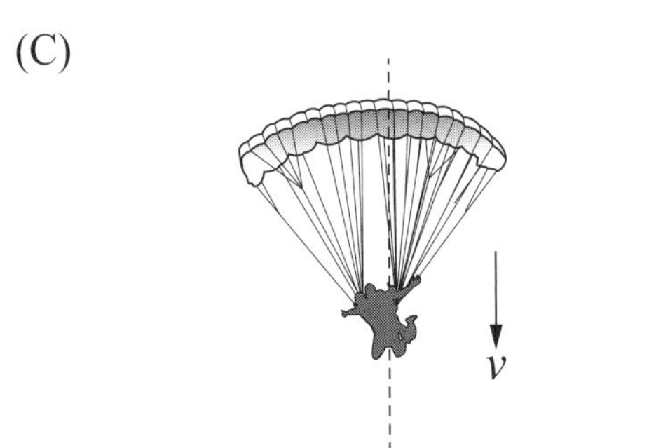

Parachute falling at a constant velocity

(D)

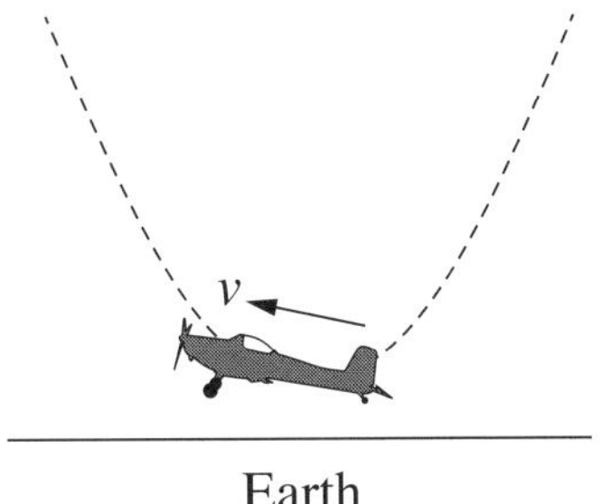

Plane pulling out of a dive

(q11, 2018 HSC)

20 A mass was hanging from the roof of a bus that was travelling forward on a horizontal road at a constant velocity.

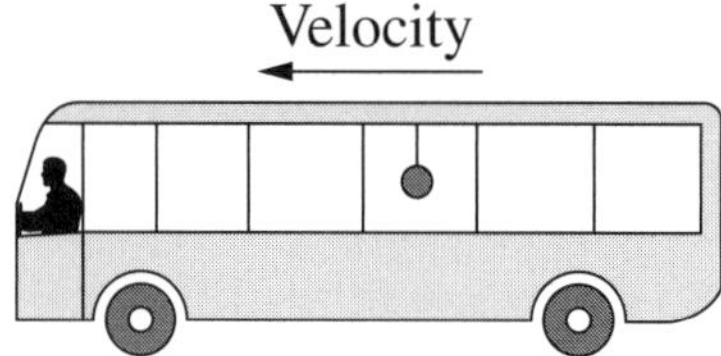

The string holding the mass was cut. At the same instant, the bus driver applied the brakes, causing the bus to slow down at a rate of 3 m s^{-2}.

To an observer outside the bus, the mass follows a parabolic trajectory.

Which statement correctly describes the resulting motion of the mass observed from within the frame of reference of the moving bus?

(A) The mass travelled in a straight line vertically downwards.

(B) The mass travelled in a straight line downwards and towards the front of the bus.

(C) The mass travelled in a parabolic path downwards and towards the back of the bus.

(D) The mass travelled in a parabolic path downwards and towards the front of the bus.

(q19, 2018 HSC)

Part B Short-answer questions

Question 21 (6 marks)

A projectile leaves the ground at point *A* with velocity components as shown in the diagram. It follows the path given by the dotted line and lands at point *B*.

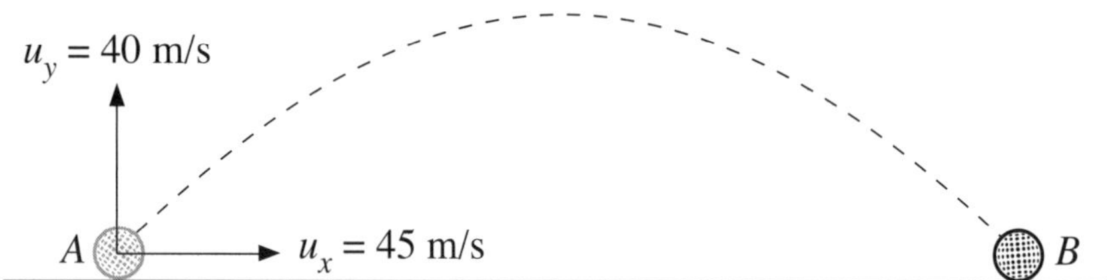

(a) State the horizontal component of the projectile's velocity when it lands. **1**
(1 line)

(b) Find the magnitude of the initial velocity of the projectile. **1**
(2 lines)

(c) Calculate the maximum height attained by the projectile. **2**
(6 lines)

(d) Calculate the range of the projectile, if it lands level with its starting position. **2**
(6 lines)

(q16, 2006 HSC)

Question 22 (4 marks)

A toy bird is launched at 60° to the horizontal, from a point 45 m away from the base of a cliff.

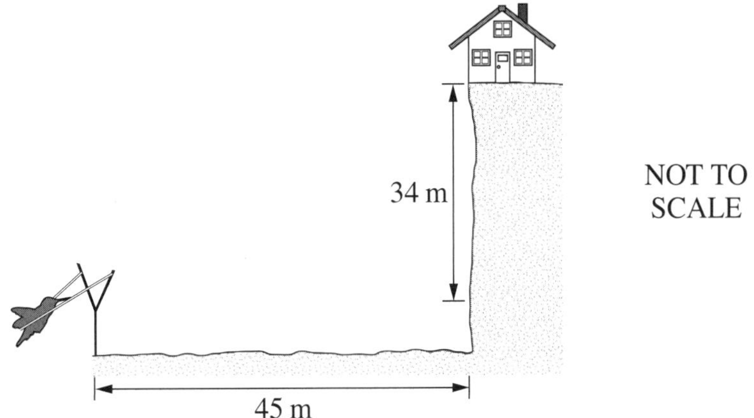

Calculate the magnitude of the required launch velocity such that the toy bird strikes the base of the wooden building at the top of the cliff, 34 m above the launch height. **4**

(8 lines) (q27, 2012 HSC)

Question 23 (3 marks)

This set of data was obtained from a motion investigation to determine the acceleration due to gravity on a planet other than Earth. **3**

Time (s)	Vertical velocity (ms^{-1})
0.60	0.02
1.00	0.09
1.20	0.12
1.40	0.17
1.80	0.23

Plot the data from the table, and then calculate the acceleration.

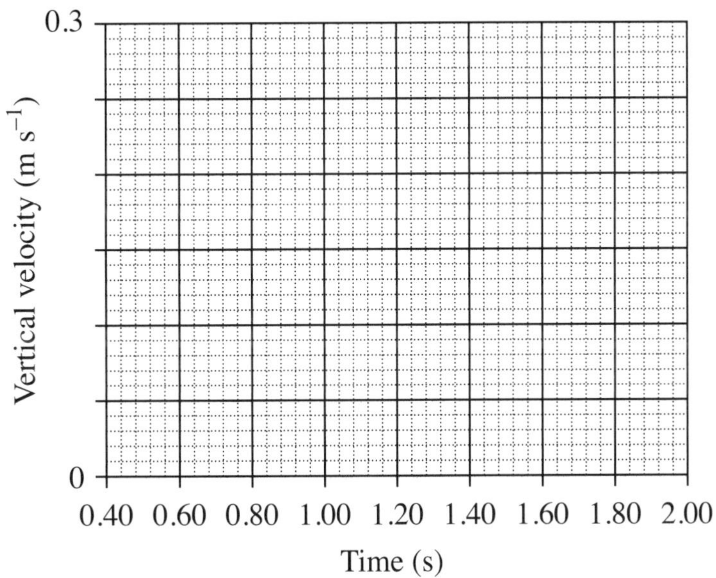

(4 lines) (q22, 2013 HSC)

Question 24 (4 marks)

Consider the following thought experiment. **4**

Two towers are built on Earth's surface. The height of each of the towers is equal to the altitude of a satellite in geostationary orbit about Earth. Tower A is built at Earth's North Pole and Tower B is built at the equator.

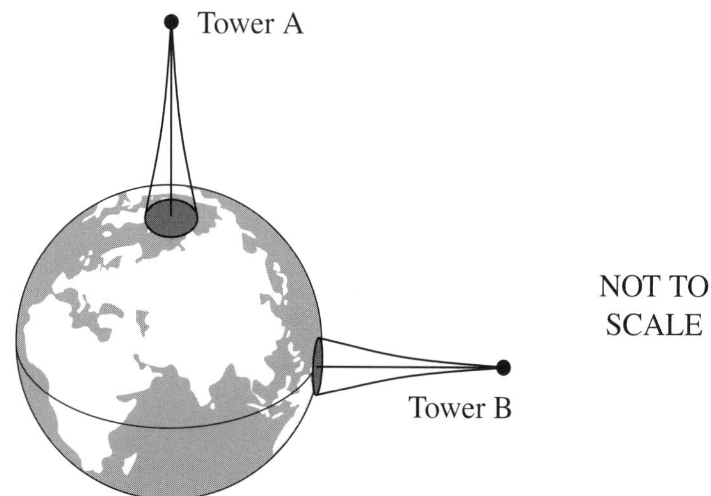

Identical masses are simultaneously released from rest from the top of each tower. Explain the motion of each of the masses after their release.

(10 lines) (q23, 2012 HSC)

Question 25 (3 marks)

In 2014 the James Webb Space Telescope (JWST) will be placed in orbit around the Sun. Earth and the JWST will follow the orbits shown, with identical orbital periods. This appears to contradict Kepler's law of periods. **3**

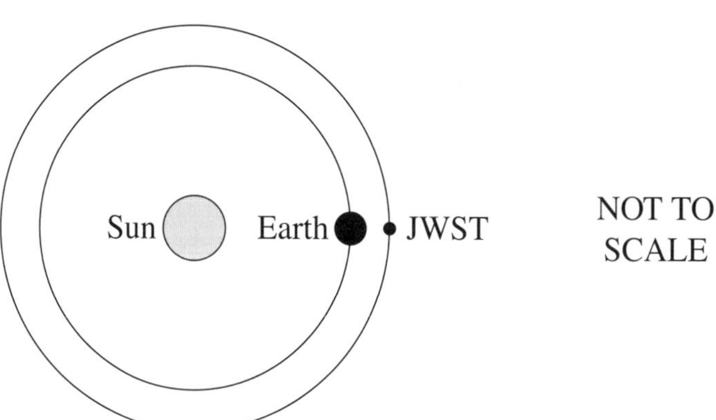

Why is it possible for the JWST to orbit the Sun with the same orbital period as Earth? In your answer, refer to Kepler's law of periods.

(8 lines) (q24, 2010 HSC)

Question 26 (4 marks)

A projectile is fired horizontally from a platform.

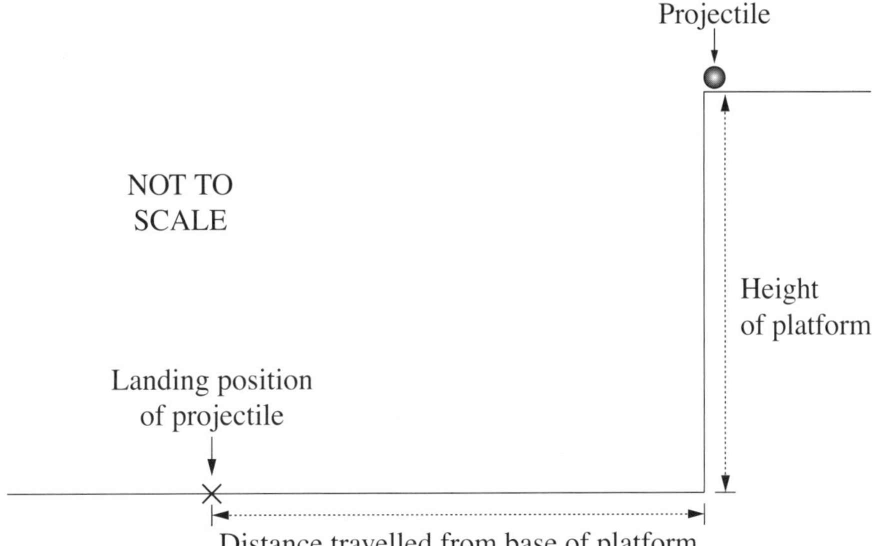

Measurements of the distance travelled by the projectile from the base of the platform are made for a range of initial velocities.

Initial velocity of projectile (m s^{-1})	*Distance travelled from base of platform* (m)
1.4	1.0
2.3	1.7
3.1	2.2
3.9	2.3
4.2	3.0

(a) Graph the data on the grid provided and draw the line of best fit. **2**

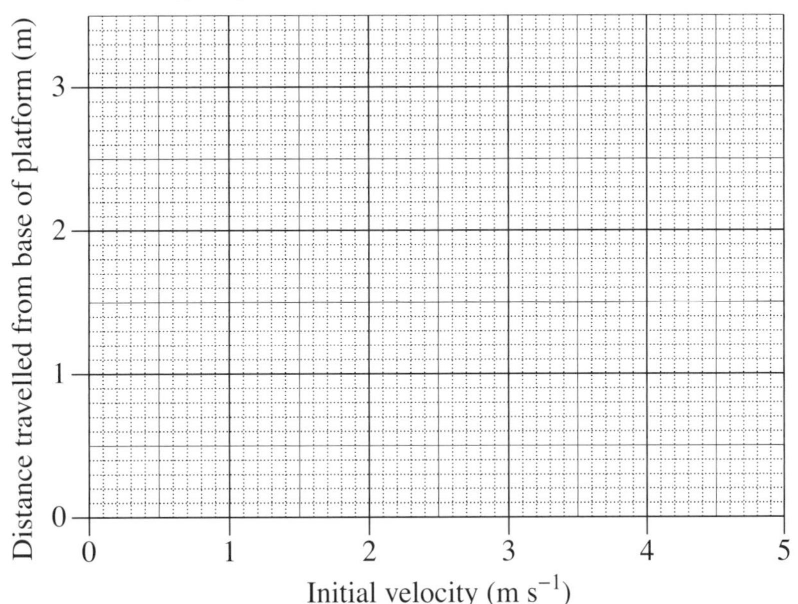

(b) Calculate the height of the platform. **2**

(6 lines)

(q21, 2015 HSC)

Question 27 (6 marks)

Consider the following two models used to calculate the work done when a 300 kg satellite is taken from Earth's surface to an altitude of 200 km.
You may assume that the calculations are correct.

Model X	*Model Y*
Data: $g = 9.8 \text{ m s}^{-2}$ $m = 300 \text{ kg}$ $\Delta h = 200 \text{ km}$ $W = Fs$ $= mg\Delta h$ $= 3 \times 10^2 \times 9.8 \times 2.0 \times 10^5$ $= 5.9 \times 10^8 \text{ J}$	Data: $G = 6.67 \times 10^{-11} \text{ N m}^2 \text{ kg}^{-2}$ $r_{\text{Earth}} = 6.38 \times 10^6 \text{ m}$ $r_{\text{orbit}} = 6.58 \times 10^6 \text{ m}$ $M = 6.0 \times 10^{24} \text{ kg}$ $m = 300 \text{ kg}$ $W = \Delta E_p$ $\Delta E_p = E_{p \text{ final}} - E_{p \text{ initial}}$ $= -\frac{GMm}{r_{\text{orbit}}} - \left(\frac{GMm}{r_{\text{Earth}}}\right)$ $= -1.824 \times 10^{10} - \left(-1.881 \times 10^{10}\right)$ $= 5.7 \times 10^8 \text{ J}$

(a) What assumptions are made about Earth's gravitational field in models *X* and *Y* that lead to the different results shown? **2**

(5 lines)

(b) Why do models *X* and *Y* produce results that, although different, are close in value? **1**

(3 lines)

(c) Calculate the orbital velocity of the satellite in a circular orbit at the altitude of 200 km. **3**

(8 lines) (q26, 2015 HSC)

Question 28 (5 marks)

(a) Why does orbital decay occur more rapidly for satellites in a low-Earth orbit than for satellites in other orbits? **2**

(4 lines)

(b) Calculate the magnitude of the gravitational force that acts on a 50 kg satellite when it is 8000 km from Earth's centre. **3**

(6 lines) (q21, 2016 HSC)

Question 29 (5 marks)

Two teams carried out independent experiments with the purpose of investigating Newton's Law of Universal Gravitation. Each team used the same procedure to accurately measure the gravitational force acting between two spherical masses over a range of distances.

The following graphs show the data collected by each team.

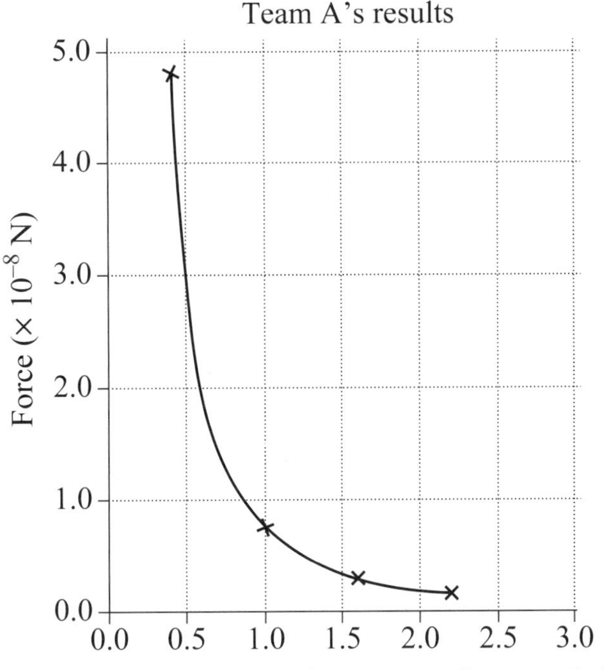

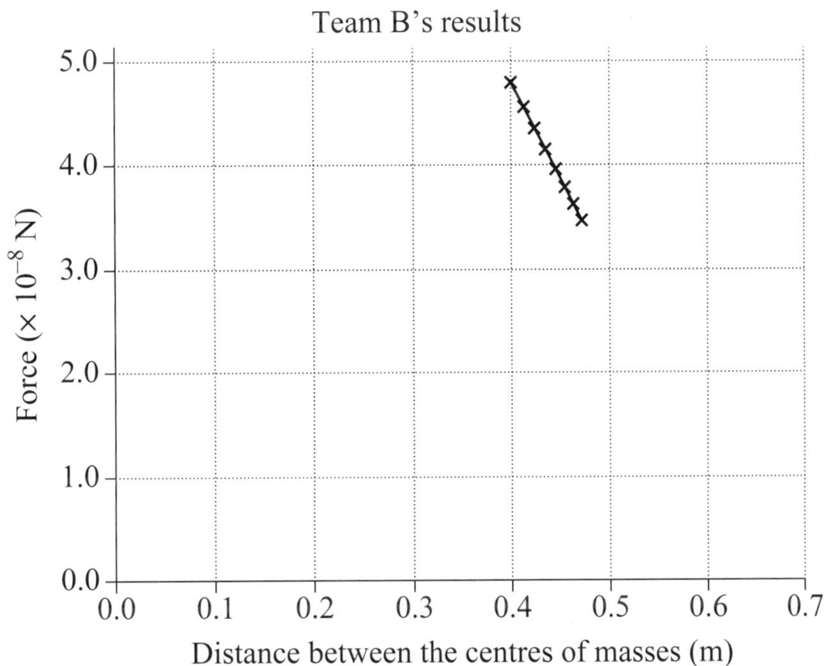

Question 29 continues on the following page

Question 29 (continued)

(a) Compare qualitatively the relationship between force and distance in the graphs. **2**

(5 lines)

(b) Assess the appropriateness of Team A's data and Team B's data in achieving the purpose of the experiments. **3**

(6 lines) (q25, 2016 HSC)

End of Question 29

Question 30 (5 marks)

The following diagram shows the acceleration of a rocket during the first stage of its launch. **5**

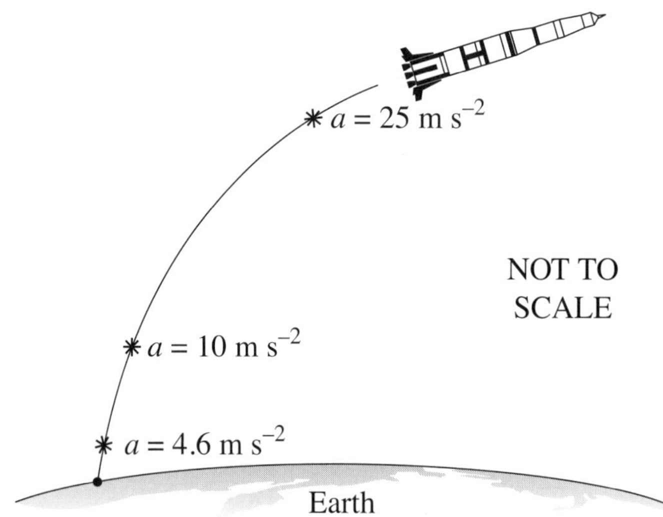

Explain the acceleration of the rocket with reference to the law of conservation of momentum.

(10 lines) (q28, 2016 HSC)

Question 31 (5 marks)

The escape velocity from a planet is given by $v = \sqrt{\frac{2GM}{r}}$.

(a) The radius of Mars is 3.39×10^6 m and its mass is 6.39×10^{23} kg. Calculate the escape velocity from the surface of Mars. **2**

(4 lines)

(b) Using the law of conservation of energy, show that the escape velocity of an object is independent of its mass. **3**

(6 lines) (q24, 2017 HSC)

Question 32 (7 marks)

A spring is used to construct a device to launch a projectile. The force (F) required to compress the spring is measured as a function of the displacement (x) by which the spring is compressed.

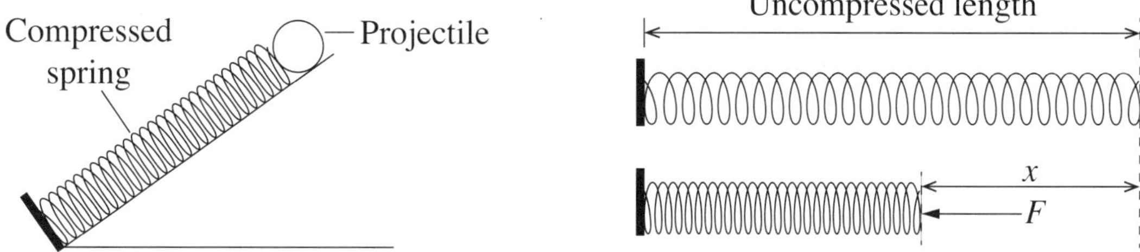

The potential energy stored in the compressed spring can be calculated from $E_p = \frac{1}{2}kx^2$, where k is the gradient of the force–displacement graph shown.

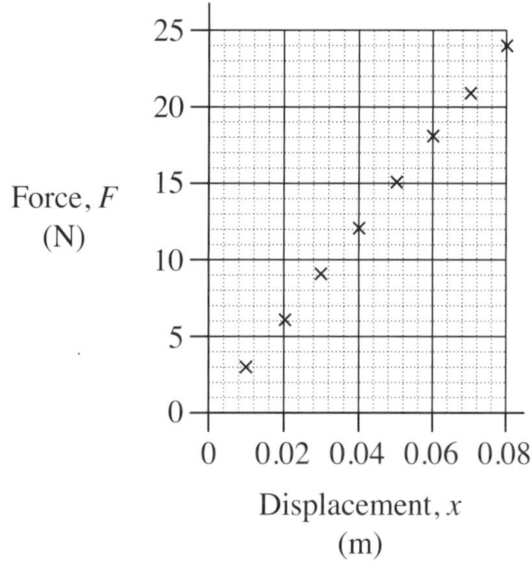

(a) A projectile of mass 0.04 kg is launched using this device with the spring compressed by 0.08 m. Calculate the launch velocity. **4**

(11 lines)

(b) Calculate the range of a projectile launched by this device from ground level at an angle of 60° to the horizontal with a velocity of 10 ms^{-1}. **3**

(9 lines) (q29, 2017 HSC)

Question 33 (4 marks)

(a) Compare the force of gravity exerted on the moon by Earth with the force of gravity exerted on Earth by the moon. **2**

(3 lines)

(b) The acceleration due to gravity on the moon is 1.6 m s^{-2} and on Earth it is 9.8 m s^{-2}. Quantitatively compare the mass and weight of a 70 kg person on the moon and on Earth. **2**

(4 lines) (q21, 2018 HSC)

Question 34 (6 marks)

(a) The diagram shows a camera and a ruler set up to obtain data about a projectile's motion along the trajectory shown. The entire trajectory is visible through the camera. **3**

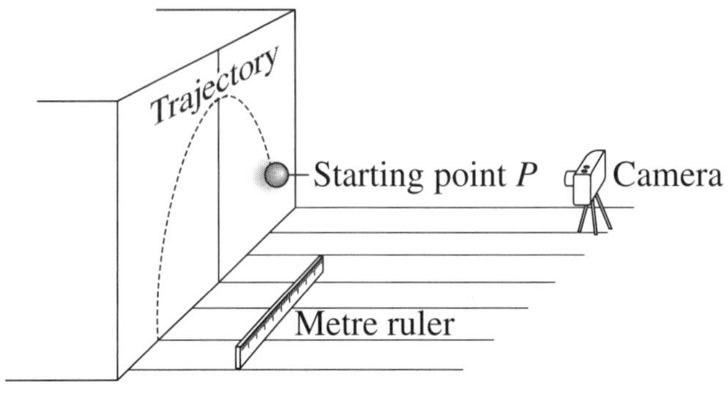

Identify ONE of the errors in this set-up and describe the effects of this error on the results.

(6 lines)

(b) An experiment was set up based on the method described in part (a), but conducted so that the data obtained were valid. **3**

The imagery shows the trajectory of the ball.

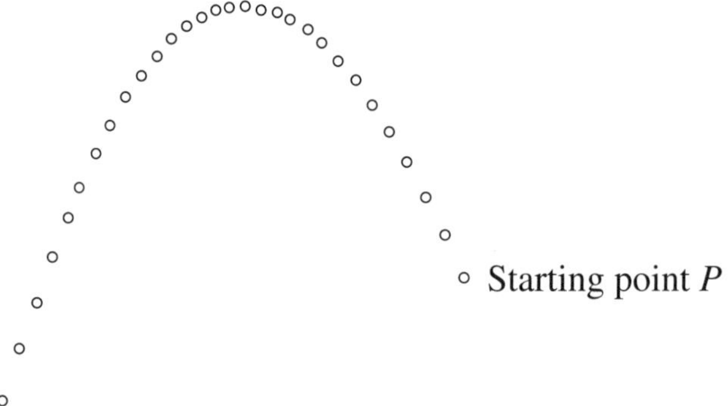

The graphs show data from this experiment.

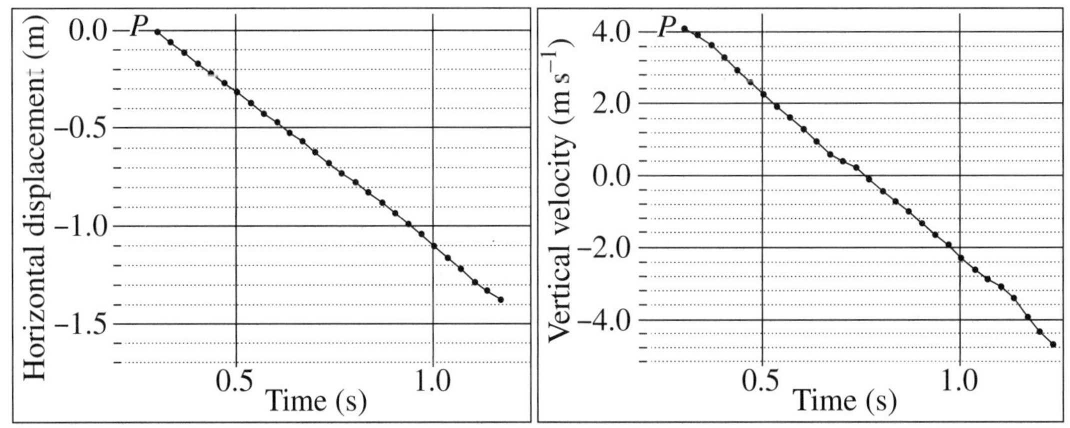

Using the graphs, describe the velocity and acceleration of the ball quantitatively and qualitatitvely.

(6 lines) (q27, 2018 HSC)

Question 35 (5 marks)

The radius of the moon is 1740 km. The moon's mass is 7.35×10^{22} kg. In this question, ignore the moon's rotational and orbital motion.

A 20 kg mass is launched vertically from the moon's surface at a velocity of 1200 m s^{-1}.

(a) Show that the change in potential energy of the mass in moving from the surface to an altitude of 500 km is 1.26×10^7 J. **2**

(8 lines)

(b) Calculate the velocity of the 20 kg mass at an altitude of 500 km. **3**

(10 lines) (q28, 2018 HSC)

Module 5
Advanced Mechanics

Sample answers

Part A Objective-response questions

1 C The horizontal component of the velocity of the cannonball does not change in direction or magnitude; its vertical velocity does change, and eventually reverses direction.

2 D The *vertical* velocity of an object initially launched horizontally is found using $v = u + at$ so after two seconds $v = 0 + 9.8 \times 2 = 19.6\ \text{ms}^{-1} \downarrow$.

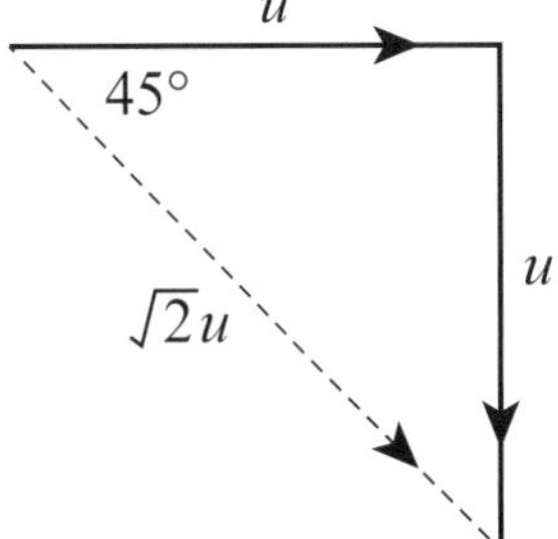

Since at this instant the object is moving at 45° below the horizontal, the values of the horizontal and vertical components are exactly equal, so $u = 19.6\ \text{ms}^{-1}$. The magnitude of its velocity at that point is therefore $\sqrt{2}u = 27.7\ \text{ms}^{-1}$

3 C Since $h = 1.0$ m and $u_v = 0\ \text{ms}^{-1}$, then $1.0 = \frac{1}{2}gt^2$; therefore, $t = 0.45$ s.

4 A The potential energy of an object in orbit is given by $E_P = \frac{Gm_E m_{obj}}{r}$ where E_P is the potential energy, m_E is Earth's mass and m_{obj} is the object's mass. Because the masses of the objects are different, their potential energy is also different. The orbital speed of any object is given by the formula $v = \frac{2\pi r}{T}$ where v is the orbital speed, r is its orbital radius and T is the orbital period. Here the mass of the orbiting object is not included, so their orbital speed is equal.

5 D Projectiles within a gravitational field move in a parabolic path. D is the only actual parabolic path shown.

6 B A is incorrect as the gravitational force is not negligible. C is incorrect as the gravitational force depends upon the masses affected, which are unequal. D is incorrect as the gravitational acceleration acting on all masses at the same point in space is equal.

7 B The horizontal component of the initial velocity is $\frac{70}{3.5} = 20\ \text{ms}^{-1}$. Its vertical component is found by $\Delta y = v_0 t + \frac{1}{2}gt^2$ so $v_0 = 17.15\ \text{ms}^{-1}$. The projectile was therefore launched at an angle of 40°.

8 B $\frac{r_1^3}{T_1^2} = \frac{GM_1}{4\pi^2}$ Equally, $\frac{r_2^3}{T_2^2} = \frac{GM_2}{4\pi^2}$ As $r_1 = r_2$ and by removing common factors

G and $4\pi^2$

$\therefore \frac{1}{T_1^2} \propto \frac{M_1}{1}$ and $\frac{1}{T_2^2} \propto \frac{M_2}{1}$ $\therefore \frac{T_1^2}{T_2^2} = \frac{M_2}{M_1} = \frac{4}{1}$ $\therefore T_2 = \frac{1}{2}T_1$

9 C Since the lift is decelerating as it is rising, in fact it is accelerating downwards, so the frame of reference within the lift is undergoing that acceleration. Because the falling projectile is observed from within that frame of reference its acceleration downwards appears to be less. Its horizontal motion is unaffected.

10 C In order to maintain the circular path the motorcycle must be in contact with the surface at all times—including when at the position in the diagram. For it to move in a circle, the motorcycle must experience a centripetal force. At the position shown its *weight* can be just enough force to apply this—in which case this will be the minimum force required to keep it moving in the circle.

$F_w \equiv F_C \therefore mg = \frac{mv^2}{r} \therefore g = \frac{v^2}{r}$ i.e. v (minimum) $= \sqrt{3.6 \times 9.8} = 5.9 \text{ ms}^{-1}$

11 B $F_w = mg$

$630 \text{ N} = 75 \text{ kg} \times g$

$g = 8.4 \text{ ms}^{-2}$

12 A The total orbital energy is proportional to $\frac{m}{r}$ and hence the energy of satellite B will be given by $E_B \left(\frac{4}{\frac{1}{2}}\right) E = 8E$.

13 A To safely negotiate a corner, friction must provide the centripetal force.

Thus $\frac{mv^2}{r} = \mu mg$ or $\mu = \frac{v^2}{rg}$ is the minimum value of μ that will ensure the car goes around the corner. Therefore provided $\mu > \frac{v^2}{rg}$, the car will safely negotiate the corner.

14 B Potential energy is greater when the distance from the Earth is greater. Since energy is conserved, the kinetic energy must be greatest when the potential energy is least.

15 B As the car enters and exits the roundabout it turns to the left, meaning acceleration towards *X*. Around the circular path there is constant acceleration in the direction of *X*′.

16 A As $F = ma$, the satellite will accelerate in the same direction as the net force on the satellite. Because the only force operating on the satellite is gravity, the satellite will experience an acceleration towards the centre of the Earth.

17 A The ISS is a low Earth satellite and, consequently, collisions with air molecules reduce the velocity of the space station, causing it to lose altitude.

18 A The gravitational acceleration for a planet is given by $g = \frac{GM}{r^2}$. If the mass and radius were doubled the new gravitational acceleration would be $g = \frac{2}{2^2} = \frac{1}{2}g$.

19 D We can assume the plane turns in an approximately circular motion at the bottom of the dive. Now because the centripetal force and centripetal acceleration of the plane are upwards, the pilot would experience an increased apparent weight force (i.e. an increased g force).

20 B The mass is initially at rest with respect to the bus driver. If the bus was moving at constant velocity the bus driver would see the mass fall straight downwards. However, as the bus was slowing down at 3 ms^{-2} the bus driver would see the mass falling downwards with an acceleration of 9.8 ms^{-2}, but the horizontal component of the motion would result in the object also accelerating forwards at 3 ms^{-2} with respect to the bus driver. The vertical distance moved in a time t would be $y = \frac{1}{2}gt^2$ and the horizontal distance would be $x = \frac{1}{2}3gt^2$. Because the ratio of the horizontal distance x to the vertical distance y is a constant (i.e. $\frac{x}{y} = \frac{\frac{3}{2}t^2}{\frac{1}{2}gt^2} = \frac{3}{g}$) the mass moves in a straight line downwards and towards the front of the bus with respect to the driver.

Part B Short-answer questions

Question 21 (Total 6 marks)

(a) The horizontal component of the projectile's velocity at landing is the same as it was at launch, i.e. 45 ms^{-1} to the right. *(1 mark)*

(b) The magnitude of initial velocity, u, is given by Pythagoras' theorem, i.e. $40^2 + 45^2 = u^2$, $\therefore u = 60.2$ ms^{-1}. *(1 mark)*

(c) At maximum height, the final vertical velocity (v_y) is zero. Using $v_y^{\,2} = u_y^{\,2} + 2a_ys_y$, where $a_y = g$ ms^{-2} down, $u_y = 40$ ms^{-1} up, and s_y = maximum height.

$s_v = \dfrac{0^2 - 40^2}{2 \times -9.8} = 81.6$ m vertically above ground. *(2 marks)*

(d) The range of the projectile is $s_x = u_xt_f$, where $u_x = 45$ ms^{-1} right. Now t_f is determined by vertical motion where s_y at landing is zero. Using $s_y = u_yt_f + \frac{1}{2}a_yt_f^2$ then,
$0 = 40t_f - 4.9t^2$, or $4.9t_f = 40$,
$\therefore t_f = 8.16$ s. Thus the range $s_x = 45 \times 8.16 = 367.2$ m to the right. *(2 marks)*

Question 22 (Total 4 marks)

Vertical displacement, $s_v = 34$ m up, horizontal displacement, $s_y = 45$ m right.

For vertical motion, $s_v = u_vt + \frac{1}{2}a_vt^2$, i.e. $34 = -u\sin 60°t + 4.9t^2$. *Equation 1*

For horizontal motion, $s_h = u_ht$, i.e. $45 = u\cos 60°t$, $\frac{45}{0.5} = ut$. *Equation 2*

Combining the two equations, $-34 = -0.866 \times 90 + 4.9t^2$.

Therefore, $t = \sqrt{\frac{77.94 - 34}{4.9}} = 2.99s$.

Substituting into Equation 2, $u = \frac{45}{2.99 \times 0.5} = 30.1 \text{ ms}^{-1}$ up at 60° to horizontal.

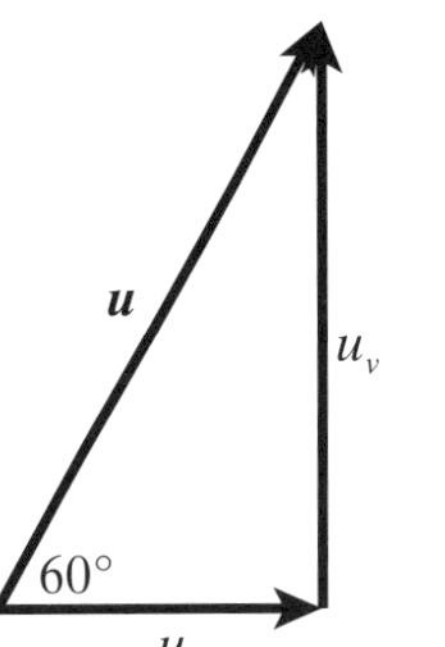

(4 marks)

Question 23 (Total 3 marks)

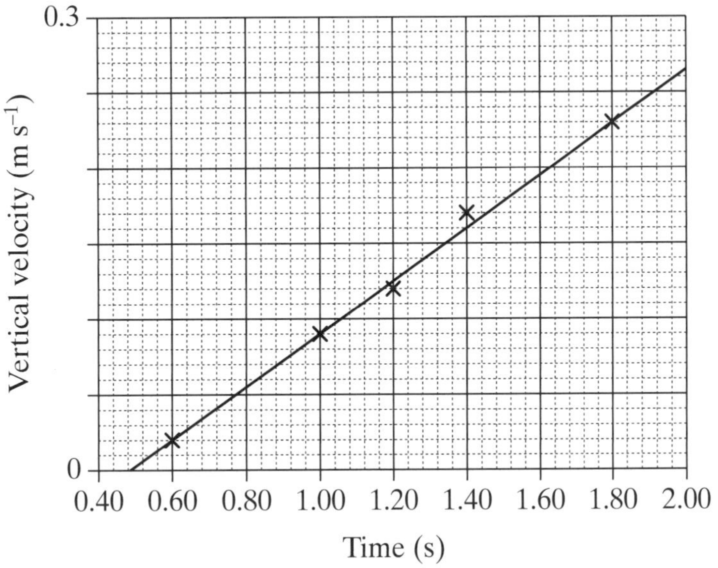

The acceleration is found from $a = \frac{\Delta v}{\Delta t}$. Here $a = \frac{0.25 - 0}{1.86 - 0.44} = 0.176 = 0.18 \text{ ms}^{-2}$.

(3 marks)

Question 24 (Total 4 marks)

Being over the pole when released, mass A has no horizontal velocity and will accelerate vertically down due to the gravitational attraction of Earth. As it falls, the acceleration will increase as it gets closer and closer to Earth, i.e. $g \alpha \frac{1}{d^2}$. The very high velocity achieved by the time it nears Earth is likely to cause mass A to burn up as it begins to enter the increasingly dense atmosphere of Earth.

Mass B, when released at this distance from Earth, will remain exactly where it is. This is due to it being directly above the equator and having the same angular velocity as the surface of Earth, i.e. it will be in a geostationary orbit with an orbital velocity, $V_B = \sqrt{\frac{Gm_E}{d}}$, where d is the distance from mass B to the centre of Earth. This velocity is just the right magnitude for mass B to remain in orbit at this distance directly above the equator. *(4 marks)*

Question 25 (Total 3 marks)

Kepler's law of periods predicts that for a larger orbital radius, the period of the orbit would also be larger, i.e. $T^2 \propto r^3$. If the JWST had only the influence of the Sun affecting its orbit this would be true, but in this case Earth is also exerting a gravitational force that is influencing the orbit of the JWST around the Sun. In order to achieve the orbit of the JWST as described, the combined forces created by the gravitational attraction to the Sun and the gravitational attraction to Earth must create a centripetal force on the JWST of just the right value when it is at this particular distance from the Sun and Earth, such that this centripetal force holds the JWST in the position shown, where it has a higher orbital velocity than Earth but remains directly behind the Earth with an orbital period around the Sun identical with Earth's. *(3 marks)*

Question 26 (Total 4 marks)

(a)

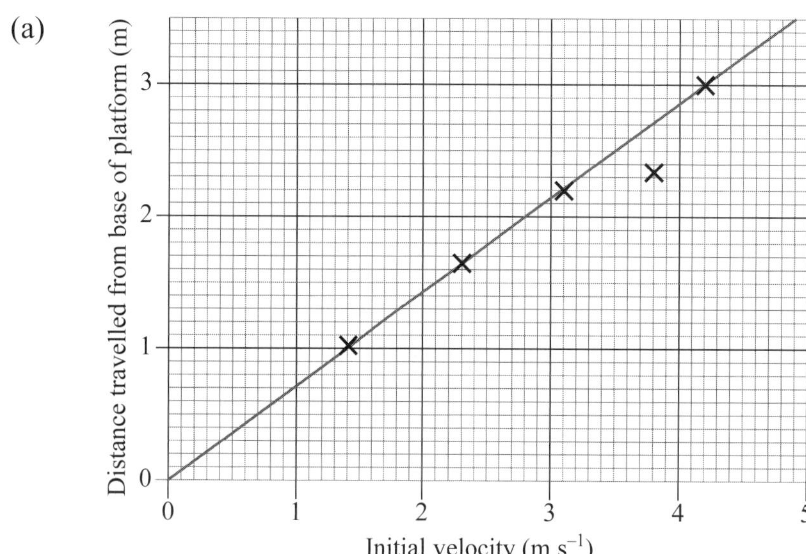

The slope of the line of best fit, $\frac{\Delta x}{\Delta v_0}$, gives the (average) time interval for the projectile to fall.

According to the line of best fit drawn for this graph,

$\therefore t = \frac{3.5 - 0.2}{4.9 - 0} = 0.67$ seconds.

Students should note there is one outlying point, which should not be taken into account because it is located well outside the variation of the other points, and should be regarded as experimental error. (Ideally that test should be repeated.) *(2 marks)*

(b) The height of the platform is found using the formula $\Delta y = u_y t + \frac{1}{2}gt^2$, where Δy is the displacement of the projectile $\therefore \Delta y = 0 - 4.9 \times 0.67^2 = 2.2$ m. *(2 marks)*

Question 27 (Total 6 marks)

(a) Model X assumes that the acceleration due to gravitation is 9.8 ms^{-2} both at sea level on Earth and at a height 200 km above. Model Y correctly assumes that it does not—rather that the strength of Earth's gravitational field decreases with height in the form $G \propto \frac{1}{d^2}$ where d is the distance from the Earth's centre. *(2 marks)*

(b) The calculation used in the incorrect Model *X* produces a result that is not greatly different from the correct one because the radius of Earth from which the satellite is to be launched, 6.38×10^6 m, is not very different from its orbital height, 6.58×10^6 m. Therefore, the actual gradual decrease in the value of gravitational acceleration over such a short range does not produce a markedly different outcome. *(1 mark)*

(c) In order for a satellite to maintain a stable circular orbit around the Earth it must experience a centripetal force that is constant in magnitude, but with its direction continually changing, always towards the centre of its orbit. This force is provided by gravitation—the gravitational force *is* the centripetal force, i.e. $F_G \equiv F_C$.

$$\therefore\ G\frac{m_E m_S}{r^2} = \frac{m_S v^2}{r}\ \therefore\ G\frac{m_E}{r} = v^2\ \therefore\ v = \sqrt{\frac{Gm_E}{r}}$$

where m_S represents the mass of the satellite, m_E represents the mass of the Earth, and v is the orbital velocity of the satellite.

$$v = \sqrt{\frac{Gm_E}{r}} = \sqrt{\frac{(6.67 \times 10^{-11}) \times (6.0 \times 10^{24})}{6.58 \times 10^6}} = 8.0 \times 10^3\ \text{ms}^{-1}$$

(3 marks)

Question 28 (Total 5 marks)

(a) Even in the emptiest regions of space between the galaxies there are some atoms scattered around; however, there are a far greater number close to planets (due to gravitational attraction) including Earth, even where craft are located in 'the vacuum of space'. The closer a craft is to the ground, the more particles are present and must be pushed aside by any moving craft, removing some of its kinetic energy and resulting in orbital decay. Then, as they lose gravitational potential energy, their speed increases, so the collisions with atmospheric particles cause even more rapid energy losses. Consequently low-Earth orbit satellites experience more rapid orbital decay than those on other orbital paths. *(2 marks)*

(b) $F_G = \dfrac{Gm_E m_{sat}}{d^2}\ \therefore\ F_G = \dfrac{(6.67 \times 10^{-11}) \times (6.0 \times 10^{24}) \times 50}{(8000 \times 10^3)^2} = 312.7 = 310\text{N}$

(3 marks)

Question 29 (Total 5 marks)

(a) Team A: Their graph indicates a hyperbolic relationship, such as $F = \dfrac{k}{d}$ or $\dfrac{k}{d^2}$ etc.

Team B: Their graph indicates a linear relationship, of the form $AF + Bd = C$. *(2 marks)*

(b) Team A: Although Team A carried out only four actual measurements, the range of its distance measurements was broad enough for the relationship to be identified $F = \dfrac{k}{d^2}$. From theory we know this is the correct relationship.

Team B: Team B did carry out eight measurements, but because the range of the readings it took was far too small, the true shape of the line of best fit appeared to be linear.

[This is a common error in investigations, notably when students are trying to determine the relationship between the length of a simple pendulum and its period—it is essential to include a few readings with the mass on a short length of cord.] *(3 marks)*

Question 30 (Total 5 marks)

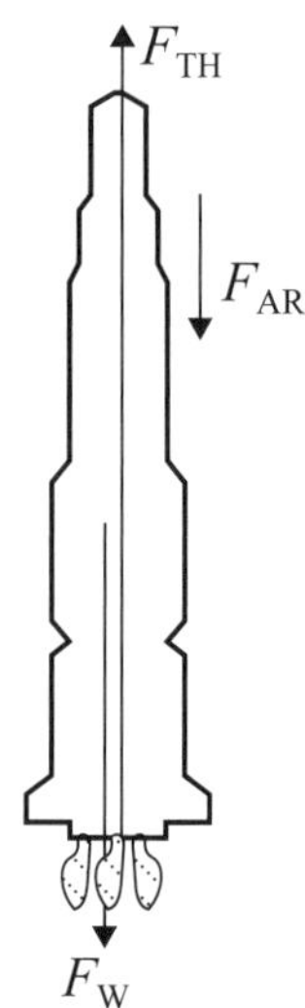

After a rocket is launched it is affected mainly by three forces: the thrust of its motors F_{TH} ↑, its weight F_W ↓ and air resistance, F_{AR} ↓.

For the launch to be successful F_R ↑= $F_{TH} - F_W - F_{AR}$, > 0.

The physics principle explaining the upward movement of the craft is that of Conservation of Momentum—fuel is burnt in the engines, the resulting gases are forced out of the base of the engine at very high speed and as the motor gives these gases downwards momentum, they apply an equal change in momentum upward onto the rocket.

According to the question, the resultant upward acceleration of the rocket, at launch, is 4.6 m s^{-2}. This low value is aimed to minimise the discomfort the astronauts aboard experience due to the '*g*-force'.

All the time fuel is being burnt the craft's mass continues to decrease. This is the main reason the acceleration rises to 10 m s^{-2} within a minute or so. It is also true that the air resistance decreases as the craft rises above the denser air nearer the ground. Meanwhile, as it moves further away from the centre of the Earth, the gravitational attraction downwards also decreases, though at this stage neither is significant relative to the weight lost by burning fuel.

By the time the craft tilts, beginning its manoeuvre to enter a parking orbit, its acceleration is identified to be 25 m s^{-2}. At this point the reduced air resistance and weaker gravitational attraction by the Earth have both become more significant, whilst the Stage-1 fuel tanks will be empty (may even have been dropped to reduce the craft's weight) to help attain orbital speed. Since the thrust force of the motor remains fairly constant the reduced mass, decreased *g* and diminished air resistance produce greater acceleration on the craft. (*5 marks*)

Question 31 (Total 5 marks)

(a) $v = \sqrt{\dfrac{2GM}{r}}$

$$v = \sqrt{\frac{2 \times 6.67 \times 10^{-11} \times 6.39 \times 10^{23}}{3.39 \times 10^{6}}}$$

$v = 5014.5 \text{ ms}^{-1} = 5.01 \times 10^{3} \text{ ms}^{-1}$ (3 sig. fig.) (*2 marks*)

(b) To escape a gravitational field the kinetic energy needs to be greater than or equal to the magnitude of the gravitational potential energy. It can be considered that the object stops just as it reaches a point an infinite distance away. Therefore $E_{kfinal} = 0$ and $E_{pfinal} = 0$. Thus $E_{totalfinal} = 0$ and as energy is conserved $E_{totalinitial} = 0$.
We can determine the velocity at which this occurs.

From above:

$E_{kinitial} = -E_{pinitial}$

$$\frac{1}{2}mv^2 = \frac{GMm}{r}$$

Where m = mass of object and M = mass of central body/planet

$$\therefore \frac{1}{2}v^2 = \frac{GM}{r}$$

Or $v = \sqrt{\frac{2GM}{r}}$

The mass of the object is not part of the equation; therefore the escape velocity is independent of this value. *(3 marks)*

Question 32 (Total 7 marks)

(a) Assuming all of the spring's potential energy is converted into kinetic energy we can make the two values equal, i.e.

$E_p = E_k$

$\frac{1}{2}kx^2 = \frac{1}{2}mv^2$

We can determine k from the gradient of the graph since $F = kx$, thus $k = \frac{F}{x}$ = gradient

$\frac{18 - 6}{0.06 \times 0.02} = 300$

$\therefore \frac{1}{2} \times 300 \times 0.08^2 = \frac{1}{2} \times 0.04 \times v^2$

$0.96 = 0.02 \times v^2$

$v = 6.93\ \text{ms}^{-1} = 6.9\ \text{ms}^{-1}$ (2 sig. fig.) *(4 marks)*

(b) Projectile launched at 10 ms^{-1} at an angle of 60°.

$\therefore\ u_x = 10 \cos 60 = 5\ \text{ms}^{-1}$

$u_y = 10 \sin 60 = 8.66\ \text{ms}^{-1}$

The time of flight can be calculated using

$\Delta y = u_y t + \frac{1}{2}at^2$

$0 = 8.66t - 4.9t^2$

$0 = t(8.66 - 4.9t)$

$\therefore t = 0$ *or* $8.66 - 4.9t = 0$

$t = \frac{8.66}{4.9}$

$t = 1.77\ s$

Then from $\Delta x = u_x t$

$\Delta x = 5 \times 1.77$

$\Delta x = 8.85\ \text{m} = 8.9\ \text{m}$ *(3 marks)*

Question 33 (Total 4 marks)

(a) The force of gravity on each object is given by $F = \frac{GMm}{r^2}$, where G is the universal constant of gravity, M is the mass of the Earth, m is the mass of the Moon and r is the distance between their centres.

The force is one of attraction between the bodies. By Newton's 3rd law, the magnitude of the gravitational force on each body due to the other is equal but it operates in the opposite direction on each body, resulting in a force of attraction between the two bodies. Thus the force exerted by the Moon on the Earth is equal but in the opposite direction to the force exerted on the Earth by the Moon. *(2 marks)*

(b) Mass is a measure of the inertia of a body and will therefore be the same (i.e. 70 kg) on the Moon and the Earth.

Weight is a measure of the force of gravity on the body given by $w = mg$. Hence on Earth the person will have a weight of $w = mg = 70 \times 9.8 = 686$ N, while on the Moon the person's weight will be $w = mg = 70 \times 1.6 = 112$ N. *(2 marks)*

Question 34 (Total 6 marks)

(a) The end of the metre rule is not placed at the start of the motion.

This will produce a zero error in measurements of the horizontal position using the ruler.

If the zero error was not taken into account, the measured horizontal position (x) of the ball from the starting point at each time would be less than the actual distance.

OR Because the camera is offset the gradations on the metre rule will not appear to be constantly spaced due to parallax error. This will result in incorrect values being measured for the horizontal position of the particle. *(3 marks)*

(b) The graphs show that the horizontal component of the velocity remains constant, while the vertical component changes due to the local acceleration due to gravity. Motion towards the left and downwards is called negative, while upwards motion is called positive.

The horizontal velocity of the ball is constant as the first graph of horizontal displacement against time is a straight line. The horizontal acceleration is therefore zero and the gradient of the first graph gives the horizontal velocity

$= v_x = \text{gradient} = \frac{(0.3 - 1.1)}{(1.0 - 0.5)} = -1.6 \text{ ms}^{-1}$ or 1.6 ms^{-1} to the left.

The vertical velocity, as shown in the graph of vertical velocity against time, changes throughout the flight. The gradient of the line of best fit of this graph will be equal to the vertical acceleration, which is constant.

$= \text{gradient} = \frac{(-2.4 - 2.3)}{(1.0 - 0.5)} = \frac{-4.8}{0.5} = -9.4 \text{ ms}^{-2}$ or 9.4 ms^{-2} downwards. *(3 marks)*

Question 35 (Total 5 marks)

(a) The change in gravitational potential energy is given by:

$$\Delta U = U_{final} - U_{initial}$$

$$= -\frac{GMm}{r_{final}} - \left(-\frac{GMm}{r_{initial}}\right)$$

$$= GMm\left(\frac{1}{r_{initial}} - \frac{1}{r_{final}}\right)$$

$$= 6.67 \times 10^{-11} \times 7.35 \times 10^{22} \times 20\left(\frac{1}{1740 \times 10^3} - \frac{1}{2240 \times 10^3}\right) = 1.26 \times 10^7 \text{J}$$

Thus moving the 20 kg mass from the surface to 500 km above the surface increases the gravitational potential energy by 1.26×10^7 J. *(2 marks)*

(b) Because the object is in free flight, the total energy of the object is conserved and the initial kinetic energy is gradually converted to gravitational potential energy as the object moves upwards.

The kinetic energy remaining at 500 km will be given by:

$$K_{\text{final}} = K_{\text{initial}} - \Delta U$$

$$= \frac{1}{2}mv^2 - \Delta U$$

$$= \frac{1}{2}(20)(1200)^2 - 1.26 \times 10^7$$

$$= 1.8 \times 10^6 \text{ J} = 1.8 \text{ MJ}$$

Now the velocity can be calculated from the expression for kinetic energy:

$$K = \frac{1}{2}mv^2$$

and hence

$$v = \sqrt{\frac{2K}{m}}$$

$$= \sqrt{\frac{2 \times 1.8 \times 10^6}{20}}$$

$$= 424.3 \text{ ms}^{-1}$$

(3 marks)

CHAPTER 2

Module 6
Electromagnetism

Part A Objective-response questions

1 Which of the following correctly shows the electric field between two parallel, charged plates?

(A)
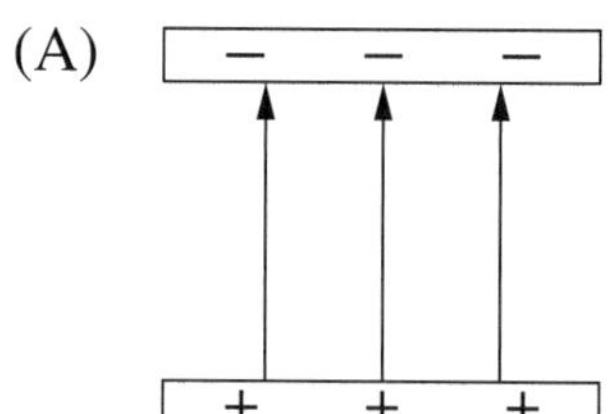

(B)
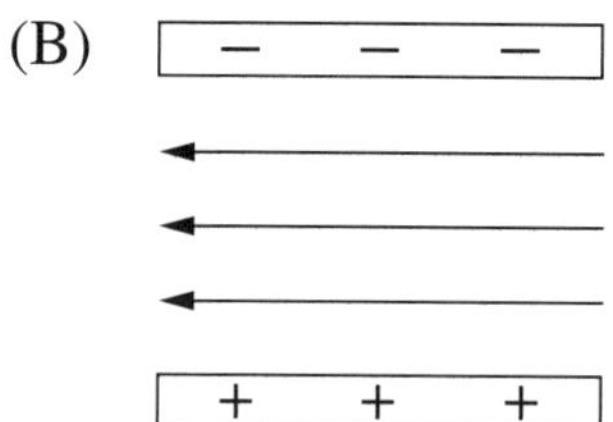

(C)
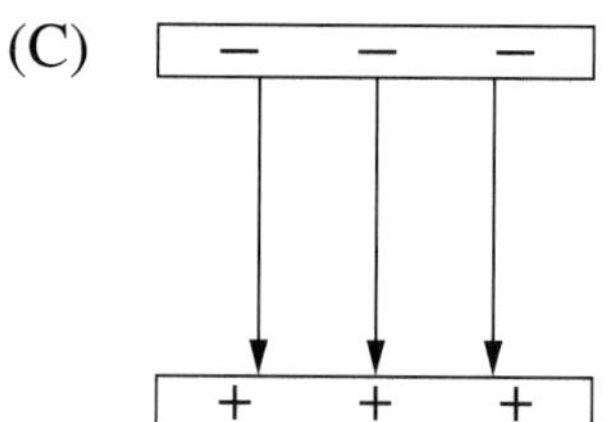

(D)
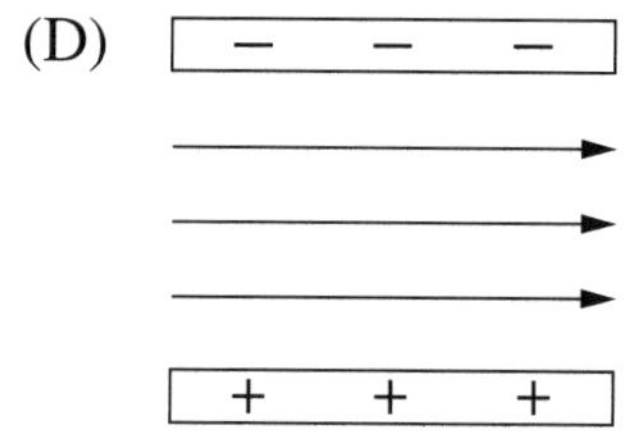

(q3, 2017 HSC)

2 A simple AC generator was connected to a cathode ray oscilloscope and the coil was rotated at a constant rate. The output is shown on this graph.

Voltage
1
0
–1
Time

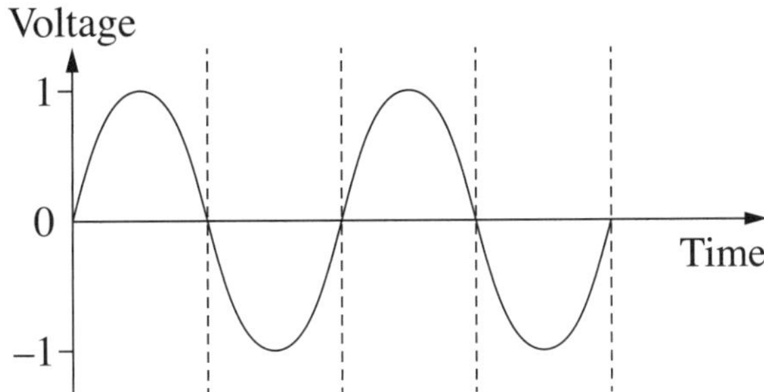

Which of the following graphs best represents the output if the rate of rotation is decreased to half of the original value?

(A)

Voltage
1
0
–1
Time

(B)

Voltage
1
0
–1
Time

(C)

Voltage
1
0
–1
Time

(D)

Voltage
1
0
–1
Time

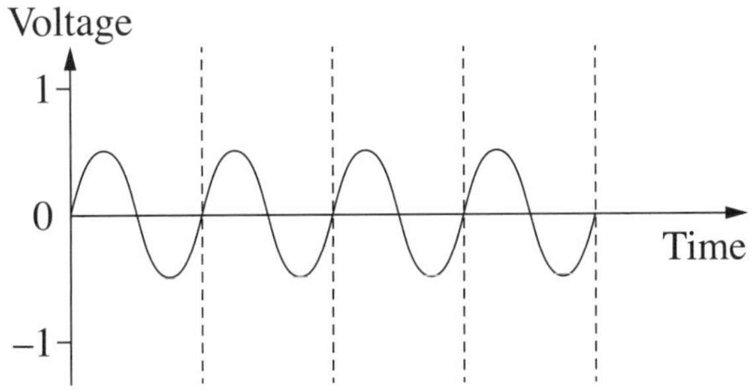

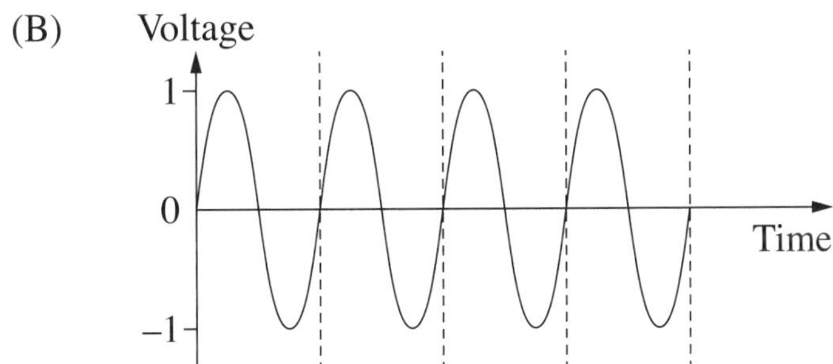

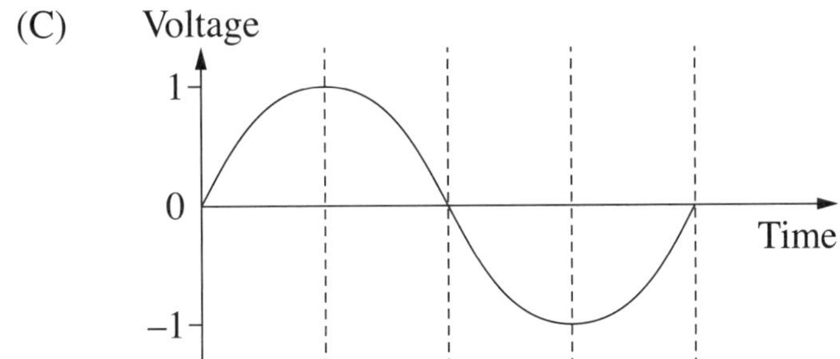

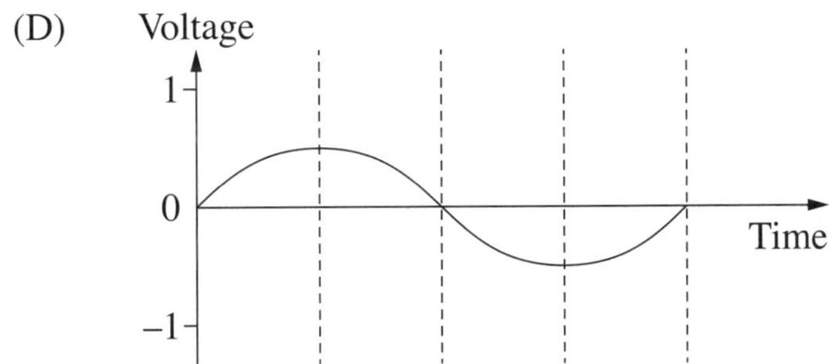

(q12, 2015 HSC)

3 A circular loop of wire is stationary in a magnetic field. The sides are then pushed together to change the shape, as shown in the diagram.

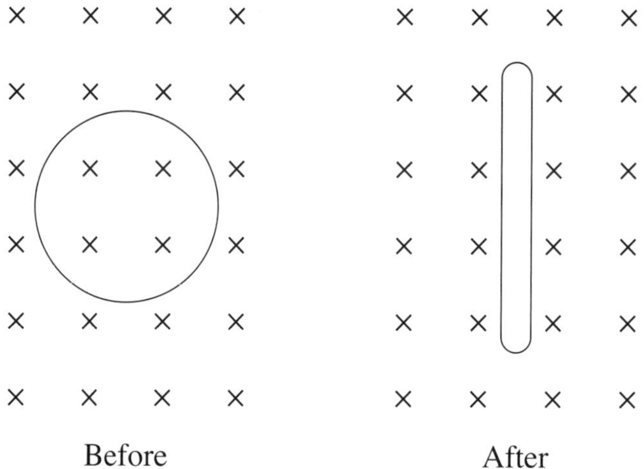

As the loop is compressed, a current is induced.

Which row of the table shows the direction of the current and explains why it is induced?

	Current direction	*Why the current is induced*
(A)	Clockwise	Change in magnetic flux
(B)	Anticlockwise	Change in magnetic flux
(C)	Clockwise	Change in magnetic flux density
(D)	Anticlockwise	Change in magnetic flux density

(q15, 2015 HSC)

4 Some mobile phones are recharged at a power point using a charger that contains a transformer.

What is the purpose of the transformer?

(A) To convert AC at the power point to DC

(B) To convert DC at the power point to AC

(C) To increase the AC voltage at the power point

(D) To decrease the AC voltage at the power point

(q1, 2016 HSC)

5 How does back emf affect a DC motor?

(A) It creates heat in the iron core.

(B) It limits the speed of the motor.

(C) It reverses the current in the coil.

(D) It increases the torque of the motor.

(q9, 2016 HSC)

6 The cone of a speaker is pushed so that the coil moves in the direction shown.

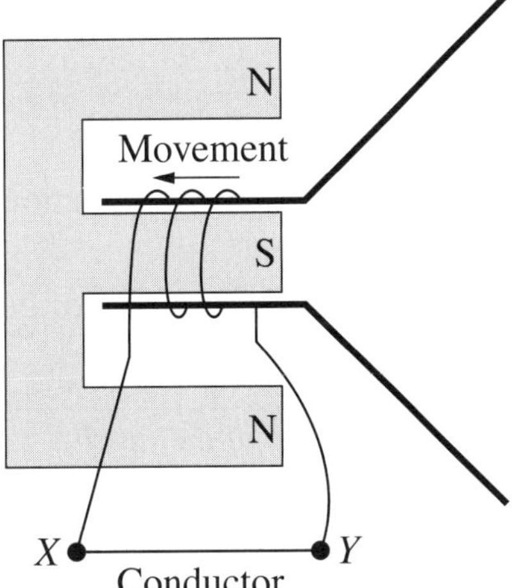

Which row of the table correctly identifies the behaviour of the speaker and the direction of the current through the conductor?

	The speaker behaves like a …	*The direction of the current is from …*
(A)	generator	*X* to *Y*
(B)	generator	*Y* to *X*
(C)	motor	*X* to *Y*
(D)	motor	*Y* to *X*

(q16, 2016 HSC)

7 The diagram shows a model of a generator connected to a galvanometer.

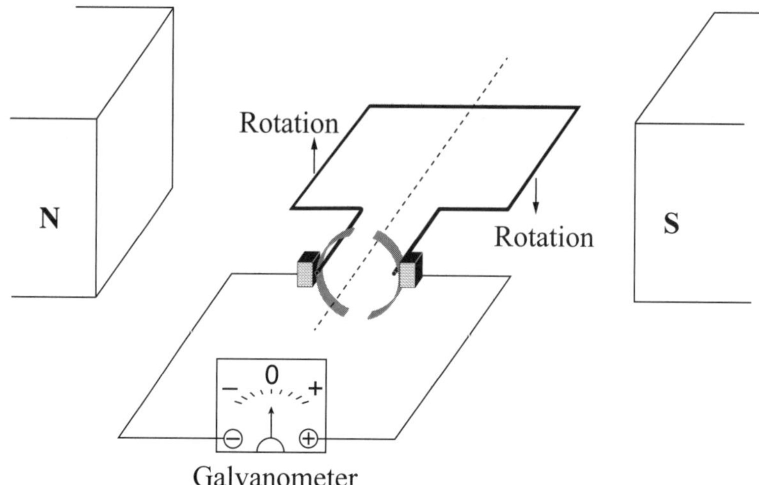

The loop is rotated continuously in a clockwise direction as viewed from the end nearest the galvanometer.

Which row of the table correctly identifies the type of generator and the movement of the needle of the galvanometer?

	Type of generator	*Movement of the needle*
(A)	DC	Swings between 0 and +
(B)	AC	Swings between – and 0
(C)	DC	Swings between + and –
(D)	AC	Swings between – and +

(q10, 2017 HSC)

8 An AC source is connected to a transformer having a primary winding of 900 turns. Connected to the secondary winding of 450 turns is a pair of parallel plates 0.01 m apart.

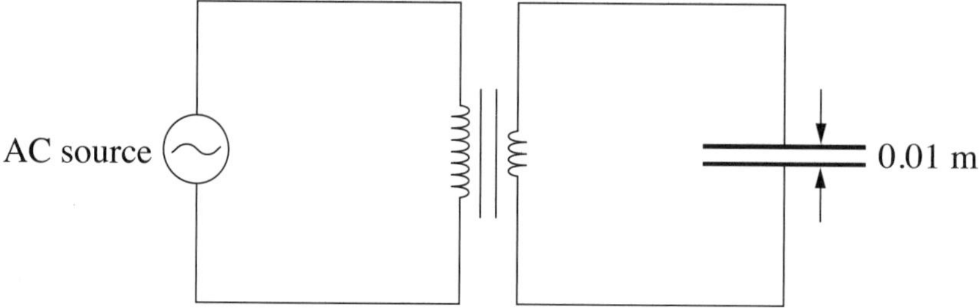

The AC input is shown in the graph.

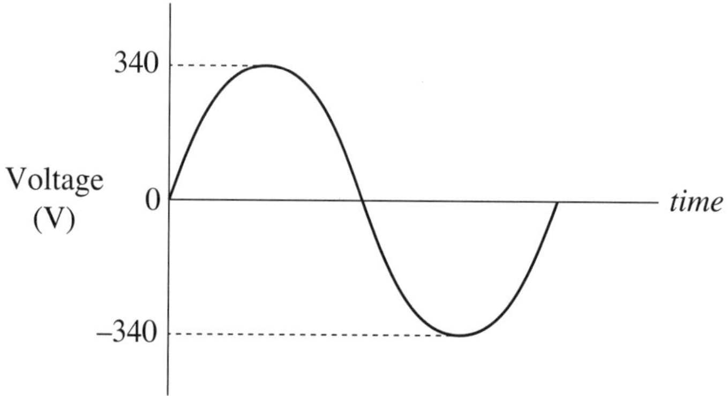

Question 8 continues on the following page

Question 8 (continued)

What is the maximum field strength (in V m^{-1}) produced between the plates?

(A) 1.7

(B) 6.8

(C) 1.7×10^4

(D) 6.8×10^4 (q11, 2017 HSC)

End of Question 8

9 A triangular piece of wire is placed in a magnetic field as shown.

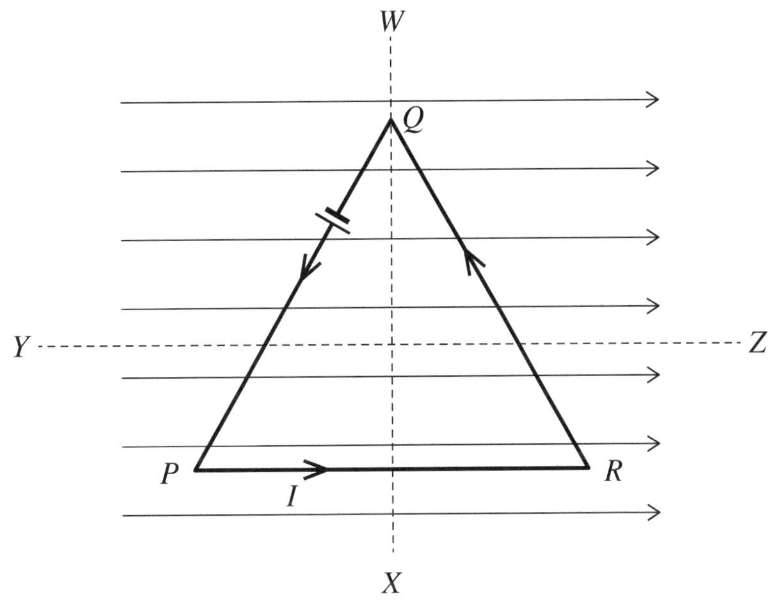

When current I is supplied as shown, how does the wire move?

	Axis of rotation	*Direction of movement*
(A)	*YZ*	*Q* into page
(B)	*YZ*	*Q* out of page
(C)	*WX*	*R* into page
(D)	*WX*	*R* out of page

(q13, 2017 HSC)

10 The diagram shows a DC circuit containing a transformer.

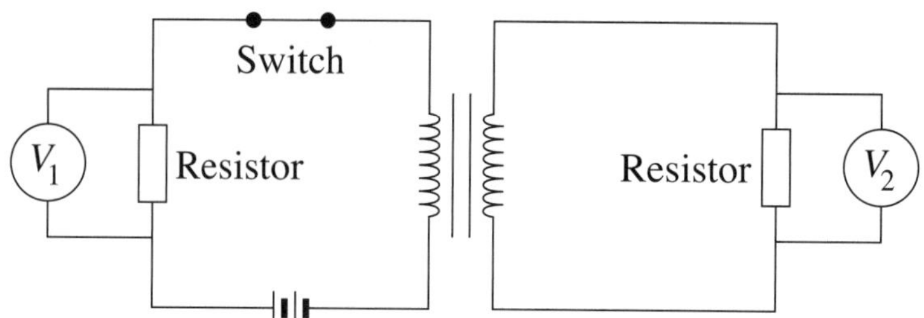

The potential differences V_1 and V_2 are measured continuously for 4 s. The switch is initially closed.

At $t = 2$ s, the switch is opened.

Which pair of graphs shows how the potential differences V_1 and V_2 vary with time over the 4-second interval?

(A) V_1 0 Time (s) 4

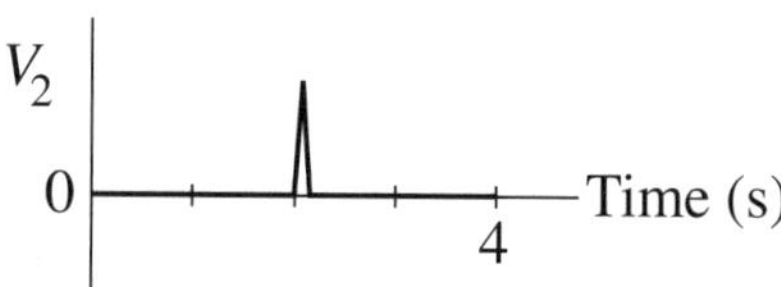

(B) V_1 0 Time (s) 4

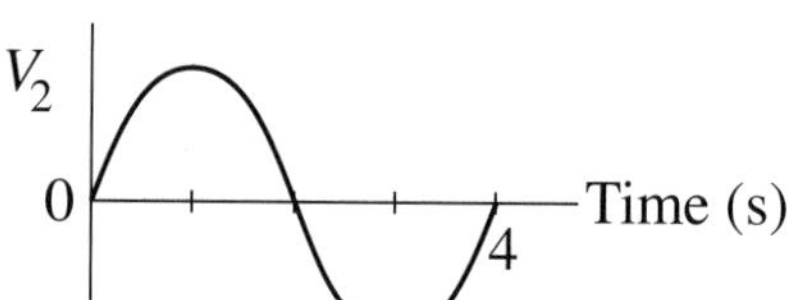

(C) V_1 0 Time (s) 4

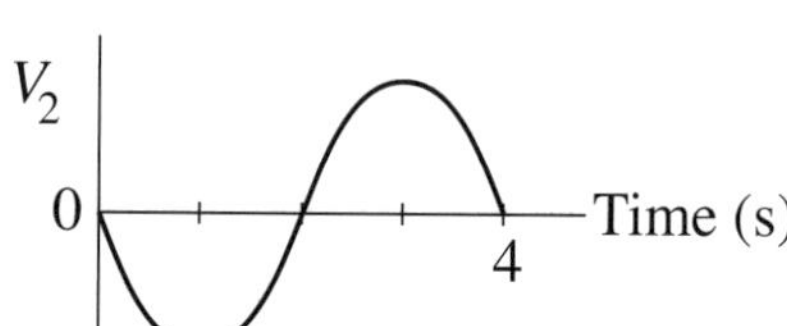

(D) V_1 0 Time (s) 4

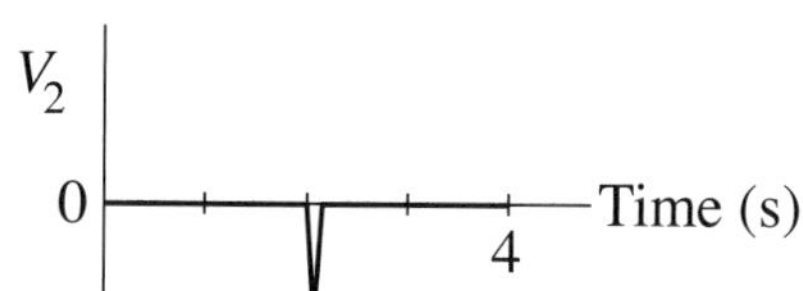

(q14, 2017 HSC)

11 An AC supply is connected to a light bulb by two long parallel conductors as shown.

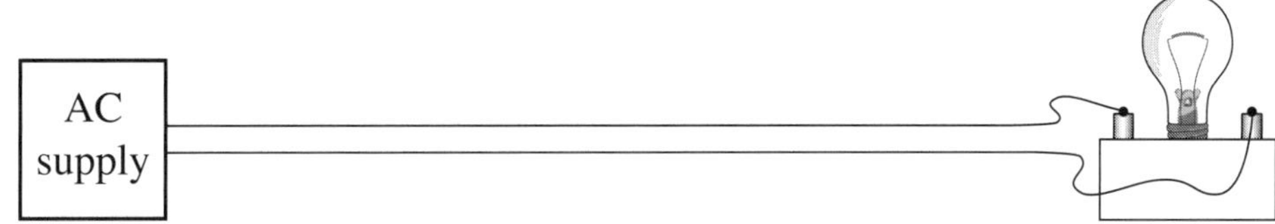

Which graph shows the variation over time of the magnetic force between the two conductors?

(A)
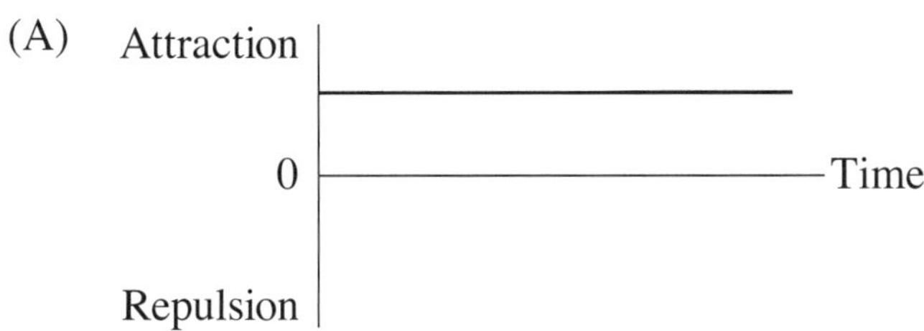

(B)
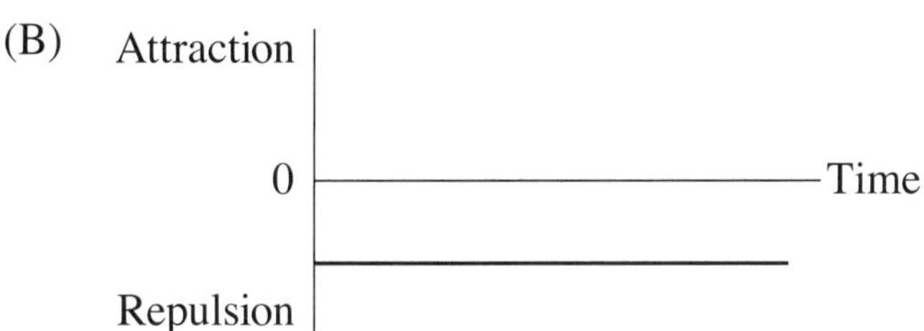

(C)
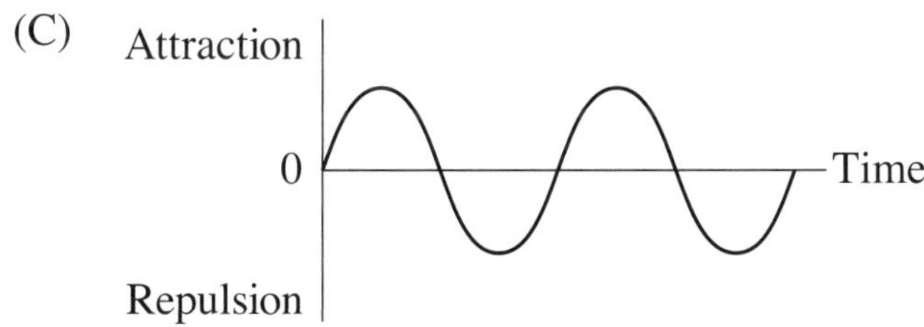

(D)
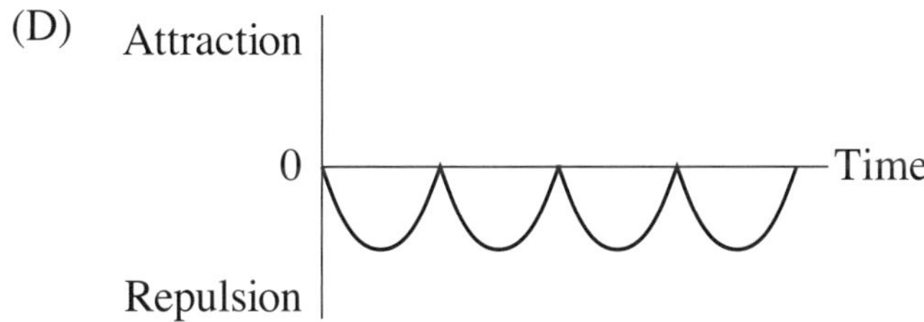

(q16, 2017 HSC)

12 A magnet rests on an electronic balance. A rigid copper rod runs horizontally through the magnet, at right angles to the magnetic field. The rod is anchored so that it cannot move.

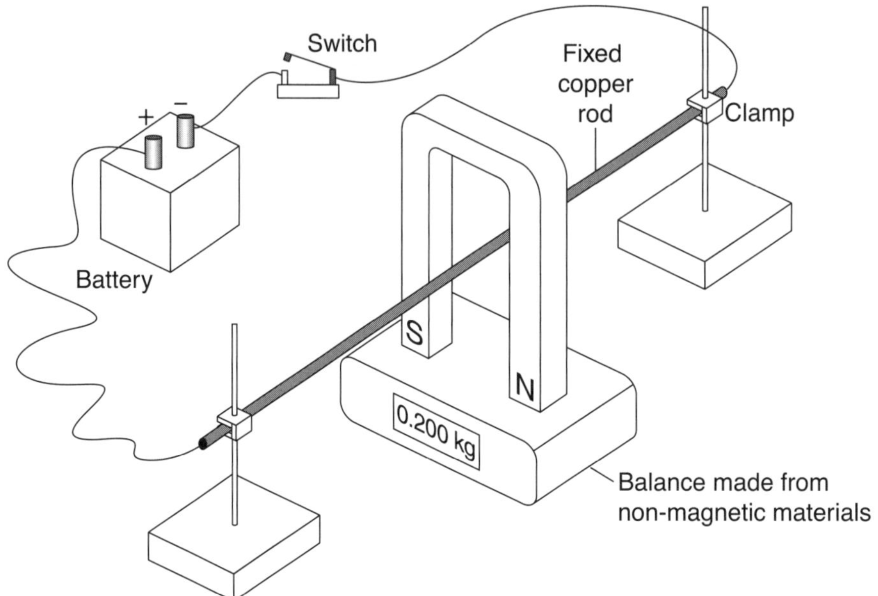

Which expression can be used to calculate the balance reading when the switch is closed?

(A) $0.200 \text{ kg} + BIl$ (B) $0.200 \text{ kg} + \dfrac{BIl}{9.8}$

(C) $0.200 \text{ kg} - BIl$ (D) $0.200 \text{ kg} - \dfrac{BIl}{9.8}$ (q17, 2017 HSC)

13 A particle of mass m and charge q travelling at velocity v enters a magnetic field of magnitude B and follows the path shown.

B

Particle

A second particle enters a magnetic field of magnitude $2B$ with a velocity of $\frac{1}{2}v$ and follows an identical path.

What is the mass and charge of the second particle?

	Mass	*Charge*
(A)	m	q
(B)	$\frac{1}{2}m$	$2q$
(C)	$4m$	q
(D)	m	$\frac{1}{2}q$

(q18, 2017 HSC)

14 Earth's magnetic field is shown in the following diagram.

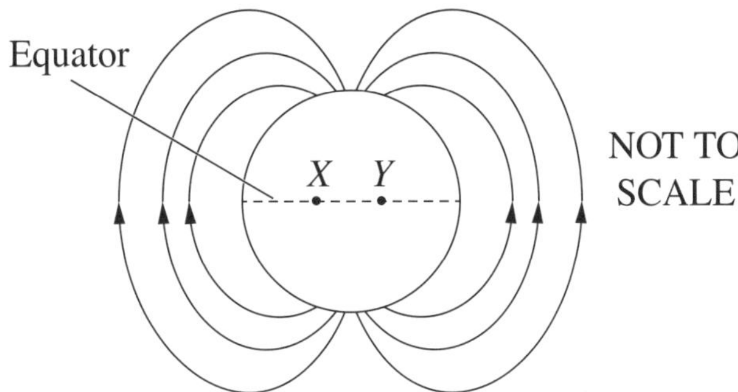

Two students standing a few metres apart on the equator at points X and Y, where Earth's magnetic field is parallel to the ground, hold a loop of copper wire between them. Part of the loop is rotated like a skipping rope as shown, while the other part remains motionless on the ground.

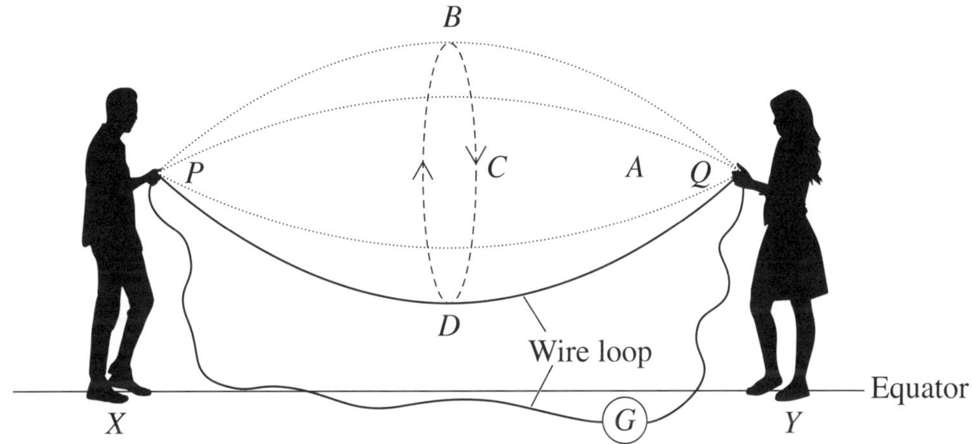

At what point during the rotation of the wire does the maximum current flow in a direction from P to Q through the moving part of the wire?

(A) A (B) B (C) C (D) D

(q19, 2017 HSC)

15 A motor, battery and ammeter are connected in series. When the motor is turning at full speed, the ammeter has a reading of 0.1 A. While the motor is spinning, a person holds the shaft of the motor to stop it.

Which row of the table correctly identifies the change in the ammeter reading and an explanation for the change?

	Reading on ammeter	*Explanation*
(A)	Decreases	Decrease in back emf
(B)	Increases	Increase in back emf
(C)	Decreases	Increase in back emf
(D)	Increases	Decrease in back emf

(q4, 2018 HSC)

16 The diagram shows a current-carrying conductor in a magnetic field.

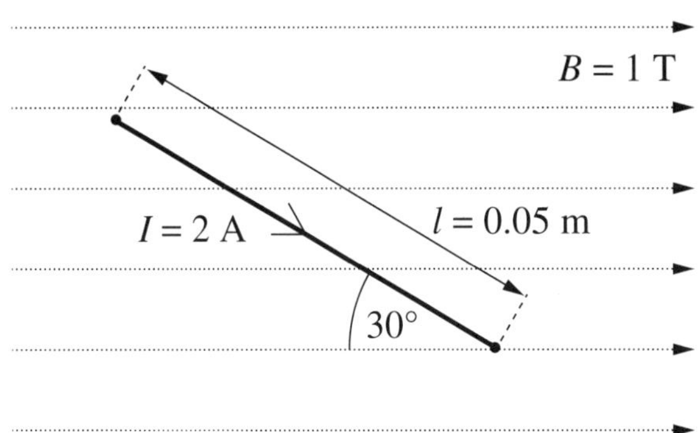

What is the magnitude of the force on the conductor?

(A) 0 N

(B) 0.05 N

(C) 0.09 N

(D) 0.10 N

(q5, 2018 HSC)

17 The diagram shows some parts of a simple DC motor.

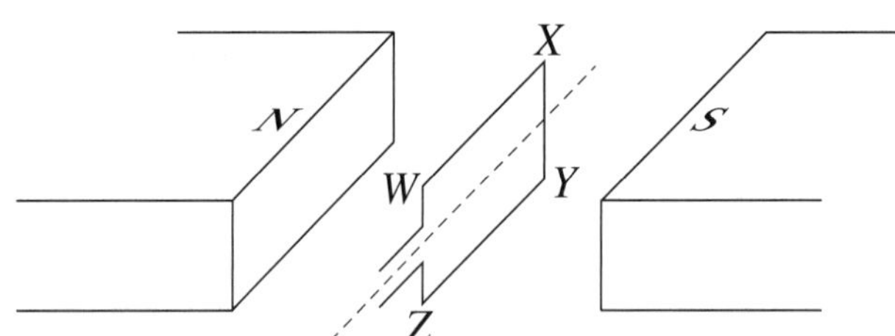

Which row of the table correctly describes the direction of the force acting on side *WX* and the direction of torque this produces on the coil?

	Direction of force acting on WX	*Direction of torque produced on the coil by the force acting on WX*
(A)	Remains constant	Remains constant
(B)	Remains constant	Reverses every 180°
(C)	Reverses every 180°	Remains constant
(D)	Reverses every 180°	Reverses every 180°

(q10, 2018 HSC)

18 The diagram shows electrons travelling in a vacuum at 2×10^6 m s^{-1} between two charged metal plates 1×10^{-3} m apart.

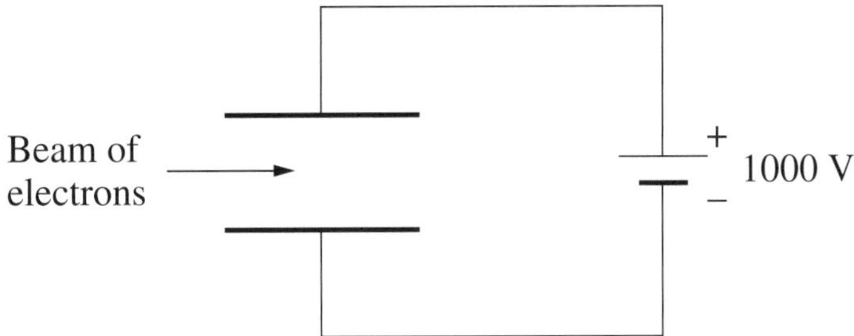

A magnetic field is to be applied to make the electrons continue to travel in a straight line.

What is the magnitude and direction of the magnetic field that is to be applied?

(A) 5×10^{-1} T into the page

(B) 5×10^{-1} T out of the page

(C) 1×10^{6} T into the page

(D) 1×10^{6} T out of the page

(q12, 2018 HSC)

19 An electron moves in a circular path with radius r in a magnetic field as shown.

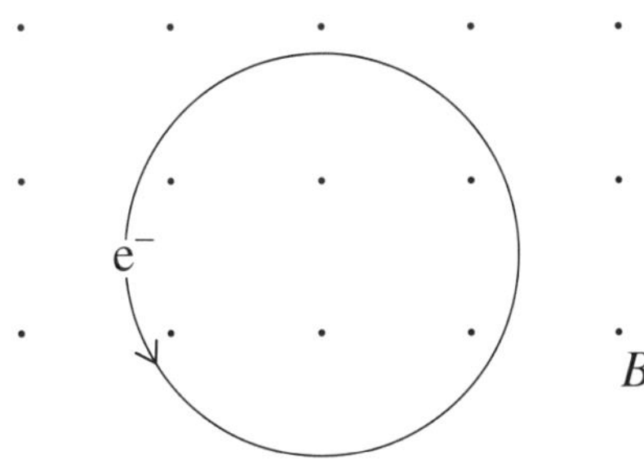

If the speed of the electron is increased, which row of the table correctly shows the effects of this change?

	Force on electron	*Radius of path*
(A)	Increases	Decreases
(B)	Increases	Increases
(C)	Decreases	Decreases
(D)	Decreases	Increases

(q13, 2018 HSC)

20 An experiment is set up as shown.

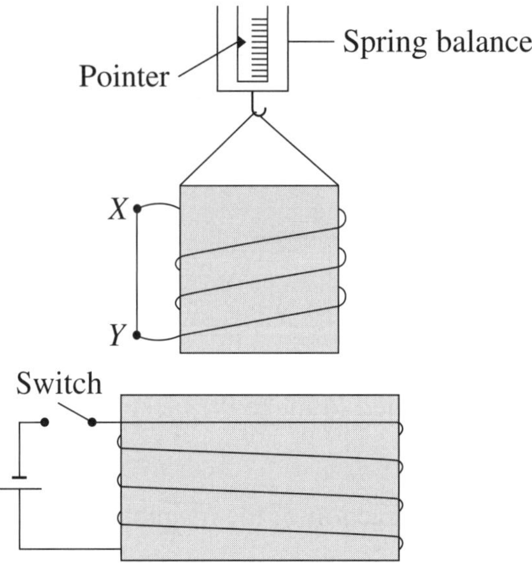

When the switch is closed, the reading on the spring balance changes immediately, then returns to the initial reading.

Which row on the table correctly shows the direction of the current through the straight conductor *XY* and the direction in which the pointer on the spring balance initially moves?

	Direction of current through the straight conductor	*Direction in which the pointer initially moves*
(A)	From *X* to *Y*	Down
(B)	From *X* to *Y*	Up
(C)	From *Y* to *X*	Down
(D)	From *Y* to *X*	Up

(q18, 2018 HSC)

Part B Short-answer questions

Question 21 (5 marks)

An electric field is produced between two charged parallel plates, *M* and *N*.

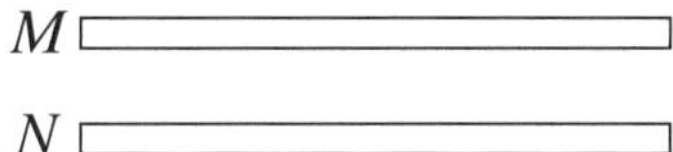

(a) The plates, *M* and *N*, are 1.0 cm apart and have an electric field of 15 V m^{-1}. **2**
Calculate the potential difference between the plates.
(4 lines)

(b) The potential difference is now changed and a magnetic field of 0.5 T is placed perpendicular to the plates, as shown in the diagram below. **3**

M
Electron
N

Determine the magnitude and direction of the electric field required to allow the electron to travel through undeflected, if the electron is moving at 1×10^4 ms^{-1}.
(8 lines) (q26, 2013 HSC)

Question 22 (5 marks)

A 0.05 kg mass is lifted at a constant speed by a DC motor. The motor has a coil of 100 turns in a 0.1 T magnetic field. The area of the coil is 0.0012 m^2. The motor shaft has a radius of 0.004 m.

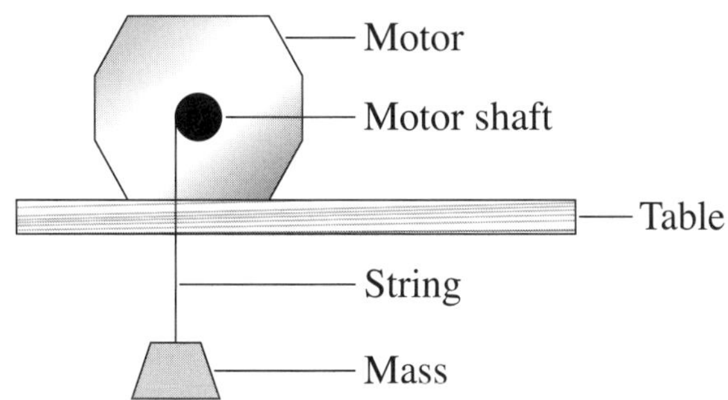

(a) Determine the force needed to lift the mass. **2**
(4 lines)

(b) Calculate the minimum current required in the coil to lift the mass. **3**
(8 lines) (q29, 2013 HSC)

Question 23 (5 marks)

The diagram shows the paths taken by two moving charged particles when they enter a region of uniform magnetic field.

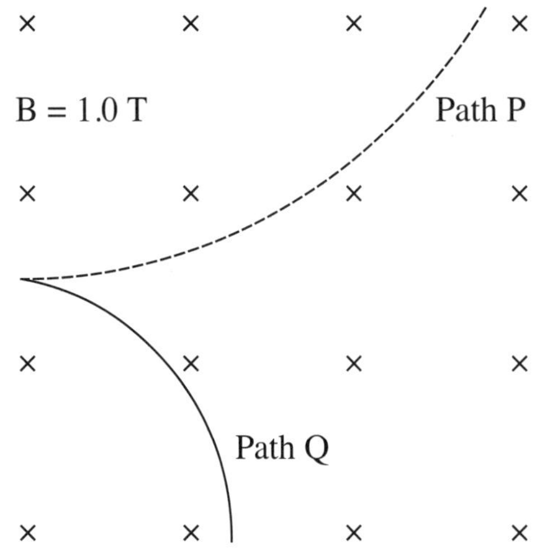

(a) Why do the paths curve in different directions? **1**

(2 lines)

(b) Why are the paths circular? **2**

(4 lines)

(c) How do the properties of a particle affect the radius of curvature of its path in a uniform magnetic field? **2**

(4 lines) (q30, 2012 HSC)

Question 24 (4 marks)

A copper rod is placed on a wooden frame, which is placed on an electronic balance. **4**

A length of 0.2 m of the rod passes at right angles to a horizontal magnetic field.

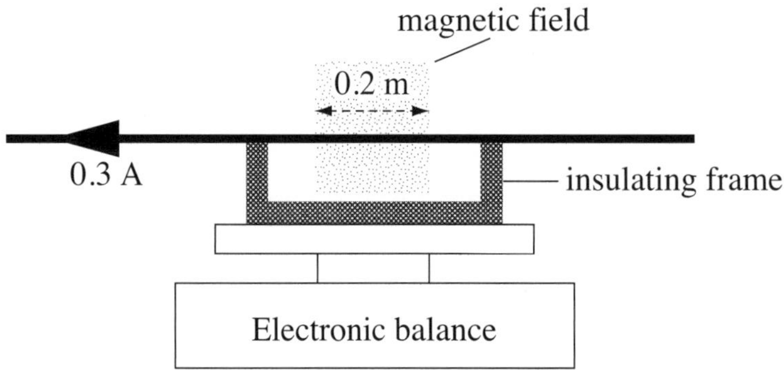

When a current of 0.3 A is passed through the rod, the reading on the balance increases by 7.5×10^{-4} kg.

What is the strength and direction of the magnetic field?

(8 lines) (q28, 2010 HSC)

Question 25 (6 marks)

An electron is emitted from a mineral sample, and travels through aperture A into a spectrometer at an angle of 60° with a speed of 6.0×10^6 ms^{-1}.

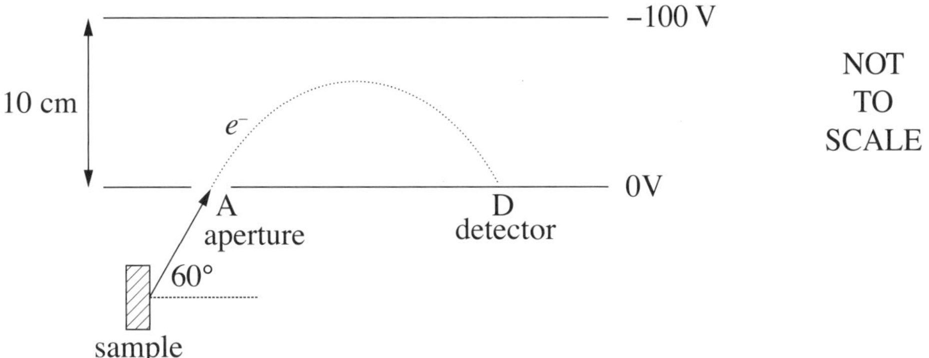

(a) Calculate the magnitude and direction of the force experienced by the electron inside the spectrometer. **3**

(6 lines)

(b) The electron experiences constant acceleration and eventually strikes the detector, D. **3**

What is the time taken for the electron to travel from A to D?

(6 lines) (q19, 2009 HSC)

Question 26 (5 marks)

In the Large Hadron Collider (LHC), the particle beams are steered using magnetic fields, as shown.

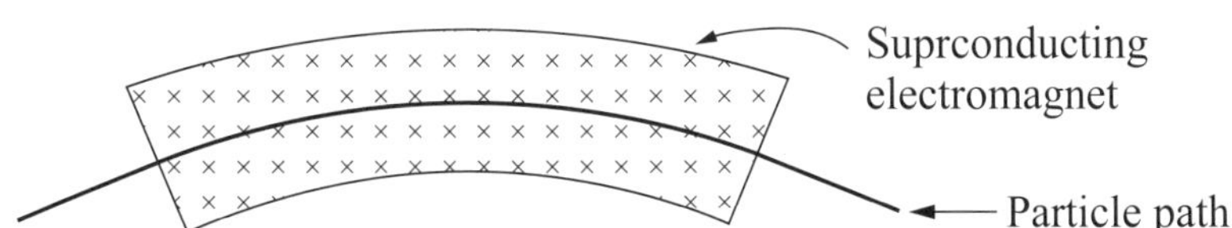

(a) Two particles with the same mass and speed are travelling through the LHC in opposite directions. **2**

What can be deduced about the charge on the particles?

(2 lines)

(b) During a test run, a proton travels with a speed of 1.0×10^7 ms^{-1} around the LHC. The radius of curvature of its path is 4.2 m. **3**

Calculate the magnetic field strength.

(6 lines) (q25, 2009 HSC)

Question 27 (5 marks)

The primary winding of a transformer contains 2000 turns. The primary AC voltage is 23 000 volts and the output voltage is 660 000 volts.

(a) Calculate the number of turns on the secondary winding. **2**

(4 lines)

(b) If the current in the primary winding of the transformer is 100 A, and the secondary winding has a resistance of 2000 Ω, what is the power loss in the secondary winding, assuming there is no power loss in the primary winding? **3**

(Show calculations.)

(6 lines)

(q24, 2014 HSC)

Question 28 (5 marks) **s**

(a) Outline an investigation that can be used to demonstrate the principle of an AC induction motor. **2**

(4 lines)

(b) Explain how the motor effect is used in an AC motor. **3**

(6 lines)

(q25, 2014 HSC)

Question 29 (5 marks)

The diagram represents a simple DC motor. A current of 1.0 A flows through a square loop *ABCD* with 5 cm sides in a magnetic field of 0.01 T.

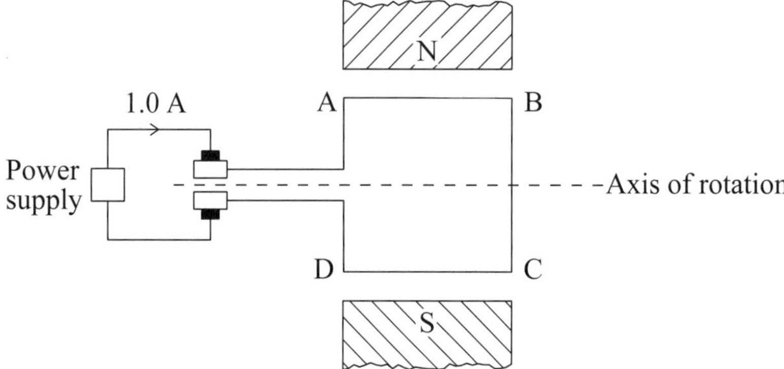

(a) Determine the force acting on section *AB* and the force acting on section *BC* due to the magnetic field, when the loop is in the position shown. **3**

(7 lines)

(b) How is the direction of the torque maintained as the loop rotates 360° from the position shown? **2**

(4 lines)

(q22, 2015 HSC)

Question 30 (7 marks)

A part of a cathode ray oscilloscope was represented on a website as shown.

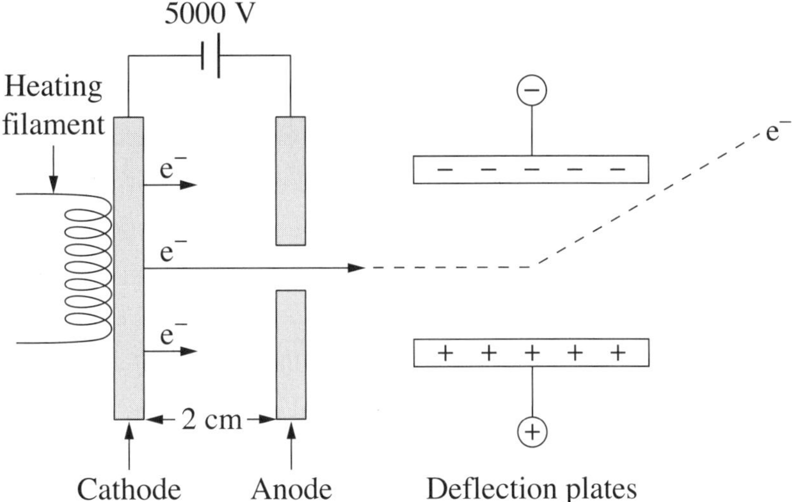

Electrons leave the cathode and are accelerated towards the anode.

(a) Explain why the representation of the path of the electron between the deflection plates is inaccurate. **3**

(7 lines)

(b) Calculate the force on an electron due to the electric field between the cathode and the anode. **2**

(6 lines)

(c) Calculate the velocity of an electron as it reaches the anode. **2**

(6 lines) (q24, 2015 HSC)

Question 31 (5 marks)

A copper plate is attached to a lightweight trolley. The trolley moves at an initial velocity, *v*, towards a strong magnet fixed to a support. **5**

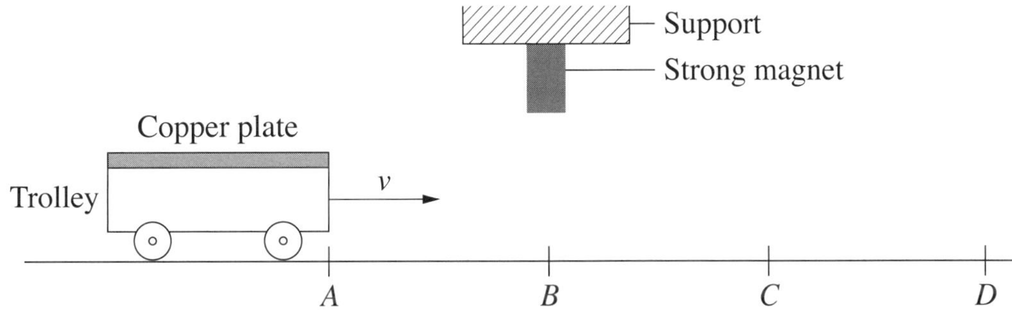

The dashed line on the graph shows the velocity of the trolley when the magnet is not present. On the axes, sketch the graph of the velocity of the trolley as it travels from *A* to *D* under the magnet, and justify your graph.

Question 31 continues on the following page

Question 31 (continued)

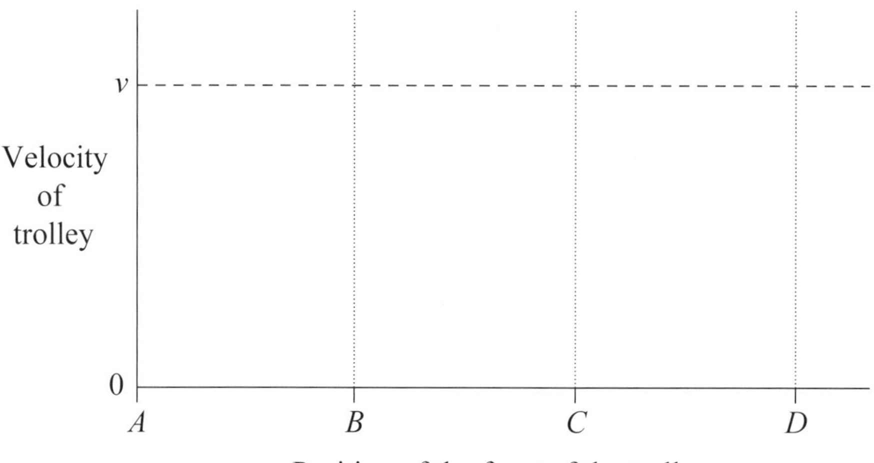

(8 lines) (q28, 2015 HSC)

End of Question 31

Question 32 (6 marks)

When an alternating current is passed through coil *A*, a voltage is observed on the oscilloscope connected to coil *B*. **2**

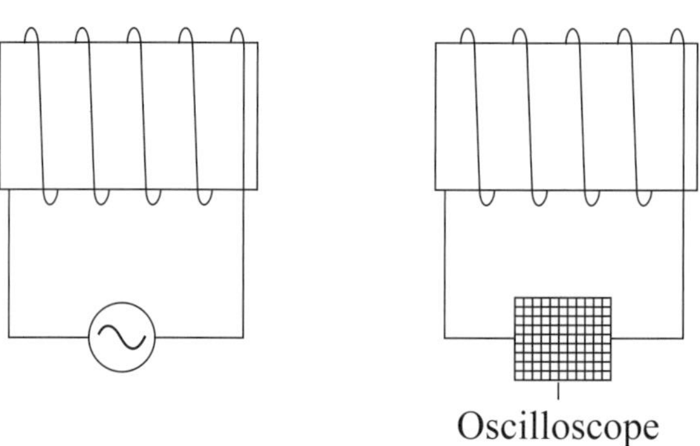

(a) How could a bar magnet be used, instead of coil *A*, to produce a similar pattern on the oscilloscope?
(3 lines)

(b) A strong magnet is at rest a few centimetres above a solid metal disc made of a non-magnetic metal. The magnet is then dropped. **4**

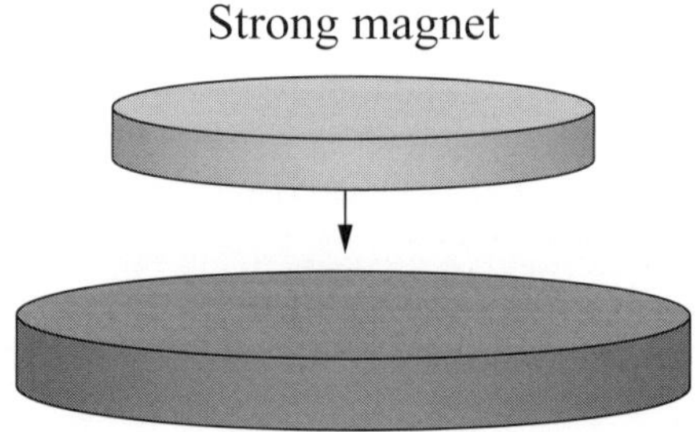

Question 32 continues on the following page

Question 32 (continued)

The velocity of the magnet is shown in this graph.

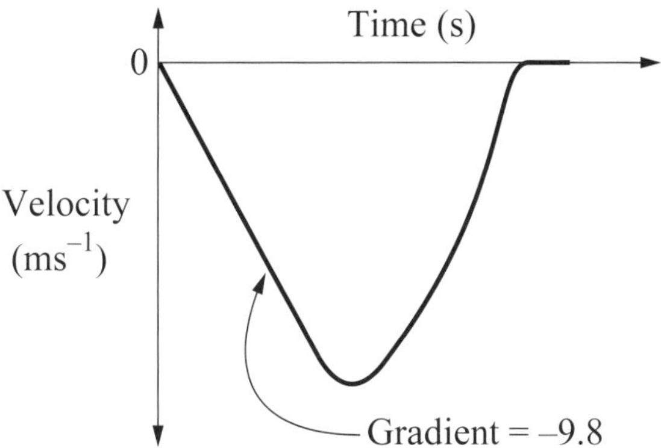

Account for the shape of the graph.

(8 lines) (q22, 2016 HSC)

End of Question 32

Question 33 (6 marks)

The following makeshift device was made to provide lighting for a stranded astronaut on Mars.

The mass of Mars is 6.39×10^{23} kg.

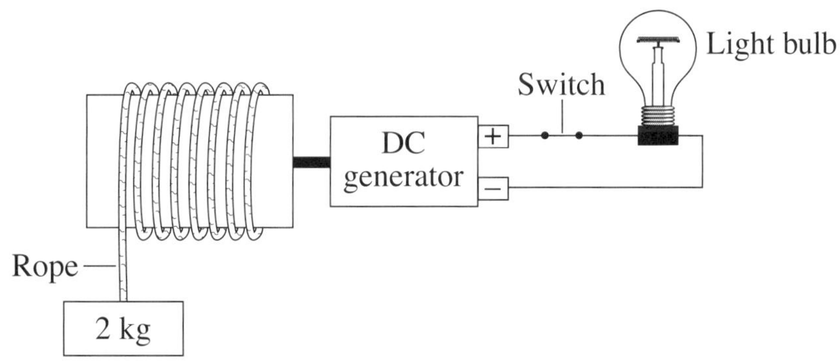

The 2 kg mass falls, turning the DC generator, which supplies energy to the light bulb. The mass falls from a point that is 3 376 204 m from the centre of Mars.

(a) Calculate the maximum possible energy released by the light bulb as the mass falls through a distance of one metre. **3**

(6 lines)

(b) Explain the difference in the behaviour of the falling mass when the switch is open. **3**

(6 lines) (q30, 2016 HSC)

Question 34 (5 marks)

(a) A torque is applied to a nut, using a wrench, as shown. **2**

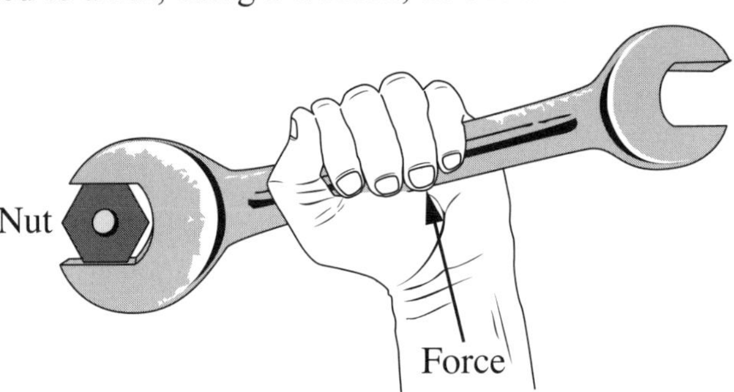

Suggest TWO ways that the applied torque could be increased.

(4 lines)

(b) A coil consisting of 15 turns is placed in a uniform 0.2 T magnetic field between two magnets. A current of 7.0 amperes flows in the direction shown. **3**

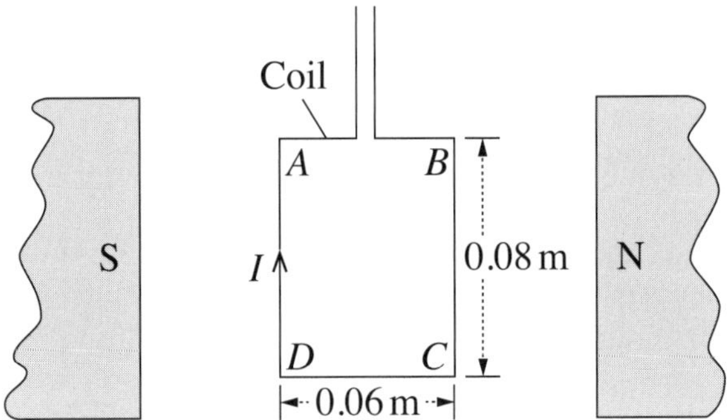

Calculate the magnitude and direction of the torque produced by the side *BC* of the 15-turn coil.

(6 lines) (q22, 2017 HSC)

Question 35 (6 marks)

Contrast the design of transformers and magnetic braking systems in terms of the effects that eddy currents have in these devices. **6**

(13 lines) (q28, 2017 HSC)

Question 36 (5 marks)

The diagram shows an electric circuit in a magnetic field directed into the page. The graph shows how the flux through the conductive loop changes over a period of 12 seconds.

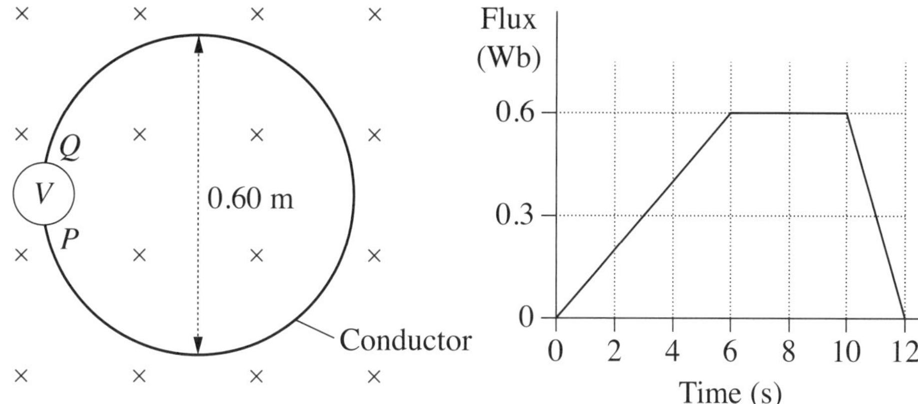

(a) Calculate the maximum magnetic field strength within the stationary loop during the 12-second interval. **2**

(4 lines)

(b) Calculate the maximum voltage generated in the circuit by the changing flux. In your answer, indicate the polarity of the terminals *P* and *Q* when this occurs. **3**

(6 lines) (q27, 2017 HSC)

Question 37 (4 marks)

In a thought experiment, a proton is travelling at a constant velocity in a vacuum with no field present. An electric field and a magnetic field are then turned on at the same time. **4**

The fields are uniform in magnitude and direction and can be considered to extend infinitely. The velocity of the proton at the instant the fields were turned on is perpendicular to the fields.

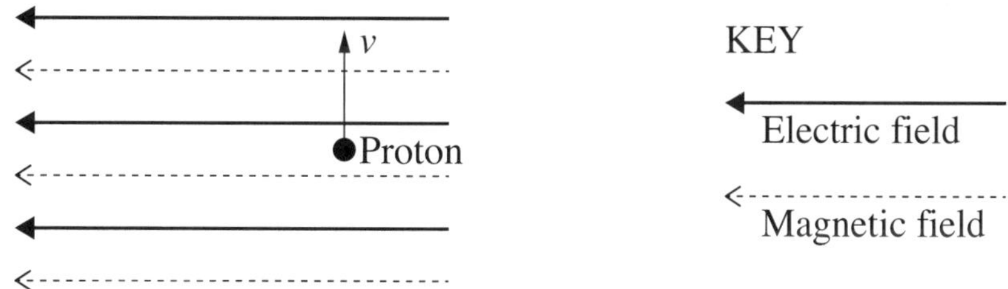

Analyse the motion of the proton after the fields have been turned on.

(8 lines) (q30, 2017 HSC)

Question 38 (6 marks)

(a) A drill spins a magnet above a non-magnetic metal disc which is free to rotate. **3**

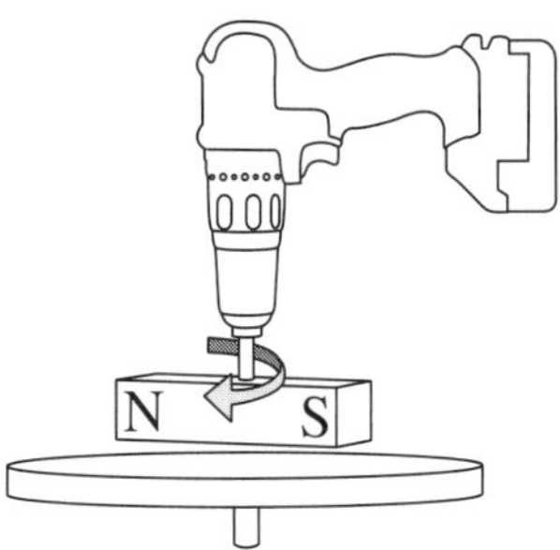

Explain the effect of the rotating magnet on the disc.

(5 lines) (q22, 2018 HSC)

(b) The diagram shows a magnet attached to an electric drill so that it can be rotated between two coils connected to a voltmeter. **3**

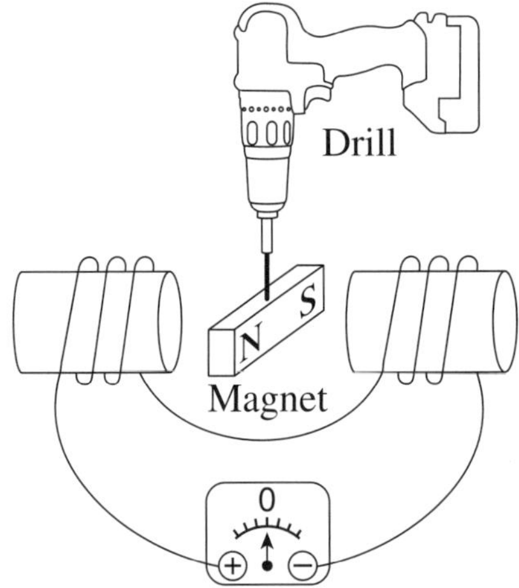

The drill starts from rest and gradually speeds up, reaching its full speed after three revolutions.

Sketch a graph showing the induced emf across the coils during the time that it takes the magnet to reach its full speed.

(q22, 2018 HSC)

Question 39 (5 marks)

(a) Explain why the eddy currents in a transformer core reduce the voltage induced in the secondary coil. **2**

(5 lines) (Sample question)

(b) Three parallel wires *X*, *Y* and *Z* all carry electric currents. A force of attraction is produced between *Y* and *Z*. There is zero net force on *Y*. **3**

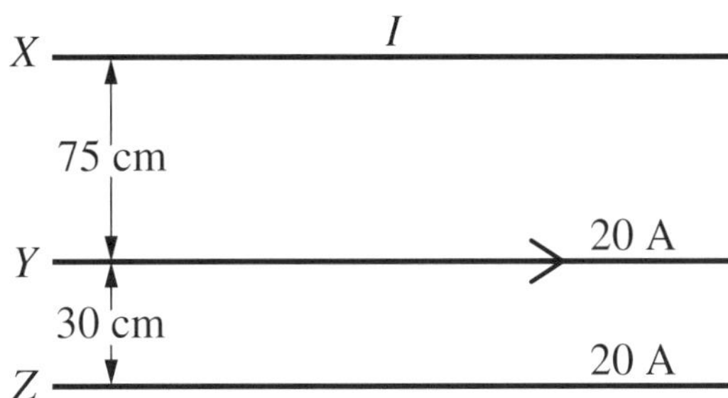

What is the magnitude and direction of the current in *X*?

(6 lines) (q24, 2018 HSC)

Question 40 (4 marks)

Outline the similarities and differences between the effects of electric fields and gravitational fields on matter. In your answer, refer to the definitions of these fields. **4**

(12 lines) (q26, 2018 HSC)

Question 41 (6 marks)

The diagram shows a model of a system used to distribute energy from a power station through transmission lines and transformers to houses. **6**

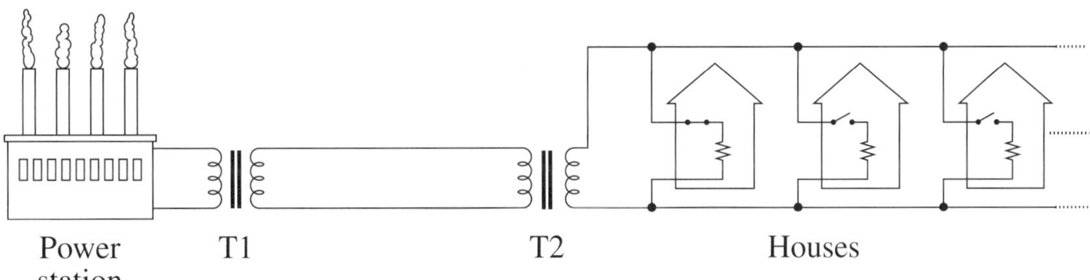

During the evening peak period there is an increase in the number of electrical appliances being turned on in houses.

Explain the effects of this increased demand on the components of the system, with reference to voltage, current and energy.

(18 lines) (q30, 2018 HSC)

Module 6
Electromagnetism

Sample answers

Part A Objective-response questions

1 A Field lines follow the direction a positive test charge would follow if placed within the field. A positive charge would move towards the negatively charged plate.

2 D If the rate of rotation of the generator is reduced, its period increases. The change in flux per rotation remains constant so as t is doubled, the induced voltage of the generator is halved.

3 A A north pole is located above the page. By Lenz's law, as the *flux* through the loop is reduced a south pole is induced above to oppose the change (i.e. clockwise emf).

4 D Although transformers can step up voltage, to charge a mobile phone does not require voltage greater than 240 volts. Transformers do not change AC to DC or vice versa.

5 B In order for the rotor in a DC motor to continue to turn, the direction of current flowing through it must reverse (or stop). Every time this occurs back-emf is induced, which creates a 'back-current'. The more rapidly the motor is rotating the more frequently the change in flux, therefore the greater the back-emf opposing the applied voltage. Once the two are equal the motor will no longer turn any faster—its rotational speed is limited.

6 A At first glance this question has two possible answers. The question reads, 'The cone of a speaker is pushed ...' as it is—by the magnetic field's action on the current-carrying coil, in which case the answer would be D since it experiences the motor effect (as identified in the syllabus). Equally, it might have been pushed by (say) a finger—in which case the answer is A as in this case, because no DC or AC source is connected between points X and Y. Had the question included '… by a finger …', or had a finger been included in the diagram provided, errors due to misinterpretation, rather than incorrect physics, might have been avoided.

7 A The generator in the diagram clearly shows a split ring commutator, making this a DC generator. In a DC generator the current only flows in one direction so the galvanometer will only record current flowing from 0 and +.

8 C In the transformer, the maximum input voltage is 340 V. The maximum voltage across the parallel plates is therefore from

$$\frac{V_P}{V_S} = \frac{n_P}{n_S}$$

$$\frac{340}{900} = \frac{Vs}{450}$$

Max. $Vs = 170$ V

So from $E = \dfrac{v}{d}$

Max. field strength $= \dfrac{170}{0.01} = 1.7 \times 10^4$ V m^{-1}

9 C Using right-hand push rule (or Fleming's left-hand rule) the direction of the force on *PQ* is out of the page, on *QR* it is into the page, and there is no force on *QP*. This will result in rotation about *WX* with *R* going into the page initially.

10 A The switch is initially closed meaning there will be a DC voltage across the resistor. When the switch is opened at 2 seconds there will no longer be a voltage, so the correct answer is A or B. There will only be a voltage across the secondary resistor when there is a changing voltage in the primary circuit. The opening of the switch in the primary circuit causes a momentary change in voltage, resulting in a brief spike of voltage in the secondary circuit.

11 D Assuming a single phase of AC passes through the wire, it will always be travelling at opposite directions along the parallel wires. This will result in a repulsive force that is intermittent as the current momentarily stops as it changes direction.

12 B When the switch is closed, the copper rod will experience a force of *BIl* upwards. However, because the rod is fixed in position the magnet will experience an equal and opposite force downward. Since the balance is calibrated for a gravitational field strength of 9.8 ms^{-2}, the reading on the balance will increase by the amount shown in answer B.

13 C The magnetic force on the particle creates a circular path. So $\dfrac{mv^2}{r} = qvB\sin\theta$.

This can be simplified to $\dfrac{mv}{r} = qB$ as $\theta = 90^0$. Or $qrB = mv$.

For r to remain the same as B is doubled, and velocity is halved, $qr2B = m\dfrac{v}{2}$

$\therefore 4qrB = mv$

$\therefore \dfrac{4rB}{v} = \dfrac{m}{q}$

$\dfrac{m}{q} = 4$ since values for r, B and v are constant.

Answer C gives this option.

14 C The direction of the magnetic field in the lower diagram will be into the page (from the information in the upper diagram). For maximum current flow the rate of change of flux needs to be a maximum. This occurs at *A* or *C* as the wire is moving perpendicularly to the field. From Lenz's law the direction of the induced current will oppose the force that caused it. At *C* the wire is moving down so the induced current will create an upwards force, meaning the induced current will flow from *P* to *Q* (from right-hand push rule or Fleming's left-hand rule).

15 D Back emf is an induced voltage that increases with the speed of rotation of the motor. If the motor is slowed down, the back emf decreases, which will increase net emf across the rotor coil and the current flowing in the coil. Hence holding the motor will decrease the back emf but increase the current flowing in the coil.

16 B The magnitude of the force on the coil is given by
$F = lIB\sin\theta = 0.05 \times 2 \times 1 \times \sin 30° = 0.05$ N.

17 C Because this is a DC motor its commutator will reverse the current direction on the side WX each half cycle, which will reverse the direction of the force on WX each half cycle. This change in the direction of the force ensures the torque on the coil remains in a constant direction.

18 A The electric field between the plates is $E = \frac{V}{d} = \frac{1000}{1 \times 10^{-3}} = 1 \times 10^6$ Vm^{-1}.
To travel straight through, the net force must be zero and, hence,
$qvB = qE$ or $B = \frac{E}{v} = \frac{1 \times 10^6}{2 \times 10^6} = 0.5$ T.
By the right-hand palm rule, the magnetic field would have to be into the page to produce a downwards force on the negative electron. This downwards force would balance the upwards force on the electron produced by the electric field.

19 B As $F = qvB$, increasing the speed will increase the force on the electron. Now as the centripetal force is produced by the force on the charge due to its motion in the magnetic field we can write $qvB = \frac{mv^2}{r}$ or $r = \frac{mv}{qB}$. Hence increasing the speed will also increase the radius (r).

20 B Switching the current on produces an increasing magnetic field downwards. By Lenz's law the induced current in the upper coil will oppose this change by producing an increasing magnetic field upwards. An upwards magnetic field would imply the current moves from XY by the right-hand grip rule. As the fields are in opposite directions, the coils will repel one another, decreasing the apparent weight of the top coil.

Part B Short-answer questions

Question 21 (Total 5 marks)

(a) $E = \frac{V}{D}$ $\therefore 15 = \frac{V}{1.0 \times 10^{-2}}$ $\therefore V = 0.15$ V *(2 marks)*

(b) The magnetic force acting on the moving electron is given by $FB = Bqv\sin\theta$. In this case θ is 90°, so $F_B = Bqv$. The electric force acting on the same moving electron is given by $F_e = qE$. For the electron to pass through the combined fields without deflection $F_B = F_e$. Hence $Bqv = qE$. Since the charge on the electron, q, features on both sides of the formula, it can be ignored.

$Bv = E$ $\therefore 0.5 \times (1 \times 10^4) = E$ $\therefore E = 5000$ Vm^{-1}

In order to counteract the magnetic field, which would tend to force the electron to bend ↓ (towards the bottom of the page), the electric field would need to be directed towards the bottom of the page ↓ so as to apply a force in the reverse direction on the negative electron. Hence plate M is positive. *(3 marks)*

Question 22 (Total 5 marks)

(a) The mass is lifted at a uniform rate, so the force applied upwards by the motor is equal to the force downwards, the weight of the mass: $F_{UP} = mg = 0.05 \times 9.8 = 0.49$ N ↑.
(2 marks)

(b) The torque required by the motor in order to raise the mass is found using $\tau = Fd$.
In this case the torque required is $\tau = 0.49 \times 0.004 = 1.96 \times 10^{-3}$ Nm. The same torque must be supplied by the motor to raise the mass, so the minimum value of the torque of the DC motor must also be 1.96×10^{-3} Nm, assuming friction is negligible.
If it is assumed that the DC motor has a radial magnetic field, so the $\cos\theta$ factor can be ignored, then the formula for the torque becomes $\tau = BIAn$.

$\therefore BIAn = 1.96 \times 10^{-3}$ Nm $\therefore 0.1 \times I \times 0.0012 \times 100 = 1.96 \times 10^{-3}$ so $I = 0.16$ A
(3 marks)

Question 23 (Total 5 marks)

(a) The particles have opposite electric charges, with *P* positively charged and *Q* negatively charged. *(1 mark)*

(b) The paths are circular because the magnetic force acts perpendicular to the initial velocity of the particles and as such it is a centripetal force which produces a circular path. The fact that the circular paths are uniform shows that the particles are experiencing a constant centripetal force with no resistance to their motion (travelling in a vacuum) and they are moving at a constant speed. *(2 marks)*

(c) The magnetic force acting on the charged particle is given by $F = Bqv$. This creates a centripetal force i.e. $F_c = \frac{mv^2}{r}$. Therefore, $Bqv = \frac{mv^2}{r}$. Rearranging gives $r = \frac{mv}{Bq}$.

As can be seen from the equation, the radius of the curving path is proportional to the mass and velocity of the particles, and inversely proportional to the charge of the particles and the strength of the external magnetic field. *(2 marks)*

Question 24 (Total 4 marks)

The balance is recording mass. When a current flows, mass increases by 7.5×10^{-4} kg. This represents an increase in the downward force of $7.5 \times 10^{-4} \times 9.8 = 7.35 \times 10^{-3}$ N.

Now this increase is due to a magnetic force that acts down in the same direction as gravity. The magnetic force is given by $FB = BIl$, i.e. $7.35 \times 10^{-3} = B \times 0.3 \times 0.2$.

Therefore $B = 0.1225$ T vertically into the page, i.e. the magnetic field created by the current-carrying wire interacts with the external magnetic field above the wire to create a high field strength that results in a force down the page. This means the external magnetic field is acting INTO the page, i.e. $B = 122.5$ mT vertically into the page. *(4 marks)*

Question 25 (Total 6 marks)

(a) $F = qE$ and $E = \dfrac{V}{D}$.

Therefore $F = 1.602 \times 10^{-19} \times \left(\dfrac{100}{0.10}\right) = 1.602 \times 10^{-16}$ N to bottom of page.

Force on electron = 1.602×10^{-16} to the bottom of page (towards plate at 0 V).

(3 marks)

(b) $a = \dfrac{F}{m} = \dfrac{1.602 \times 10^{-16}}{9.109 \times 10^{-31}} = 1.7587 \times 10^{14}$ ms^{-2} downwards.

u_v and v_v, are equal and opposite = $6.0 \times 10^6 \sin 60° = 5.196 \times 10^6$ ms^{-1}, and a_v is in the same direction as v_v.

Now $v_v = u_v + a_v t$, therefore $t = \dfrac{-5.196 \times 10^6 - 5.196 \times 10^6}{-1.7587 \times 10^{14}}$

$= 5.91 \times 10^{-8}$ seconds. *(3 marks)*

Question 26 (Total 5 marks)

(a) Given that their paths are identical and the fact that the particles have the same mass and are travelling at the same speed in opposite directions, this means the particles must have an EQUAL but OPPOSITE charge. *(2 marks)*

(b) Force magnetic = Centripetal force, i.e.

$$Bqv = \frac{mv^2}{r}$$

$$\text{Therefore } B = \frac{mv}{qr}$$

$$= \frac{1.673 \times 10^{-27} \times 1.0 \times 10^7}{1.602 \times 10^{-19} \times 4.2}$$

$$= 0.0249$$

Magnetic field strength = 25 mT. *(3 marks)*

Question 27 (Total 5 marks)

(a) $\dfrac{n_P}{n_S} = \dfrac{V_P}{V_S} \therefore \dfrac{2000}{n_S} \quad \dfrac{23\,000}{660\,000} \quad \therefore n_S = 57\,391$ (57 000) turns *(2 marks)*

(b) According to the principle of conservation of energy, the power input is equal to the total power output (useful power plus power loss) $P_{IN} = P_{OUT}$

$P = VI$, so $V_P I_P = V_S I_S \therefore 23\,000 \times 100 = 660\,000 \times I_S \therefore I_S = 3.485$ A

$P_{LOSS} = I_S^2 R_S = 3.485^2 \times 2000 = 24\,290$ J (24 kW) *(3 marks)*

Question 28 (Total 5 marks)

(a)

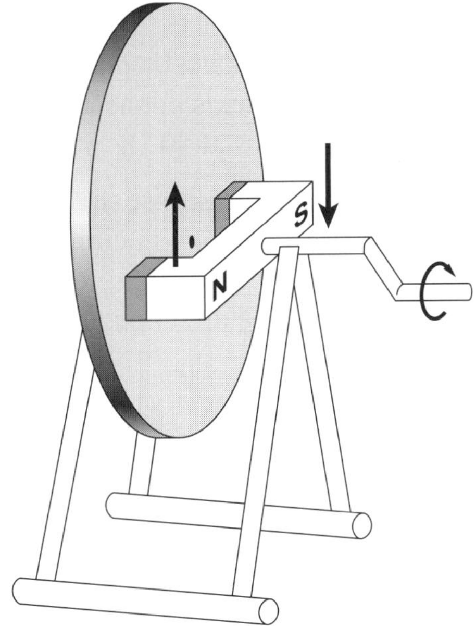

This device is called an Aragó disc. A circular plate made of aluminium or another non-magnetic metal is free to rotate on a crank handle. A long horseshoe magnet fixed to the crank handle is forced to rotate as the handle is turned, so the magnetic poles at opposite ends of the magnet move across the disc's surface. This changes the magnetic flux penetrating it, generating an emf opposing this change, in accordance with Lenz's law. The plate is a bulk metal, so eddy currents occur in the disc. As long as the crank handle is turned, the disc is forced to turn in the same direction. This demonstration shows that a rotating magnetic field causes a conductor to turn in the same direction without direct contact or supplied current to the rotor, which is the principle of the AC induction motor. *(2 marks)*

(b)

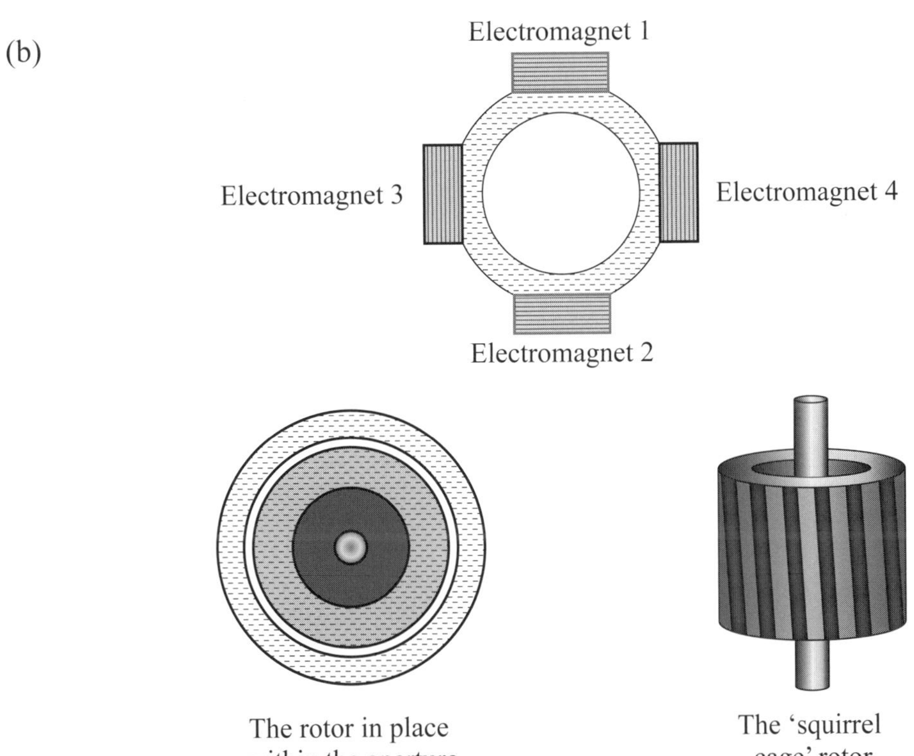

The rotor in place within the aperture

The 'squirrel cage' rotor

The motor effect states that a current-carrying conductor within a magnetic field experiences a force.

In an AC induction motor, alternating current is connected to electromagnets 1 and 2, and a second alternating current, 90° out of phase, is connected to electromagnets 3 and 4. This results in a magnetic field that rotates at 50 Hz within the aperture.

The rotor is designed to have low resistance, to maximise the currents induced within it when it is located within the aperture. Currents flow in circuits created by bars of the squirrel cage and conducting top and bottom plates. As demonstrated by the Aragó disc, the rotor is then forced to rotate in the same direction as the rotating magnetic field created by the electromagnets. *(3 marks)*

Question 29 (Total 5 marks)

(a) The formula for the force acting on a straight current-carrying conductor within a uniform magnetic field is $F_B = BIl \sin\theta$, where θ is the angle between the directions of the wire and the magnetic field.

Wire *AB*: $F_b = BIl \sin\theta = 0.01 \times 1.0 \times (5 \times 10^{-2}) \sin 90° = 5 \times 10^{-4}$ N directed into the page

Wire *BC*: $F_b = BIl \sin\theta = 0.01 \times 1.0 \times (5 \times 10^{-2}) \sin 0° \times 0$ N *(3 marks)*

(b) The situation shown in the diagram is the point of maximum torque on the coil. When it has rotated through 90° the torque drops to zero, and beyond 90° its direction is reversed. In order to maintain the direction of the torque so the motor actually functions, the current direction through loop *ABCD* is reversed. This is achieved using a split-ring commutator and brushes: as the coil rotates, the section of the split-ring attached to wire *AB* turns as well, until having rotated through 180° it comes in contact with the other brush (i.e. is at an angle 270° from the original position). The opposite happens with the section of the split-ring attached to wire *DC*. With torque in the same direction, the loop returns to its initial position. The current again changes direction when the split in the commutator passes under the brushes. *(2 marks)*

Question 30 (Total 7 marks)

(a) There are two problems associated with the given diagram.

The most obvious is that it shows the electrons being deflected towards the negative plate instead of away from it. Being negatively charged, electrons experience a force within an electric field that is applied in the direction opposite to that of the field.

The second error is that the deflection should be parabolic, identical to that of a projectile within a gravitational field, and not a path in the form of two straight lines with a sharp angle as shown. *(3 marks)*

(b) The electric field strength between the cathode and the anode is given by $E = \frac{V}{d}$.

Here, $E = \frac{V}{d} = \frac{5000}{2 \times 10^{-2}} = 2.5 \times 10^{5}\ \text{V m}^{-1}$. This field is directed towards the cathode.

The force on an electron within the electric field is determined by $F = qE$.

In this case the force is $(1.602 \times 10^{-19}) \times (2.5 \times 10^{5}) = 4.0 \times 10^{-14}$ N (towards the anode). *(2 marks)*

(c) The initial velocity (hence kinetic energy) of the electron on the cathode is considered to be effectively zero. The acceleration of an electron within a uniform field is found by:

$F = ma \therefore 4.0 \times 10^{-14} = 9.109 \times 10^{-31}a \therefore a = 4.38 \times 10^{16}\ \text{ms}^{-2}$

$v^2 = u^2 + 2as \therefore v^2 = 0 + 2 \times (4.38 \times 10^{16}) \times (2 \times 10^{-2})$

$\therefore v = 4.2 \times 10^7\ \text{ms}^{-1}$ towards the anode. *(2 marks)*

(Note: although it is obvious, since the question asks for the velocity of the electron and not its speed, students should guarantee their full marks by including its direction.)

Question 31 (Total 5 marks)

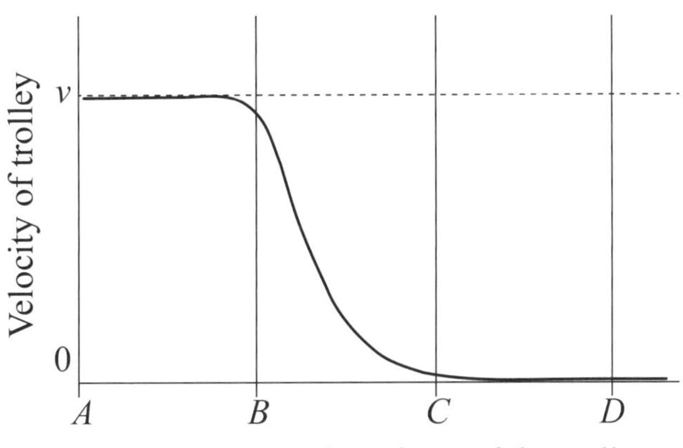

The trolley's velocity remains constant from point *A* until the trolley is close enough to point *B* for the strong magnet to begin to affect the copper plate—any form of friction is negligible. As the copper plate penetrates the magnetic field close to point *B*, an emf is generated within it. In accordance with Lenz's law, this emf opposes the motion of the trolley, causing it to lose kinetic energy which is converted into eddy currents and hence heat. As the trolley slows, the rate of change in magnetic flux reduces, so the rate of deceleration of the trolley also decreases. As the front of the trolley reaches point *C*, its rear continues passing through the magnetic field of the strong magnet. Although the direction of the eddy currents within the copper plate has reversed, they are still converting the kinetic energy of the trolley into heat, slowing it down, until at last the field is left behind. Its speed would then remain constant if friction really was negligible. *(5 marks)*

Question 32 (Total 6 marks)

(a) The pattern observed on the oscilloscope when AC is switched on in coil *A* on the left would be a regular sinusoidal wave pattern having the identical frequency as the AC input into coil *A*. (Its amplitude depends upon the number of loops in each coil, which appear to be equal.)

If coil *A* is replaced by a bar magnet, a similar pattern would appear on the oscilloscope if either pole of the bar magnet were moved rapidly back and forth towards either end of coil *B*. The pattern would again appear wave-like. The frequency would depend on the rate of back-and-forth motion of the magnet. Amplitude would be determined by the strength of the magnet, the rate of motion and the number of coils. *(2 marks)*

(b) At time $t = 0$ s the magnet falls, having an acceleration vertically downwards of 9.8 m s^{-2}. The graph shows that its downward acceleration decreases rapidly to zero when it is about halfway down. It then experiences a slowing descent until it stops as indicated by the slope of the graph changing from –9.8 to 0 and rapidly to positive until reaching $v = 0$. This is caused by the falling magnet inducing an emf in the metal solid underneath (Faraday's Law), which in accordance with Lenz's Law causes an opposing force. The metal is not magnetic, but the change in magnetic flux as the magnet approaches from above induces a like pole in the surface of the metal opposing the flux change, hence an eddy current flows. This flow of current in the conductor requires energy, and the only energy in this action is the potential and kinetic energy of the falling magnet, which is therefore reduced.

It is important to note that because the graph shows velocity/time, the bottom of the curve does *not* identify that the magnet has stopped, merely that the downward acceleration has become upward. It is still moving downwards, but slowing rapidly to zero because the flux change increases while the magnet moves downwards, maintaining the induced current (as shown by the upward curve) even while the magnet rapidly slows down, therefore continuing to remove mechanical energy from the magnet. *(4 marks)*

Question 33 (Total 6 marks)

(a) The initial gravitational potential energy of the mass would be:

$$E_P = -\frac{Gm_M m}{d} = -\frac{Gm_M \times 2}{d}$$

$$\therefore E_P = -\frac{(6.67 \times 10^{-11}) \times (6.39 \times 10^{23}) \times 2}{(3\,376\,204)} = -25\,248\,059.66 \text{ J}$$

After falling 1 metre, its new potential energy would become:

$$E'_P = -\frac{Gm_M \times 2}{d'} \quad \therefore E'_P = -\frac{(6.67 \times 10^{-11}) \times (6.39 \times 10^{23}) \times 2}{(3\,376\,203)} = -25\,248\,067.13 \text{ J}$$

The loss in GPE is therefore 7.47 J. The maximum possible energy released by the bulb is 7.47 J. *(3 marks)*

(b) If the switch to the lamp is left open, there would be no current flowing through the light bulb, so it would stay off, and the 2-kg mass would fall as normally on Mars, with a gravitational acceleration of 3.9 m s^{-2}, its gravitational PE being changed only to KE.

More interesting is what occurs if the switch is closed: the 2-kg mass would have to fall downwards, otherwise there would be no change in its GPE so the bulb would not light up. In fact, initially it would fall quickly, but decelerate rapidly to a uniform speed, after which the light bulb would maintain a uniform brightness until the mass stopped. *(3 marks)*

Question 34 (Total 5 marks)

(a) Since $\tau = Fd$

The τ will be increased if either F or d is increased; that is, applying more force in the direction shown on the diagram, or applying the force at a greater distance from the nut. *(2 marks)*

(b) Students needed to be careful to only calculate the torque on side *BC* as asked by the question.

To calculate the torque on side *BC*, we need to apply the torque formula from part (a) using the force due to the motor effect on side *BC* only and multiplying by the number of turns.

$$\begin{aligned}\tau &= nFd \\ &= nBIl\sin\theta \times d \\ &= 15 \times 0.2\text{ T} \times 7\text{ A} \times 0.08 \times 1 \times 0.03 \\ &= 0.0504\text{ Nm}\end{aligned}$$

From the RH push rule, the force on BC will be into the page. *(3 marks)*

Question 35 (Total 6 marks)

Transformers use a soft iron core to improve the magnetic flux linkage between the primary and secondary coils. However, changes in magnetic flux due to the applied AC current induce eddy currents within the soft iron core that create resistive heating. This heat leads to energy loss and reduces the efficiency of the transformer. The iron core is laminated with non-conductive material to reduce the size of the eddy currents and therefore minimise their heating effects.

Magnetic braking, however, relies on eddy currents in order to function so magnetic braking systems are designed to increase the formation of eddy currents. The braking systems rely on a wheel moving through a strong magnetic field (usually produced by electromagnets). This motion of a conductor through the magnetic field creates a change in flux. That induces eddy currents (due to Faraday's Law). The eddy currents flow so the magnetic field they produce opposes the motion that caused them (due to Lenz's Law) and therefore slows down the moving conductor. To increase the braking effect, solid conductive materials (such as copper and aluminium wheels) are used with strong magnetic fields. *(6 marks)*

Question 36 (Total 5 marks)

(a) Magnetic flux is the field strength × the area. This means the maximum magnetic field strength can be calculated from the maximum magnetic flux divided by the area of the conductive loop.

$$B = \frac{\varphi}{A} = \frac{\varphi}{\pi r^2}$$

$$B = \frac{0.6}{\pi \times 0.3^2}$$

$$B = 2.1\text{ T}$$

(2 marks)

(b) The emf (voltage) produced is equal to the rate of change of flux. The flux is changing most rapidly between 10 and 12 seconds. The rate of change is given by the slope of the graph.

$$\begin{aligned}V &= \frac{0.6}{2} \\ &= 0.3\text{ V}\end{aligned}$$

From Lenz's law, the current in the loop will flow so as to oppose the change that caused it. From 10–12 seconds the flux is reducing, so the current will flow so as to increase the flux into the page. From the right-hand grip rule, the current will flow from P to Q (clockwise through the loop). *(3 marks)*

Question 37 (Total 4 marks)

The force on the proton due to the electric field will result in acceleration towards the left of the page, $F = Eq$ so $a = \frac{Eq}{m}$. This acceleration will continue indefinitely since the field continues indefinitely.

The force due to the magnetic field will produce circular motion, with the proton initially accelerating out of the page. $F = qvB$ so $a = \frac{qvB}{m}$ (as initially the motion is perpendicular to the field).

The combination of the two forces will result in a helical path initially out of the page and then continuing towards the left. As the horizontal component of velocity increases due to the electric field, the helix will become more stretched out. The component perpendicular to the magnetic field remains constant so the radius of the spiral will also remain constant.

(4 marks)

Question 38 (Total 6 marks)

(a) The conducting disc will experience a changing magnetic field which will induce eddy currents in the disc.

By Lenz's law the induced currents will be in a direction that will produce magnetic fields which will oppose the change that produced the currents.

Hence the eddy currents will produce magnetic fields that will exert a force on the magnet in the opposite direction to the magnet's motion. By Newton's 3rd law an equal and opposite force will be exerted on the disc in the same direction as the magnet's motion. This force will cause the disc to rotate in the same direction as the magnet is rotating. *(3 marks)*

(b) The magnitude of the induced emf will increase as the magnet spins faster as, by Faraday's law of electromagnetic induction, the emf is proportional to the rate of change of magnetic flux linking the coils (i.e. $\text{emf} = \frac{-n\Delta BA}{\Delta t}$).

The emf produced will be AC and the period of the first three cycles will decrease until the motor reaches operating speed.

At operating speed, the period will be constant and the peak magnitude of the induced voltage will remain constant. *(3 marks)*

The induced emf is shown in Figure E2.

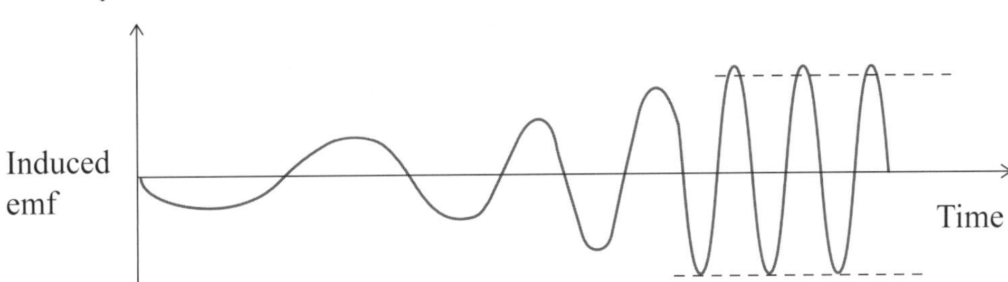

Figure E2. Induced emf as a function of time

Question 39 (Total 5 marks)

(a) The AC current on the primary coil produces a changing magnetic field in the transformer core that induces eddy currents in the core. The direction of these currents by Lenz's law will produce magnetic fields to oppose the change that created them.

Thus eddy currents reduce the magnetic field in the core. By Faraday's law the induced emf on the secondary coil is proportional to the rate of change of magnetic flux and as the eddy currents reduce the rate of change of magnetic flux, less emf will be induced on the secondary coil. *(2 marks)*

(b) If the net force on wire *Y* is zero, the force on wire *Y* due to the current in wire *Z* must be equal and opposite to the force on wire *Y* due to the current in wire *X*.

$$F_{xy} = F_{zy}$$

$$\frac{kI_x I_y l}{r_{xy}} = \frac{kI_z I_y l}{r_{zy}}$$

$$I_x = \frac{I_z r_{xy}}{r_{zy}}$$

$$= \frac{20 \times 0.75}{0.3} = 50\text{A}$$

in the same direction as the current in wire *Y*. *(3 marks)*

Question 40 (Total 4 marks)

Gravitational fields exert a force on all objects with mass. The gravitational field (g) at a point in space is defined as the gravitational force per unit mass placed at the point ($g = \frac{F}{m}$) and hence has units of N kg^{-1} = ms^{-2}. The gravitational field strength at a point is equal to the acceleration due to gravity at that point. The gravitational force on a body is always in the direction of the gravitational field at that point. The gravitational field a distance r from a planet of mass M would be given by $g = \frac{GM}{r^2}$.

Electric fields exert a force on charged objects. The electric field (E) at a point is defined as the force that would appear on a unit, with a positive test charge placed at the point ($E = \frac{F}{q}$). The units of the electric field are NC^{-1}. Positively charged objects experience a force in the direction of the electric field while negatively charged objects experience a force in the opposite direction to the charge. The electric field a distance r from a point charge q is given by $E = \frac{kq}{r^2}$.

Objects with mass are surrounded by gravitational fields, while charged objects are surrounded by electric fields.

The force on a charged particle in an electric field is given by $F = qE$, while the force on a charged object in an electric field is given by $F = mg$. Both electric and magnetic fields are vector fields as both have a magnitude and direction at each point in the field. Both types of fields can be represented by lines of force. Gravitational fields exert a force on all matter, while electric fields only exert a force on charged objects. In addition, the gravitational field exerts a force in the direction of the field only, while the force exerted by an electric field depends on the sign of the charge on the body.

OR Students could represent this information in a table of similarities and differences between the two types of fields. *(4 marks)*

Question 41 (Total 6 marks)

First, in terms of energy, during peak electrical use periods, more power is used by consumers and hence more power is drawn from the system.

To supply the increased power being drawn from each transformer, more power must be put into the first transformer T1 by the generator.

Because energy is conserved, drawing more power from the system will mean more mechanical power must be supplied to the generator. As the load is increased, the generator will become harder to turn at the same rate of rotation and hence will require more mechanical power to be put into the generator to maintain the rate of rotation.

If the system was 100% efficient, the electrical power drawn from the system would be equal to the mechanical power used to turn the generator. In reality, some energy is lost to heat in the lines and in the transformers. The loss in the lines and the windings of the transformers increase when the current is increased as $P_{loss} = I^2R$. More current will also heat the wire, increasing the resistance and further increasing the power lost in the lines and transformer windings.

Now, in terms of voltage and current, more electrical power (current) is used when householders switch on more parallel circuits. This will increase the current drawn from the transformers T1 and T2 (as they operate at specific voltages) and from the generator.

This increased current in the generator will, by Lenz's law, make the generator harder to turn.

We can understand this by remembering that the induced current in the generator produces magnetic fields that will oppose the change that created the current. That is, the magnetic fields will oppose the generator being turned. If the mechanical input power was not increased when the load was increased, the speed of rotation of the generator would decrease and the AC frequency and peak voltage produced by the generator would decrease. This drop in voltage would prevent any extra power being drawn from the system. *(6 marks)*

CHAPTER 3

Module 7

The Nature of Light

Part A Objective-response questions

1 The spectrum of a star was found to have strong ionic absorption lines and weak atomic absorption lines. When compared with spectra obtained in the laboratory, all the absorption lines in the star's spectrum were found to have higher frequencies than the laboratory lines. What does the stellar spectrum tell us about the star?

(A) The star is moving away from the Earth and has a comparatively low surface temperature.

(B) The star is moving towards the Earth and has a comparatively high surface temperature.

(C) The star is moving away from the Earth and has a comparatively high surface temperature.

(D) The star is moving towards the Earth and has a comparatively low surface temperature.

(Sample question)

2 The graph shows the maximum kinetic energy (E) with which photoelectrons are emitted as a function of frequency (f) for two different metals X and Y.

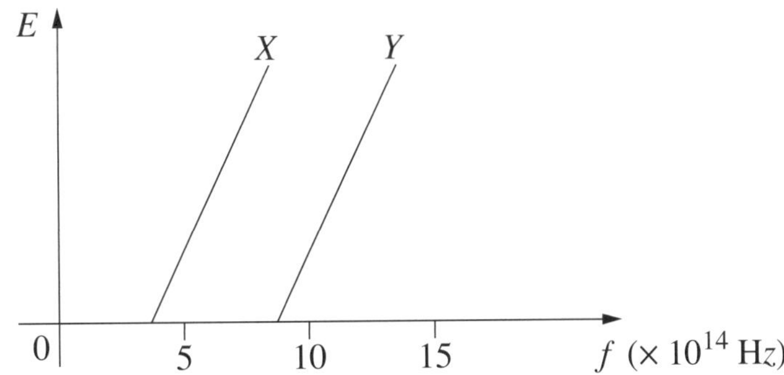

The metals are illuminated with light of wavelength 450 nm.

What would be the effect of doubling the intensity of this light without changing the wavelength?

(A) For metal X, the number of photoelectrons emitted would not change but the maximum kinetic energy would increase.

(B) For metal X, the number of photoelectrons emitted would increase but the maximum kinetic energy would remain unchanged.

(C) For both metals X and Y, the number of photoelectrons emitted would not change but the maximum kinetic energy would increase.

(D) For both metals X and Y, the number of photoelectrons emitted would increase but the maximum kinetic energy would remain unchanged.

(q20, 2013 HSC)

3 Which of the following is a true statement about scientific theories, such as Einstein's theory of special relativity?

(A) They are valid but unreliable ideas.

(B) They are useful in making predictions.

(C) They are concepts that lack an experimental basis.

(D) They are ideas that can't be accepted until they have been tested.

(q6, 2015 HSC)

4 Astronauts travel at a velocity of 0.9 *c to* Alpha Centauri. Newtonian physics predicts that this journey would take 4.86 years.

How many years will the journey take in the frame of reference of the astronauts?

(A) 0.923

(B) 1.54

(C) 2.12

(D) 11.1

(q16, 2015 HSC)

5 Which row of the table correctly shows ideas that Planck and Einstein contributed to quantum theory?

	Planck	*Einstein*
(A)	Hot objects emit radiation in discrete amounts.	Light consists of packets of energy with specific values.
(B)	Planck's constant determines the energy of photons.	Objects emit energy that increases exponentially with frequency.
(C)	No energy is lost from black body radiators.	Energy is absorbed if the band gap is less than the photon energy.
(D)	The energy of photons decreases as the wavelength increases.	Photons have energy proportional to their frequency.

(q17, 2015 HSC)

6 In a thought experiment, a jet is travelling at 0.5 *c* relative to the ground, towards a train that is travelling at 0.1 *c* relative to the ground, as shown.

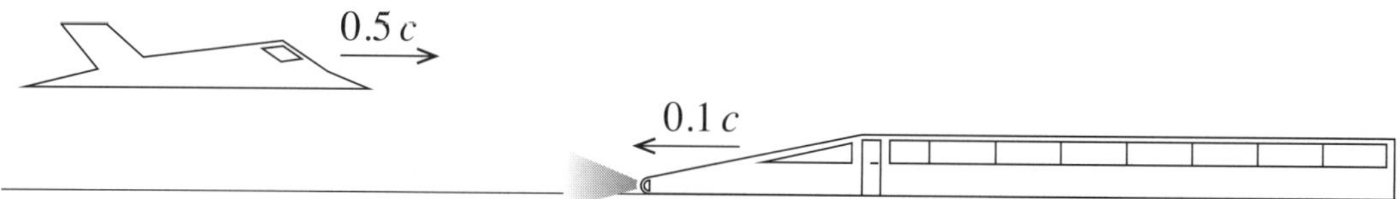

What is the speed of the light emitted from the train's headlight, as measured by a pilot in the jet?

(A) 0.1 *c* (B) 0.4 *c* (C) 0.6 *c* (D) 1.0 *c*

(q6, 2016 HSC)

7 In a thought experiment, a train is moving at a constant speed of $0.8\ c$. A lamp is located at the midpoint of a carriage. There are doors W and Z at each end of the carriage which open automatically when light from the lamp reaches them.

The passenger standing at the midpoint of the carriage switches on the lamp. Which statement best explains what the passenger observes about the doors?

(A) Z opens before W because the lamp is moving towards Z.

(B) W opens before Z because W is moving towards the lamp.

(C) W and Z open simultaneously because the lamp is placed at an equal distance from both.

(D) W and Z open simultaneously because the distance from the lamp to each door has contracted by the same amount.

(q10, 2016 HSC)

8 What is the wavelength, in metres, of a photon with an energy of 3.5 eV?

(A) 1.2×10^{-6} (B) 3.5×10^{-7}

(C) 1.18×10^{-15} (D) 5.67×10^{-26}

(q11, 2016 HSC)

9 When light of a specific frequency strikes a metal surface, photoelectrons are emitted.

If the light intensity is increased but the frequency remains the same, which row of the table is correct?

	Number of photoelectrons emitted	*Maximum kinetic energy of the photoelectrons*
(A)	Remains the same	Remains the same
(B)	Remains the same	Increases
(C)	Increases	Remains the same
(D)	Increases	Increases

(q13, 2016 HSC)

10 Muons are subatomic particles which at rest have a lifetime of 2.2 microseconds (μs). When they are produced in Earth's upper atmosphere, they travel at $0.9999\ c$.

Using classical physics, the distance travelled by a muon in its lifetime can be calculated as follows:

$$\begin{aligned} x &= vt \\ &= 660 \text{ m} \end{aligned}$$

Question 10 continues on the following page

Question 10 (continued)

Which row of the table correctly summarises the behaviour of these muons?

	Muon's reference frame		Earth's reference frame	
	Distance travelled (m)	*Lifetime* (μs)	*Distance travelled* (m)	*Lifetime* (μs)
(A)	660	2.2	> 660	> 2.2
(B)	> 660	> 2.2	660	2.2
(C)	660	2.2	< 660	< 2.2
(D)	< 660	< 2.2	660	2.2

(q19, 2016 HSC)

End of Question 10

11 The spectrum of electromagnetic waves emitted from a black body could not be explained by classical physics. What assumption did Planck have to make to explain black-body radiation?

(A) The radiation is quantised with energy $E = hf$.

(B) The electron orbitals in atoms are quantised.

(C) Electrons have an associated wavelength given by $\lambda = \frac{h}{mv}$.

(D) Radiation can be emitted or absorbed only in whole-number multiples of energy $E = hf$.

(Sample question)

12 Which of the following is an inertial frame of reference?

(A) A rocket during launch

(B) A train travelling at a constant velocity

(C) A car turning a corner at a constant speed

(D) A lift slowing down as it approaches the ground floor

(q2, 2017 HSC)

13 Unpolarised light with an intensity of $1\ \text{kWm}^{-2}$ is incident on a pair of polaroid filters with their polarisation axes at 30°, as illustrated.

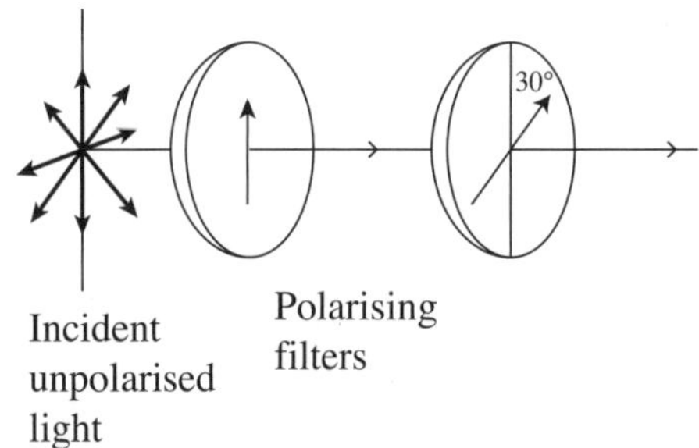

What is the light intensity after passing through the two polarising filters?

(A) 375 W (B) 866 W (C) 433 W (D) 25° W

(Sample question)

14 The length of a spaceship is measured by an observer to be 3.57 m as the spaceship passes with a velocity of $0.7c$.

At what velocity would the spaceship be moving relative to the observer if its measured length was 2.5 m?

(A) $0.490c$

(B) $0.707c$

(C) $0.714c$

(D) $0.866c$

(q20, 2017 HSC)

15 Which graph is consistent with predictions resulting from Planck's hypothesis regarding radiation from hot objects?

(A)

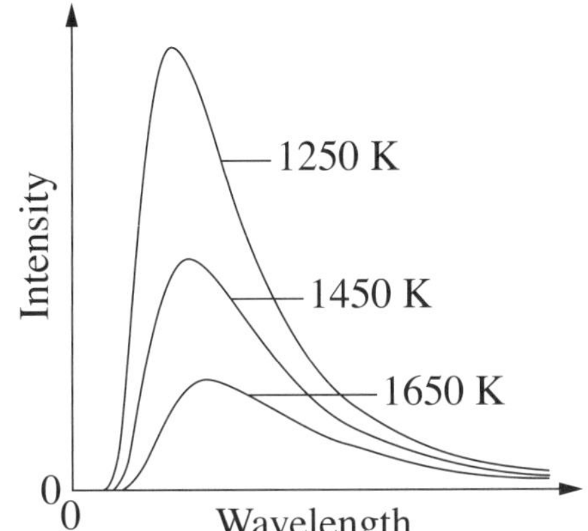

(B)

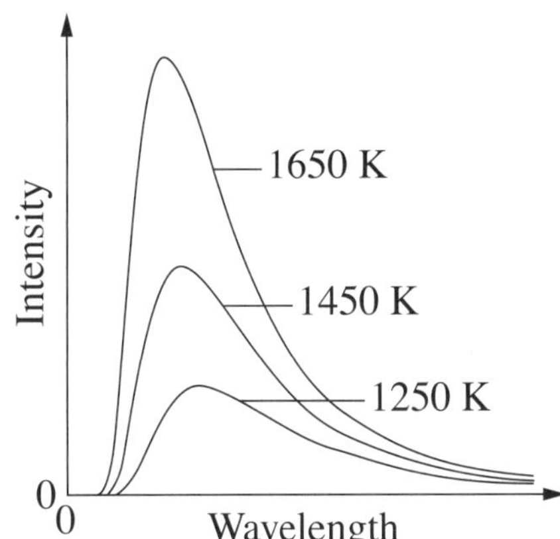

(C)

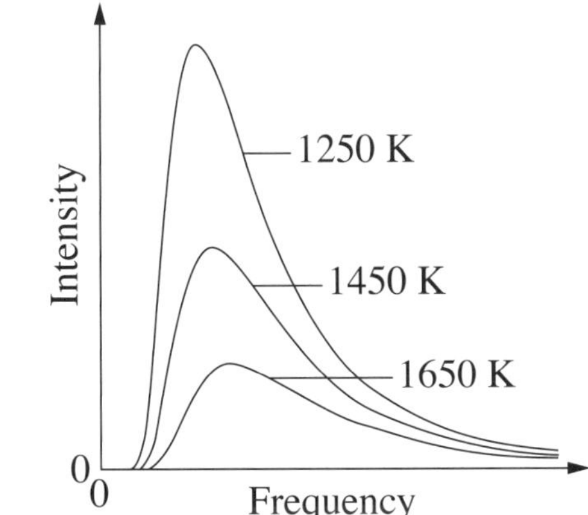

(D)

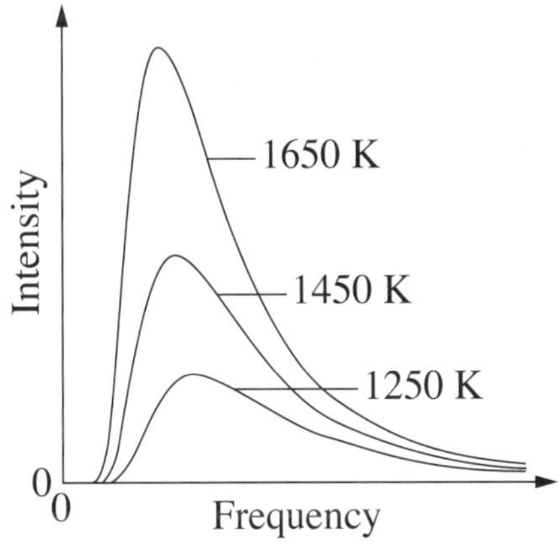

(q2, 2018 HSC)

16 Passing coherent, monochromatic light through a pair of closely spaced slits produces an interference pattern on a distant screen like the one shown in Figure E1. If 20 slits were used rather than two slits and the separation between each pair of slits remained constant, how would the interference pattern on the screen change?

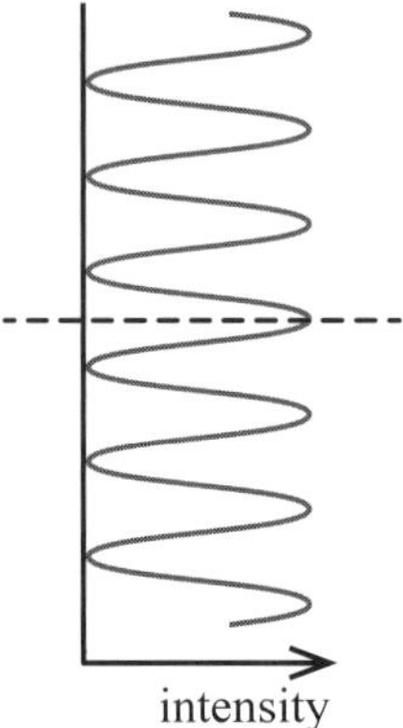

Figure E1. Double slit interference pattern

(A) The maxima would become sharper and less intense but the separation between the maxima would not change.

(B) The maxima would become sharper but the intensity of the maxima and the separation between the maxima would not change.

(C) The maxima would become sharper and more intense but the separation between the maxima would not change.

(D) The maxima would become sharper and less intense and the separation between the maxima would decrease.

(Sample question)

17 A hypothetical journey to a distant star might be accomplished in an astronaut's life span by travelling at relativistic speeds.
What is the key concept which underpins such a hypothetical journey?

(A) From the frame of reference of the spacecraft making the journey, the distance to the star is less.

(B) Clocks on the spacecraft run slower and hence the rate at which fuel is used for the journey is decreased.

(C) Relativistic effects on the spacecraft reduce its mass, making it possible to accelerate to the speeds needed for such a journey.

(D) The relativistic increase in the mass of the fuel on board makes it possible to complete a longer journey with less fuel.

(q9, 2018 HSC)

18 The diagram shows a simplified model of the Michelson–Morley experiment. It can be assumed that distances L_1 and L_2 are equal without affecting the outcome.

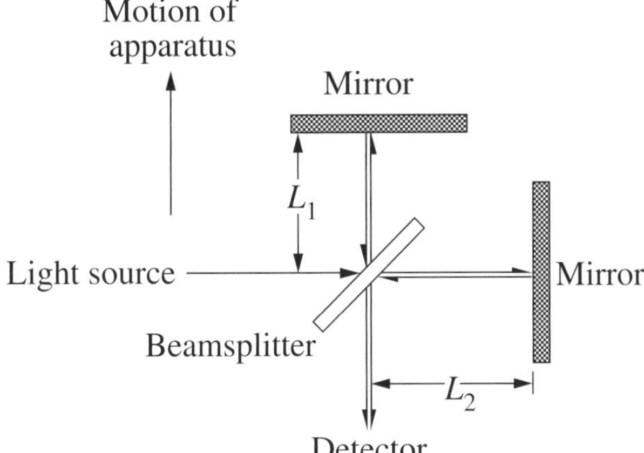

The times taken for the light to travel in both directions along the lengths L_1 and L_2 are t_1 and t_2 respectively.

What was Michelson attempting to demonstrate in this experiemnt?

(A) The relativistic contraction of L_1would cause t_1 to be less than t_2.

(B) The ether is carried through space with Earth, which would cause t_1 to be equal to t_2.

(C) The times t_1 and t_2 would be the same because the velocity of the apparatus was much less than the speed of light.

(D) The motion of the apparatus resulting from Earth's orbital motion around the sun would cause t_1 to be greater than t_2.

(q15, 2018 HSC)

19 When a train is at rest in a tunnel, the train is slightly longer than the tunnel.

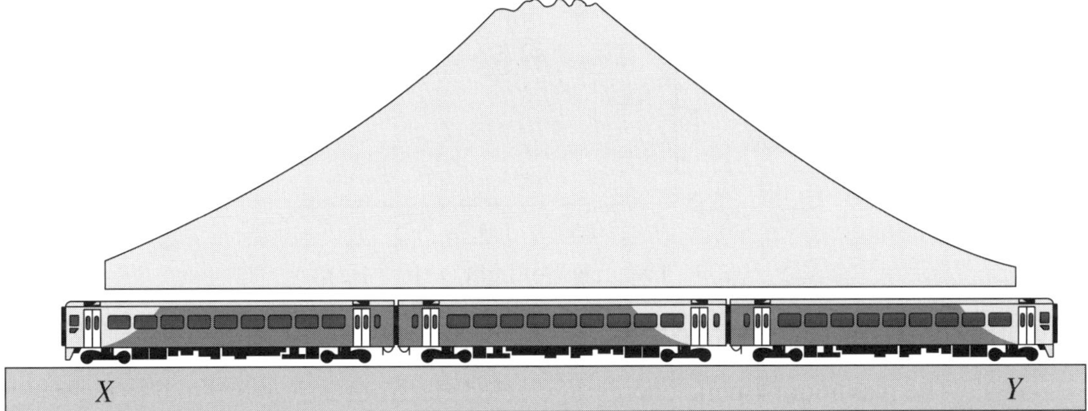

In a thought experiment, the train is travelling from left to right fast enough relative to the tunnel that its length contracts and it fits inside the tunnel.

An observer on the ground sets up two cameras, at X and Y, to take photos at exactly the same time. The photos show that both ends of the train are inside the tunnel.

Question 19 continues on the following page

Question 19 (continued)

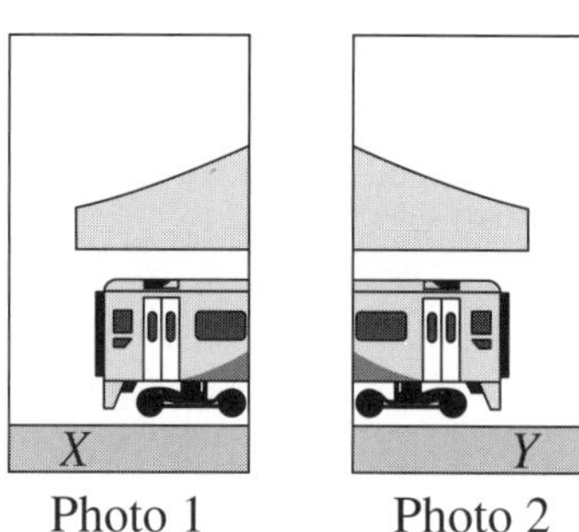

Photo 1 Photo 2

A passenger travelling on the train at its centre can see both ends of the tunnel and is later shown the photos.

From the point of view of the passenger, what is observed and what can be deduced about the photos?

(A) The tunnel's length contracts so the train does not fit, and photo 2 is taken before photo 1.

(B) The tunnel's length contracts so the train does not fit, and photos 1 and 2 are taken at the same time.

(C) The tunnel appears to expand due to length contraction of the train, allowing it to fit in the tunnel, and photo 1 is taken before photo 2.

(D) The tunnel appears to expand due to length contraction of the train, allowing it to fit in the tunnel, and photos 1 and 2 are taken at the same time.

(q16, 2018 HSC)

End of Question 19

20 The graph shows the maximum kinetic energy of electrons ejected from different metals as a function of the frequency of the incident light.

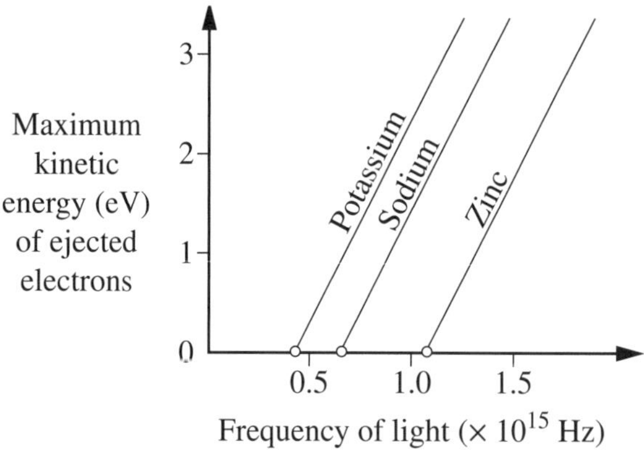

What can be deduced from this graph?

(A) The maximum kinetic energy of ejected electrons is proportional to the number of photons incident on the metal surface.

(B) More photons are required to cause an electron to be ejected from zinc than from potassium.

(C) Any photon that can eject an electron from the surface of zinc must also be able to cause an electron to be ejected from potassium.

(D) For any given frequency that causes electrons to be ejected from all three metals, the number of electrons ejected is always greatest for potassium.

(q17, 2018 HSC)

Part B Short-answer questions

Question 21 (6 marks)
The speed of light was first measured in 1676 and has been measured more accurately using a range of different techniques over the past 350 years.

(a) Why was the speed of light not measured until 1676? **1**
(2 lines)

(b) Outline one experiment that was used to measure the speed of light. **4**
(5 lines)

(c) Why have we now fixed the speed of light? **1**
(2 lines) (Sample question)

Question 22 (8 marks)
Light occurs as continuous spectra, emission spectra and absorption spectra.

(a) Describe how the spectral distribution in a beam of light could be determined. **2**
(3 lines)

(b) Outline the differences between each type of spectra and how each type could be produced. **6**
(8 lines) (Sample question)

Question 23 (6 marks)
In the double slit experiment shown in figure, the slit separation was $d = 30\ \mu m$ and the angle between the central and the second-order interference maximum was $\theta = 2°$.

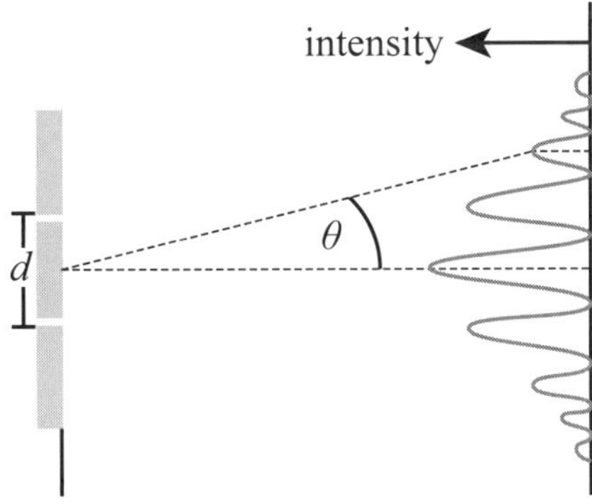

Double slit experiment (not drawn to scale)

(a) Find the wavelength of light used in the experiment. **2**
(3 lines)

(b) Explain why the intensity of the second-order interference maximum is less than the intensity of the first order maximum. **2**
(2 lines)

(c) Explain how the interference pattern would change if the double slit was replaced with a diffraction grating that had a slit separation that was greater than the separation between the double slits. **2**
(4 lines) (Sample question)

Question 24 (4 marks)

An interference device is used to produce the interference pattern on a distant screen, as shown in the figure.

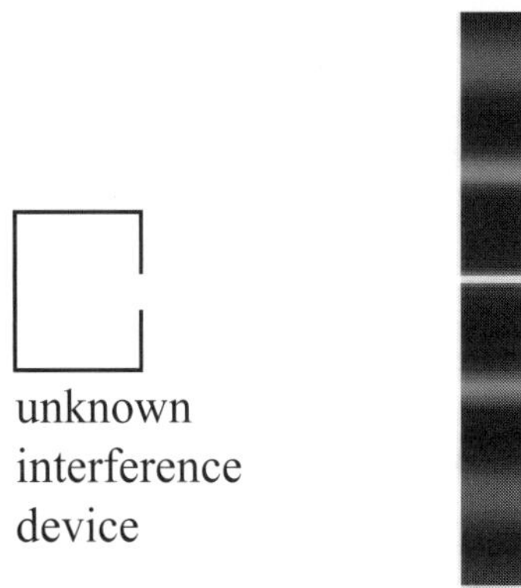

Interference pattern produced by an unseen device

(a) Suggest a method that could be used in the school laboratory to produce an interference pattern of this type. **1**

(1 line)

(b) Explain how the method you suggested would produce the pattern shown in the figure. **3**

(5 lines) (Sample question)

Question 25 (6 marks)

(a) Explain the difference between polarised and unpolarised electromagnetic waves. **2**

(3 lines)

(b) Describe how you could determine if a beam of light was polarised or unpolarised. **2**

(4 lines)

(c) Outline why polarisation was important to the development of the wave model of light. **2**

(4 lines) (Sample question)

Question 26 (5 marks)

(a) Calculate the number of photons, $\lambda = 450$ nm, which are required to transfer 1.0×10^{-3} J of energy. **3**

(7 lines)

(b) A 1 W beam of light transfers 1 J per second from one point to another. **2**

With reference to the particle model of light, contrast a 1 W beam of red light and a 1 W beam of blue light.

(5 lines) (q29, 2011 HSC)

Question 27 (4 marks)

The nearest galaxy to ours is the Large Magellanic Cloud, with its centre located 1.70 × 105 light years from Earth. Assume you are in a spacecraft travelling at a speed of 0.999 99 c toward the Large Magellanic Cloud.

(a) In your frame of reference, what is the distance between Earth and the Large Magellanic Cloud? **2**

(4 lines)

(b) In your frame of reference, how long will it take you to travel from Earth to the Large Magellanic Cloud? **2**

(4 lines) (q18, 2009 HSC)

Question 28 (6 marks)

How did Einstein's theory of special relativity and his explanation of the photoelectric effect lead to the reconceptualisation of the model of light? **6**

(21 lines) (q24, 2008 HSC)

Question 29 (7 marks)

(a) Outline ONE piece of evidence supporting Einstein's theory of relativity. **2**

(2 lines)

(b) What criteria are used to test and validate a theory? **3**

(6 lines)

(c) The distance between the cathode and screen in a cathode ray tube is 40 cm. **2**

If an electron travels through the tube at 3.0×10^7 ms^{-1}, what is the apparent distance from the cathode to the screen in the electron's frame of reference?

(4 lines) (q28, 2012 HSC)

Question 30 (4 marks)

Consider the following 'thought experiment'. **4**

A scientist on board a spaceship wishes to synchronise two clocks. To achieve this, beams of light from a source placed midway between the clocks activate photocells, turning on both clocks.

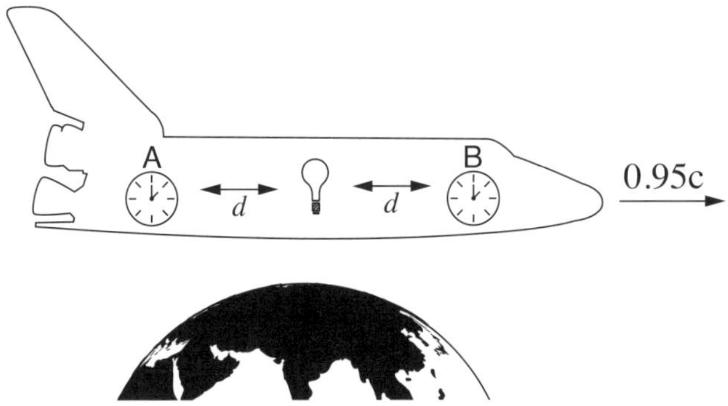

The scientist observes the synchronisation of the clocks as the rocket flies past Earth at 0.95c. A person on Earth observes that the clocks are not synchronised. Account for these observations.

(8 lines) (q24, 2011 HSC)

Question 31 (6 marks)

Explain how different discoveries in physics led to the development of THREE technologies, including the electric generator. **6**

(15 lines) (q29, 2016 HSC)

Question 32 (5 marks)

A laser emits 0.2 W of monochromatic red light with a wavelength of 630 nm.

(a) Determine the frequency of the electromagnetic waves emitted by the laser. **1**

(2 lines)

(b) Determine the energy of each photon emitted by the laser. **1**

(2 lines)

(c) Calculate the number of photons emitted by the laser per second. **1**

(2 lines)

(d) If the frequency of the laser was halved and the power of the laser was doubled, how many photons would the laser emit per second? Justify your answer. **2**

(4 lines) (Sample question)

Question 33 (6 marks)

A spacecraft has length of 72 m and a mass of 12 000 kg when at rest on the Earth, but when the craft moves at high speed past the Earth, an Earth-bound observer measures the length of the craft to be 50 m. While moving past the Earth the space traveller measures the time it takes for a laser beam to travel from the back to the front of the spacecraft.

(a) Determine the time it will take for the laser beam to travel from the back to front of the spaceship as measured by an observer on the spacecraft. **2**

(2 lines)

(b) What is the velocity of the laser beam as measured by an observer on the ground? **2**

Justify your answer.

(4 lines)

(c) Calculate the velocity of the spacecraft with respect to the Earth. **2**

(5 lines) (Sample question)

Question 34 (6 marks)

Light has been described with a variety of scientific models that have been changed and refined over time.

(a) Outline two pieces of evidence that led scientists to abandon Newton's corpuscular model of light in favour of a wave model of light. **2**

(6 lines)

(b) Describe Maxwell's wave theory of light and give an example of how electromagnetic waves in one region of the electromagnetic spectrum can be produced and detected. **4**

(12 lines) (Sample question)

Question 35 (5 marks)

A laser emits light of wavelength 550 nm.

(a) Calculate the frequency of this light. **2**

(4 lines)

(b) The electrons in a specific metal must absorb a minimum of 5×10^{-19} J in order to be ejected from its surface. **3**

Explain why electrons will not be ejected from this metal when photons of wavelength 550 nm strike its surface. Support your answer with relevant calculations.

(5 lines) (q21, 2017 HSC)

Question 36 (5 marks)

Using examples from special relativity, explain how theories in science are validated in different ways. **5**

(8 lines) (q23, 2017 HSC)

Question 37 (5 marks)

In an experiment involving polarised light, a polarising filter was observed to reduce the intensity of a beam of polarised light to 25% of the incident intensity.

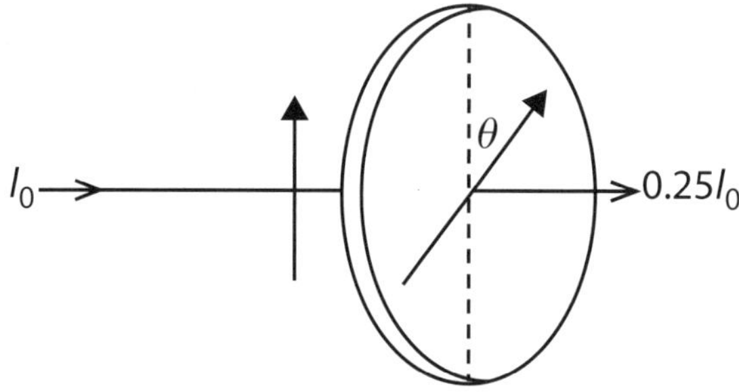

(a) With the aid of a diagram outline the difference between a polarised electromagnetic wave and an unpolarised electromagnetic wave. **3**

(5 lines)

(b) Find the angle ∠ between the direction of polarisation of the incident light and the polarising axis of the filter in the experiment illustrated above. **2**

(4 lines) (Sample question)

Question 38 (3 marks)

Explain how the light that reaches the Earth from a star can be used to determine the chemical composition of the star's atmosphere. **3**

(8 lines) (Sample question)

Module 7
The Nature of Light

Sample answers

Part A Objective-response questions

1 B The increased frequency is caused by the star moving towards the Earth (blueshift) and the intensity of ionic lines implies a high surface temperature (i.e. high-speed collisions causing ionisation).

2 B The threshold frequency of metal X is about 3.5×10^{14} Hz; that of Y is about 8.5×10^{14} Hz. The illuminating light has a wavelength of 450 nm, corresponding to a frequency of 6.67×10^{14} Hz. This means that only metal X releases photoelectrons when illuminated by this light. Increasing the light intensity increases the number of photoelectrons emitted.

3 B Theories, notably relativity, certainly are useful in making predictions (but are not always correct). D is a strong distractor; strictly it applies to hypotheses, not theories.

4 C The applicable relativity formula is $t_v = \dfrac{t_0}{\sqrt{1 - \frac{v^2}{c^2}}} = \dfrac{4.86}{\sqrt{1 - \frac{(0.9\,c)^2}{c^2}}} = 2.12$ years

5 A In B and C statements attributed to both scientists are incorrect. D is incorrect because Planck's observations were not directed at photons—that was Einstein.

6 D According to Einstein's Special Relativity Theory the speed of light through empty space is a definite speed irrespective of the speed of the source or observer. That speed is 1.0 c.

7 C All three of the alternatives are incorrect because the observer, the lamp and both doors are all in the same inertial frame of reference.

8 B $E = hf \quad \therefore \; 3.5 \times \left(1.062 \times 10^{-19}\right) = \left(6.63 \times 10^{-34}\right) f$

$$\lambda = \frac{c}{f} \quad \therefore \; \lambda = \frac{\left(3.0 \times 10^{8}\right) \times \left(6.63 \times 10^{-34}\right)}{3.5 \times \left(1.062 \times 10^{-19}\right)} = 3.5 \times 10^{-7} \text{ m}$$

9 C Since photons are being emitted the light frequency equals or exceeds the threshold frequency for the metal surface. When the intensity of the light is increased, the number of photons striking the surface increases and, since every photon is able to release one electron, the number of photoelectrons is greater. However, since the frequency of the photons remains the same, so does the energy of each photon. Since the Work Function of the metal is uniform, the maximum kinetic energy of every released photoelectron remains the same.

10 A Alternatives B and D are incorrect because within the frame of reference of the actual muons there can be no relativistic effects. C is incorrect because from outside the muons' frame of reference they are observed to travel much further because their time to travel has been extended relativistically.

11 D By making this assumption Planck was able to write an equation which fitted the black-body radiation curves perfectly.

12 B An inertial frame of reference is one that is not accelerating. B is the only choice that has no change of velocity.

13 A The intensity will halve when the unpolarised light passes through the first polaroid filter and, by applying Malus's law to the second filter, $I = \left(\frac{I_0}{2}\right)\cos^2\theta = 500\cos^2 30° = 375\ \text{Wm}^{-2}$.

14 D At a velocity of $0.7c$, the length is 3.57. Using $l_v = l_0\sqrt{1 - \frac{v^2}{c^2}}$ the rest length l_0 can be calculated to be 5.00 m. Therefore to have a length of 2.5 m we can calculate the required velocity:

$$2.5 = 5\sqrt{1 - \frac{v^2}{c^2}}$$

$$0.25 = 1 - \frac{v^2}{c^2}$$

$$\frac{v^2}{c^2} = 1 - 0.25$$

$$v = 1\ 0.25c$$

$$v = 0.866c$$

15 B The intensity of the radiation emitted from a black body increases with the temperature but the wavelength at which the greatest intensity of radiation is emitted decreases (by Wien's Law $\lambda = b/T$).

16 C This is similar to replacing the double slits with a diffraction grating. Increasing the number of slits increases the number of minima between each maximum, which sharpens the maxima in the pattern. Because each maxima is made up of light from many slits, rather than from two slits, increasing the number of slits increases the intensity of each maxima. The distance between maxima is related to the slit separation and as the slit separation remains constant in this example the separation between maxima will not change.

17 A From the spacecraft's frame of reference, the stars are moving rapidly towards the ship and hence the distance between the stars is contracted in accordance with $l = l_0\sqrt{\left(1 - \frac{v^2}{c^2}\right)}$. This means that an observer in the spacecraft measures the distance to the star to be shorter than the distance measured by an observer on Earth.

18 D Michelson and Morley believed the speed of light would change if the source or observer moved with respect to the ether. They reasoned that light would take longer to move back and forth when travelling parallel to the ether wind than when moving back and forth across the ether wind. Their experiment produced a null result because the speed of light is an absolute constant that is independent of the speed of the source or observer.

19 A The tunnel is moving with respect to the observer in the train and hence an observer in the train sees the length of the tunnel contracted. Because the observer on the train is moving with respect to the outside observer, the observer in the train will not agree with the outside observer's observation that the photographs were taken simultaneously. The observer in the train will see photo 2 taken before photo 1 because the light from this event will reach them first because the train has moved to the right in the time it takes for the light from the event to reach the observer.

20 C As the threshold frequency to remove an electron from zinc is greater than the threshold frequency for the other two metals, a photon that can eject an electron from zinc will have enough energy to eject an electron from either of the other two metals.

Part B Short-answer questions

Question 21 (Total 6 marks)

(a) Attempts were made to measure the speed of light prior to 1676 but the speed was too fast for any of the available technologies to measure. *(1 mark)*

(b) Students could describe any of the methods used to measure the speed of light; for example, that of Fizeau.

Fizeau measured the speed of light in 1849 using the experiment shown in the figure. Fizeau passed light through a gap in a spinning, toothed wheel and then, after reflecting the light from a mirror 8 km away, he adjusted the speed of his wheel until the reflected light passed through the next gap in the wheel. He then used the speed of the toothed wheel to determine the time it took for the light to reach the mirror and return. This enabled him to calculate the speed of light. *(2 marks)*

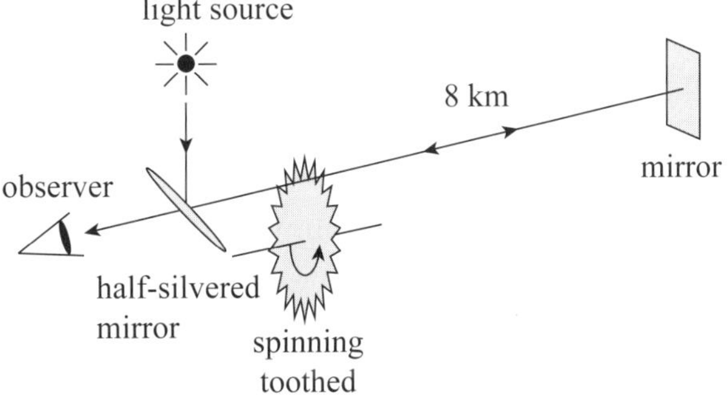

(2 marks)

Fizeau's experiment to measure the speed of light

(c) By the early 1970s the speed of light was known more accurately than the metre could be measured. In 1983 scientists around the world agreed to fix the speed of light in a vacuum (to 299 792 458 ms^{-1}) and use this fixed speed to define the metre. The standard metre is now the distance light travels in 1/c seconds. *(1 mark)*

Question 22 (Total 8 marks)

(a) We can determine the spectral distribution in a beam of light by passing the beam of light through a slit and then through a glass prism. The different wavelengths are refracted by different amounts and the intensity of the light at different wavelengths can be recorded using electronic detectors. *(2 marks)*

(b) (i) A continuous spectrum consists of all wavelengths. A continuous spectrum is emitted from hot bodies, such as the filament of an incandescent lamp. *(2 marks)*

(ii) When an elemental gas is excited by a flame of electric discharge, it emits light at specific wavelengths characteristic of the specific element being excited. Emission spectra can be produced by elemental-gas discharge tubes. *(2 marks)*

(iii) When a continuous spectrum of light is passed through an unexcited gas, specific wavelengths are absorbed by the gas. The resulting spectrum is continuous but has dark lines where specific wavelengths are missing. *(2 marks)*

Question 23 (Total 6 marks)

(a) Applying Young's double slit equation with $m = 2$:

$d\sin\theta = m\lambda$ and hence $\lambda = \dfrac{d\sin\theta}{m} = \dfrac{30 \times 10^{-6}\sin 2^\circ}{2} = 523$ nm *(2 marks)*

(b) The light is diffracted from each slit and the greater the angle of diffraction, the lower the intensity. The second maximum occurs at a greater angle of diffraction and hence will be less intense than the first maximum. *(2 marks)*

(c) The maxima would become much brighter and sharper, but the angular separation between maximum would be reduced. That is, the interference pattern on the screen would be compressed but the maxima would appear as sharp, bright lines on the screen. *(2 marks)*

Question 24 (Total 4 marks)

(a) Shining a narrow beam of white light on a diffraction grating would produce the image shown. *(1 mark)*

(b) Maxima occur when $d\sin\theta = m\lambda$, where m is an integer that refers to the order of interference. Hence in each order of interference the wavelengths will occur at slightly different angles, producing multiple spectra as shown. All wavelengths produce a central maximum and hence the central maximum will be white because it is made up of the superposition of all the colours. *(3 marks)*

Question 25 (Total 6 marks)

(a) Unpolarised light has the electric field changing in all planes that are perpendicular to the wave velocity, but polarised light has an electric field that changes in only one plane. *(2 marks)*

(b) A polarising filter could be used to determine if the light was polarised because the intensity of polarised light that passes through a polarising filter depends on the angle between the electric field of the incident light and the polarisation axis of the filter (i.e. Malus's law $I = I_{max}\cos^2\theta$). If unpolarised light is passed through a polarising filter, rotating the filter will not change the intensity of the transmitted light. However, if polarised light is passed through a polarising filter, rotating the filter will change the intensity of the transmitted light between a maximum and zero each 90° of rotation.

(2 marks)

(c) 'Polarisation' was difficult to explain using Newton's particle model and impossible to explain using Huygen's compressional wave model of light. Malus's investigation of polarisation was important because it led to Huygen's compressional wave model of light being modified to a transverse wave model by Young. *(2 marks)*

Question 26 (Total 5 marks)

(a) Energy of a single photon is given by

$$E = hf \;\therefore\; E = \frac{hc}{\lambda} = 6.626 \times 10^{-34} \times \frac{3 \times 10^{8}}{4.5 \times 10^{-7}} = 4.417 \times 10^{-19} \text{ joules per photon.}$$

(3 marks)

Now, $E_{Total} = n(E_{photon}) = 1.0 \times 10^{-3}$ J

$$\therefore n = \frac{1.0 \times 10^{-3}}{4.417 \times 10^{-19}} = 2.264 \times 10^{15} \text{ photons per mJ.}$$

(b) A photon represents a particle of light carrying a quantum of light energy related to its frequency. Red light is lower frequency, longer wavelength than blue light. As $E = hf$, the energy carried per red photon is less than the energy carried per blue photon. This means that in order to transfer 1.0 W (1 J.s2[1]) in a beam of red light, it would have to contain more red photons per second than would be required to transfer the same amount of energy per second using a beam of blue photons. 1 W equals 1 J per second = the number of photons arriving per second, each with an energy, E_{photon} and $E_{red} < E_{blue}$.

(2 marks)

Question 27 (Total 4 marks)

(a) Distance measured in the moving frame of reference is:

$$l_v = l_0\sqrt{1 - \frac{v^2}{c^2}} = 1.7 \times 10^5 \sqrt{1 - 0.99999^2}$$

Therefore, l_v according to the passenger = 760.261 light years. *(2 marks)*

(b) Time for trip, $t = \frac{l_v}{v} = \frac{760.261c}{0.99999c} = 760.269$ years. *(2 marks)*

Question 28 (Total 6 marks)

In 1905 Einstein published his first major works and in the process produced a new view of the nature of light. With his paper on the Special theory of relativity, Einstein suggested that light's velocity, *c*, was the 'terminal' velocity with nothing being able to travel faster than *c*. Einstein went further by proposing an entirely new concept when he said that c would remain constant, irrespective of the frame of reference where measured, meaning that mass, length, and time would now become variable. Einstein also incorporated *c* into his mass-energy equivalence equation, i.e. $E = mc^2$.

In a second paper Einstein proposed an explanation for the photo-electric effect and put forward the idea that light energy came in little bundles, photons, that behaved like tiny particles with the energy of the photon given using Planck's quantum idea, i.e. $E = hf$. His explanation suggested that the kinetic energy of the emitted photoelectron would be equal to the 'photon energy' minus the 'energy required to remove the electron' from an atom, i.e. $E_k = hf - \phi$. This suggested a particle property for light. As experimental evidence for Einstein's ideas was found, it led to a reconceptualization of light and it was now seen as being a wave-particle duality, travelling as electromagnetic waves but with the energy existing in the tiny, quantized bundles called 'photons'. This incorporation of Planck's quantum idea provided the impetus for others to look more closely at quantum ideas which led to the development of 'Quantum theory'. *(6 marks)*

Question 29 (Total 7 marks)

(a) The following are two examples of evidence for relativity.

Particle accelerators must be constructed to allow for the relativistic effects for the very high velocity particles being studied, i.e. increased masses and length contraction.

GPS atomic clocks must be corrected for time dilation (Special theory) because of their high speed and time changes due to the variation in the gravitational field (General theory). *(2 marks)*

(b) The theory must be testable by experiment and make predictions which can be verified by measurements. Suitable experiments must be designed which can explore whether the predictions of the theory are observed and supported by the results. The experiments will involve the use of accurate measuring devices which are sensitive enough for the required measurements. For the theory to be validated, the experiments must follow the scientific method with suitable controls and when repeated be found to always produce results that are consistent with the predictions of the theory. *(3 marks)*

(c) $l_y = l_0\sqrt{1 - \frac{v^2}{c^2}}$. The electron considers the tube to be moving relative to it.

Therefore

$$l_y = 0.4\sqrt{1 - \frac{10^{14}}{10^{16}}}.$$

$l_y = 0.398$ m. According to the electron, the distance is **39.8** cm. *(2 marks)*

Question 30 (Total 4 marks)

These observations are produced due to the Relativity of Simultaneity. Based on Einstein's Special Theory of Relativity, the speed of light is always the same value irrespective of the frame of reference where it is measured, and simultaneous events in one frame of reference may not be simultaneous when observed from a different frame of reference. The scientist and the observer are in different inertial frames of reference. The scientist onboard the spacecraft is moving at 0.95c relative to the observer on Earth but is at rest relative to the light source and the clocks. In their frame of reference the distance from the light source to each clock is exactly the same. This explains the scientist's observation of the clocks being triggered simultaneously. For the observer on Earth, in a different frame of reference where the spacecraft is moving at 0.95c relative to them, the result is different because they will observe the light reaching clock A first as, relative to the observer, clock A is moving at 0.95c towards the position of the light source where it originally emitted the light. The observer sees clock B being triggered slightly after A because, from their frame of reference, clock B is moving at 0.95c away from the original position of the light source where it produced the light. This explains the observer in a stationary frame of reference on Earth recording that the starting of the clocks was NOT simultaneous and reporting clock A started before B. *(4 marks)*

Question 31 (Total 6 marks)

(I) Fusion

There is a reasonable possibility that just at the right time in its history our Earth may be saved from the worst consequences of human-induced climate change by something found on the Moon. It is helium-3, a rare isotope of helium rarely found on Earth because it has only 75% of the mass of helium-4, and the denser gas of Earth's atmosphere allows very little to be found. In addition, being an inert element it forms no compounds. But it is found on the Moon, evidenced by its presence in the samples brought back from the four Lunar Missions. Since the Moon has almost no atmosphere, the tiny helium-3 'bullets' have been arriving daily from the Sun over the past 4.5×10^9 years since it formed, and there is a lot of it trapped in the rocks.

The point is that fusion of four hydrogen atoms to form helium-4 works well in the cores of stars, but although temperatures of well over 100 million degrees can now be produced on Earth, enough to fuse hydrogen into helium [theoretically, $4\,{}^1H_1 \rightarrow {}^4He_2$] the pressure of the Sun's core is not available to us. Instead, practical fusion reactions on Earth are carried out within a doughnut-shaped tokamak, where the hydrogen we have to use is hydrogen-2 (deuterium). The hydrogen plasma at extremely high temperatures is trapped within the tokamak by a powerful superconducting magnet requiring a temperature of just above 0 K, –273 °C. When deuterium fuses it does produce helium-4, but also two neutrons, and since these are uncharged they cannot be trapped within a magnetic field—and as they escape they take most of the energy produced in the reaction with them (being the least massive products). Consequently, current fusion reactions don't reach 'break-even point'.

But when two helium-3 nuclei fuse they release two *protons* instead, which *are* charged, hence *can* be contained. The nuclear equation is $2\,{}^3He_2 \rightarrow {}^4He_2 + 2\,{}^1H_1$.

(2 marks)

(II) Fission powers reactors

The discovery that some large nuclei such as uranium-235 would split (fission) when they absorbed a neutron and release energy led to the development of fission power plants. Fission can be self-sustaining as the fission process releases neutrons that can be used to initiate further fission reactions. Reactors use controlled-fission chain reactions where the process runs at the constant rate to produce heat to turn water into steam to drive turbine-driven electric generators. There are now approximately 500 fission power reactors operating around the world. *(2 marks)*

(III) Electric generator

Michael Faraday was able to demonstrate that electric current could be induced in a circuit by moving a magnet nearby. Following further investigation he explained the phenomenon with Faraday's Law, where he showed that changing the magnetic flux passing into a coil caused electric current to flow—provided there was a path for it. Subsequently Heinrich Lenz amended the Law to define the direction of current flow.

Nowadays, in every country in the world, electric generators are used to produce electricity for the populace using this method, but energy cannot be created—it must be converted from some source. Though the potential energy of water in a dam can provide that energy, most commonly the source is heat, whether from the Sun, burning fuel or by nuclear fission. Heat boils water—the steam is passed through a turbine attached to a huge wire coil causing it to rotate within a powerful magnetic field, and the AC electricity generated is ready for transmission. *(2 marks)*

Question 32 (Total 5 marks)

(a) For all electromagnetic waves:

$c = f\lambda$ and hence $f = \frac{c}{\lambda} = \frac{3 \times 10^8}{630 \times 10^{-9}} = 4.76 \times 10^{14}$ Hz *(1 mark)*

(b) $E = hf = \frac{hc}{\lambda} = \frac{6.626 \times 10^{-34} \times 3 \times 10^8}{630 \times 10^{-9}} = 3.2 \times 10^{-19}$ J *(1 mark)*

(c) Power = energy/time and hence the energy of the photons passing a point each second will be 0.2 J. The number of photons required to produce this energy will be given by:

Number of photons = beam energy/energy of each photon

$= \frac{0.2}{3.2 \times 10^{-19}} = 6.25 \times 10^{17}$ photons per second. *(1 mark)*

(d) If the frequency was halved, the energy of each photon would be halved and hence twice as many photons would be emitted per second if the power of the laser remained constant. To double the power, twice as many photons would be required. Now as the photon energy was halved and the power was doubled, $2 \times 2 = 4$ times as many photons would be emitted per second.

Hence there would be $4 \times 6.25 \times 10^{17} = 2.5 \times 10^{18}$ photons per second. *(2 marks)*

Question 33 (Total 6 marks)

(a) The observer in the spacecraft is at rest with respect to the craft and hence will measure the length of the spacecraft to be 72 m. Hence the time for the beam to travel the length of the craft will be:

$t = \frac{d}{v} = \frac{72}{3 \times 10^8} \times 10^8 = 2.4 \times 10^{-7}$ s *(2 marks)*

(b) Because the speed of light is an absolute constant (independent of the speed of the source or observer) both observers will measure the speed of the beam to be the speed of light *c*. *(2 marks)*

(c) Applying Einstein's equation of length contraction:

$l = l_0\sqrt{\left(1-\frac{v^2}{c^2}\right)}$ and, making v the subject of the equation:

$v = c\sqrt{1-\frac{l^2}{l_0^2}} = 3 \times 10^8\sqrt{1-\frac{50^2}{72^2}} = 2.16 \times 10^8 \text{ ms}^{-1}$ *(2 marks)*

Question 34 (Total 6 marks)

(a) The multiple bright and dark regions produced by double-slit interference experiments could be modelled numerically by assuming the patterns were produced by the interference of waves emitted from the slits.

Observations of double refraction could be explained by assuming light was a transverse wave and each component of polarisation was refracted by a different angle.

Both phenomena could be readily explained using wave theory, but neither could be explained satisfactorily by Newton's particle theory. *(2 marks)*

(b) Maxwell proposed that light was a transverse electromagnetic wave, made up of mutually perpendicular, changing electric and magnetic fields, as shown in the figure below.

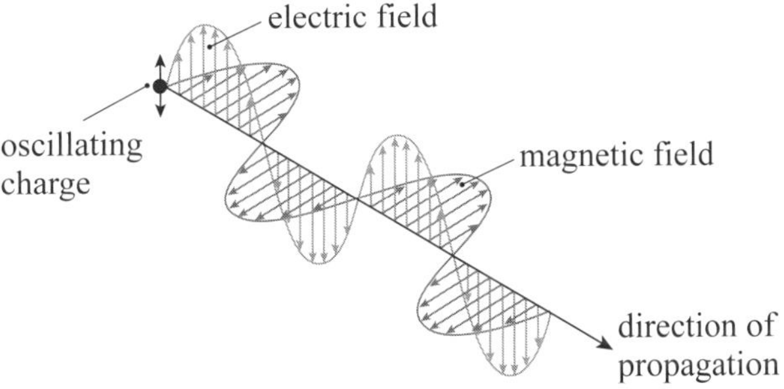

Maxwell showed that oscillating charges would produce a changing electric field which would induce a changing magnetic field, which would induce a changing electric field, etc. Hence producing a self-sustaining electromagnetic wave that would propagate away from the charge. Maxwell also showed from electromagnetic theory that the velocity of electromagnetic waves was the same as the speed of light. These electromagnetic waves could exist over a broad range of frequencies called the electromagnetic spectrum. One region of the electromagnetic spectrum is called the radio wave region. Waves with radio frequencies can be produced by rapidly oscillating charges back and forth in a wire. Radio waves can be detected by amplifying the oscillation of electrons produced by the radio waves when they pass another wire (i.e. an antenna). *(4 marks)*

Question 35 (Total 5 marks)

(a) Using $v = f\lambda$

$3 \times 10^8 \text{ ms}^{-1} = f \times 550 \times 10^{-9} \text{ m}$

$f = 5.45 \times 10^{14}$ Hz (3 sig. fig.) *(2 marks)*

(b) Photon energy is calculated using $E = hf$.

$E = 6.626 \times 10^{-34} \times 5.45 \times 10^{14}$ Hz

$= 3.61 \times 10^{-19}$ J

Therefore each photon has 3.61×10^{-19} J. Each electron within the metal can only gain this much energy (one photon per electron). However, 5×10^{-19} J is needed to escape from the metal (the work function). More energy is needed for electrons to be ejected from the surface. *(3 marks)*

Question 36 (Total 5 marks)

Theories in science can be validated by experiments and observations, or through predictions made by them. Einstein's theory of time dilation was tested and validated experimentally through the use of synchronised atomic clocks. One of the clocks was sent up in an aeroplane and, when it returned, it had 'experienced' less time than the clock that had stayed on Earth (time dilation).

Predictions made using $E = mc^2$ have been verified through measuring the amount of energy produced in nuclear transformations involving mass changes.

Muons are subatomic particles that are produced in the upper atmosphere and move at relativistic speeds. They are unstable and have a very short half-life. They should not exist for long enough to reach Earth's surface, yet they are detected at ground level. This observation could be explained through time dilation (or length contraction).

Einstein's theories were made before technology was available to test them. Since then they have been validated in many different ways and are used in designing technologies such as GPS systems, nuclear power stations and particle accelerators (where mass dilation is also observed). *(5 marks)*

Question 37 (Total 5 marks)

(a) Light is a transverse electromagnetic wave, which is made up of changing electric fields and magnetic fields that are perpendicular to each other and to the direction of travel of the wave. As shown in the figure below, a polarised light wave (i) has the electric field in one plane only, while an unpolarised wave (ii) has the electric field in all planes. Note that both diagrams in the figure show the electric field orientation for an electromagnetic wave moving towards the observer (out of the page).

Electric field

(i) polarised light (ii) unpolarised light

(3 marks)

(b) Applying Malus's law,

$I = I_{max} \cos^2\theta$ and hence,

$\theta = \cos^{-1}\left(\sqrt{\frac{I}{I_{max}}}\right) = \cos^{-1}\left(\sqrt{\frac{0.25}{1}}\right) = 60°$ *(2 marks)*

Question 38 (Total 3 marks)

Light emitted from the photosphere of the star passes through the star's atmosphere. The atoms, ions and molecules in the star's atmosphere absorb specific wavelengths of light that are characteristic of the type of atom (i.e. element), ion or molecule.

A spectrometer connected to a telescope can be used to examine the spectrum of light produced by the star. By comparing the wavelengths missing from the star's spectrum to the known absorption spectra of known atoms, ions and molecules (obtained from experiments in the laboratory) the chemical composition of the star's atmosphere can be determined.

In addition the intensity of the absorption lines can be used to estimate the proportion of different atoms, ions and molecules in the star's atmosphere. *(3 marks)*

CHAPTER 4

Module 8

From the Universe to the Atom

Part A Objective-response questions

1 If its luminosity decreased and the surface temperature of a star increased, in what direction would it move on a Hertzsprung–Russell diagram?

(A) Towards the upper right

(B) Towards the upper left

(C) Towards the lower right

(D) Towards the lower left

(Sample question)

2 Shortly after the Big Bang, subatomic matter and antimatter particles condensed from radiation. Given that matter and antimatter particles annihilate each other to produce energy, where did the matter in our universe come from?

(A) It was formed from energy after all the matter and antimatter had disappeared.

(B) It was formed much later by fusion in stars.

(C) For some reason, more matter condensed from the initial radiation than antimatter.

(D) It was produced later in supernova explosions.

(Sample question)

3 JJ Thomson determined the charge/mass ratio of the electron by constructing a device which contained

(A) perpendicular magnetic fields.

(B) perpendicular electric fields.

(C) parallel electric and magnetic fields.

(D) perpendicular electric and magnetic fields.

(q17, 2010 HSC)

4 When cathode rays were first discovered there was a debate about what the rays were. Physicist JJ Thomson ended the debate with a series of famous experiments in 1896. Which of the following was NOT determined by Thomson?

(A) Cathode rays travel much slower than light.

(B) Cathode rays can pass through thin metal foils.

(C) Cathode rays have a charge–mass ratio that is much larger than the charge–mass ratio of a hydrogen ion.

(D) Cathode rays can be deflected by electric fields.

(Sample question)

5 In a Millikan-style oil-drop experiment, an oil droplet with 2×10^6 electron charges and a mass of $2\mu g$ was levitated between two charged parallel plates 1.0 cm apart. What potential difference would be required to be placed across the plates to levitate the charge?

(A) 610 V

(B) 610 kV

(C) 61 V

(D) 6.1 V

(Sample question)

6 If a specific isotope has a half-life of 20 minutes, what percentage of the isotopes would have decayed after 2 hours?

(A) 96.9%

(B) 98.4%

(C) 1.56%

(D) 93.75%

(Sample question)

7 A fast-moving electron is found to have a de Broglie wavelength of $\lambda = 7$ nm. If the kinetic energy of the electron was doubled, what would the new de Broglie wavelength of the electron be?

(A) 5 nm (B) 14 nm

(C) 3.5 nm (D) 10 nm

(Sample question)

8 What would the daughter-product be when a carbon-14 isotope underwent a beta-minus decay?

(A) $^{10}_{4}Be$ (B) $^{14}_{5}B$

(C) $^{14}_{7}N$ (D) $^{13}_{7}N$

(Sample question)

9 Bohr's atomic model enabled the Balmer series to be explained. How did Bohr's model explain the limiting minimum wavelength of the Balmer series?

(A) There is a fixed energy between the $n = 2$ and $n = 3$ levels.

(B) There are an infinite number of orbitals an electron can occupy in the hydrogen atom.

(C) The energy difference between adjacent orbitals approaches zero for very large values of the principal quantum number (n).

(D) The principal quantum number has a specific limit for the Balmer series.

(Sample question)

10 How is the rate of reaction in a fission power reactor controlled?

(A) The amount of enrichment of the fuel is continually adjusted.

(B) A moderator is used to slow the high-speed neutrons.

(C) Neutrons are absorbed by metal rods.

(D) A coolant is continually circulated past the core to prevent it from overheating.

(Sample question)

11 What fundamental particles make up a deuterium $\left({}^{3}_{1}\text{H}\right)$?

(A) 5 up-quarks, 4 down-quarks and 1 electron

(B) 4 up-quarks, 5 down-quarks and 1 electron

(C) 2 up-quarks, 4 down-quarks and 1 electron

(D) 4 down-quarks, 2 up-quarks and 1 electron

(Sample question)

12 Which of the following is a limitation of the standard model?

(A) The standard model deals with only fundamental particles.

(B) The standard model does not include the electromagnetic force.

(C) The standard model does not include the gravitational force.

(D) The standard model cannot be used in cosmology.

(Sample question)

13 Within the nucleus, which of the following inequalities correctly shows the relative strengths of the four fundamental forces? Note that:

F_g = force of gravity

F_s = strong nuclear force

F_w = weak nuclear force

F_e = electromagnetic force

(A) $F_g < F_w < F_e < F_s$

(B) $F_w < F_g < F_e < F_s$

(C) $F_g < F_e < F_w < F_s$

(D) $F_g < F_w < F_s < F_e$

(Sample question)

14 Which of the following best describes a neutron?

(A) A neutron is a lepton made of two down-quarks and an up-quark.

(B) A neutron is a lepton made of two up-quarks and a down-quark.

(C) A neutron is a hadron made of two down-quarks and an up-quark.

(D) A neutron is a hadron made up of two up-quarks and a down-quark.

(Sample question)

15 Which of the following statements best describes why the following reaction releases energy?

$$^{2}_{1}H + ^{2}_{1}H \rightarrow ^{4}_{2}He$$

(A) The binding energy of $^{4}_{2}He$ is greater than the binding energy of $^{2}_{1}H$.

(B) The binding energy per nucleon of $^{4}_{2}He$ is greater than the binding energy per nucleon of $^{2}_{1}H$.

(C) The mass defect of $^{4}_{2}He$ is less than the mass defect of $^{2}_{1}H$.

(D) The binding energy per nucleon of $^{4}_{2}He$ is less than the binding energy per nucleon of $^{2}_{1}H$.

(Sample question)

16 Millikan measured the charge on the electron by suspending charged oil droplets between parallel charged plates. To find the charge on a droplet, what variables did Millikan need to determine?

(A) The voltage needed to suspend the droplet and the mass of the droplet

(B) The voltage needed to suspend the droplet, the separation between the parallel plates and the mass of the droplet

(C) The separation between the parallel plates, the number of electrons on the droplet, the voltage needed to suspend the droplet and the mass of the droplet

(D) The terminal velocity of the droplet and the separation between the parallel plates

(Sample question)

17 The radioactive decay series for $^{238}_{92}U$ is shown in the graph. Which of the isotopes (marked *p* to *q*) in the decay series would have the highest-binding energy per nucleon?

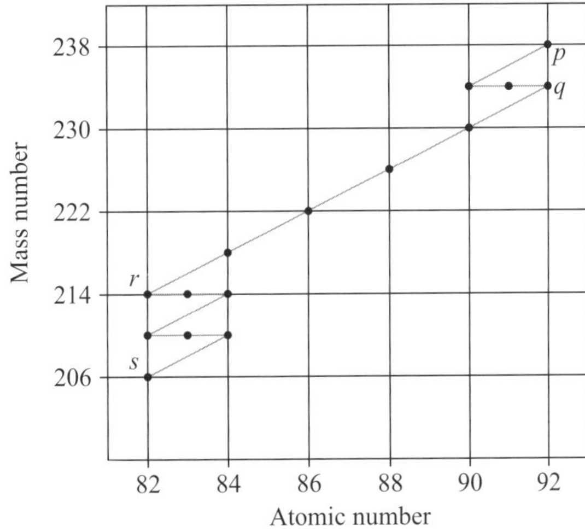

(A) *p* (B) *q*

(C) *r* (D) *s*

(Sample question)

18 How is energy produced in a red dwarf main sequence star?

(A) By hydrogen fusing to helium in the core via the CNO cycle

(B) By hydrogen fusing to helium in the core via the proton–proton (P–P) chain reaction

(C) By helium fusing to carbon in the core via the triple alpha reaction

(D) By the continual gravitational compression of the star causing gravitational potential energy to be converted to internal kinetic energy

(Sample question)

19 Two stars *X* and *Y* were observed to have a similar pattern of absorption lines in their spectra. All the absorption lines in star *Y*'s spectrum were, however, shifted to a slightly higher frequency and were broader than the absorption lines in the spectrum from star *X*. What could cause the difference in the observed spectra?

(A) Star *Y* could be moving away from the Earth faster and rotating faster than star *X*.

(B) Star *Y* could be rotating faster but moving towards the Earth more slowly than star *X*.

(C) Star *Y* could be rotating faster but moving away from the Earth more slowly than star *X*.

(D) Star *Y* could be moving towards the Earth faster than star *X* and be less dense than star *X*.

(Sample question)

20 How is Rydberg's constant (R) related to the longest wavelength (λ_α) emitted from the Balmer series of the hydrogen atom?

(A) $R = 6\lambda_\alpha$

(B) $R = 36\lambda_\alpha/5$

(C) $R = 36/5\lambda_\alpha$

(D) $R = 6/\lambda_\alpha$

(Sample question)

Part B Short-answer questions

Question 21 (6 marks)

(a) Thomson's experiment measures the charge/mass ratio of an electron. **3**

Use an annotated diagram to show how Thomson's experiment can be performed.

(b) An electron is projected at 90° into a magnetic field of 9×10^{-4} T, at a speed of 1×10^{7} ms^{-1}. This causes the electron to undergo uniform circular motion. **3**

Calculate the radius of the electron's path.

(6 lines) (q28, 2014 HSC)

Question 22 (4 marks)

(a) An atom of Carbon-12 has 6 protons and 6 neutrons in its nucleus. The mass of a Carbon-12 atom is 12.000 atomic mass unit. Show that the mass defect of one Carbon-12 atom is 0.097 atomic mass unit. **3**

(9 lines)

(b) How much energy is this mass defect equivalent to? **1**

(2 lines) (q31(c)(i) and (ii), 2008 HSC)

Question 23 (5 marks)

(a) A photon is emitted when an electron in a hydrogen atom transitions from the $n = 3$ excited state to the ground state. **2**

Calculate the wavelength of the photon.

(6 lines)

(b) A photon is incident on a hydrogen atom in the ground state. **3**

Explain, using de Broglie's hypothesis, why the photon is not absorbed by the hydrogen atom.

(q35(a)(i) and (ii), 2014 HSC)

Question 24 (6 marks)

Assess the impact of THREE advances in knowledge about particles and forces on the understanding of the atomic nucleus. **6**

(11 lines) (q34e, 2015 HSC)

Question 25 (5 marks)

(a) Identify the TWO types of nucleon and state ONE difference between them. **2**

(4 lines)

(b) Explain the stability of ${}^{4}_{2}$He nuclei in terms of TWO forces. **3**

(6 lines) (q26, 2016 HSC)

Question 26 (8 marks)

Particle accelerators have proved invaluable in understanding the fundamental building blocks and forces of nature.

(a) With the aid of a diagram explain how one type of particle accelerator produces high-energy particles. **4**

(6 lines) (Sample question)

(b) Explain how evidence from experiments involving particle accelerators and detectors has provided support for the standard model of matter. **4**

(11 lines) (q31, 2016 HSC)

Question 27 (7 marks)

(a) The diagrams show features of the hydrogen emission spectrum. **4**

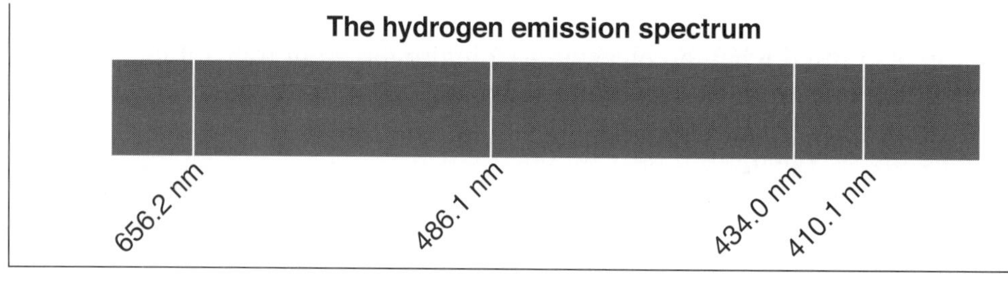

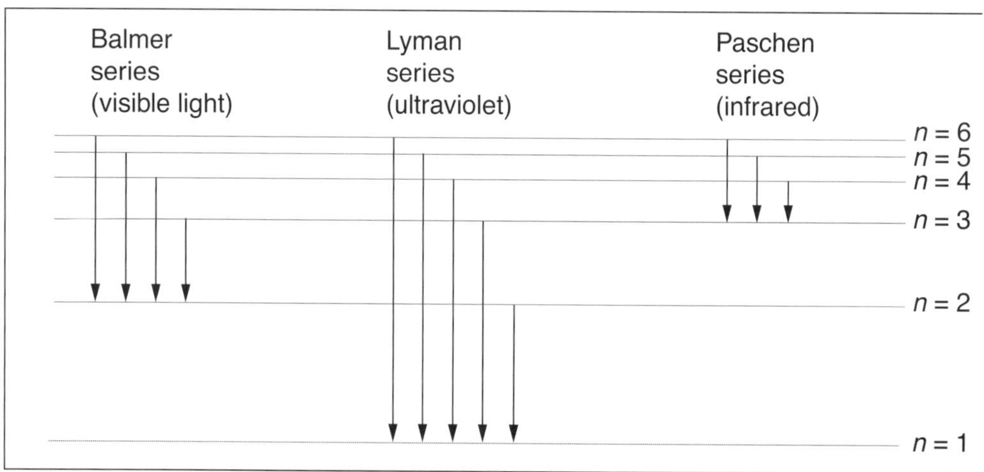

With reference to Bohr's postulates, explain how the line at 434.0 nm in the hydrogen emission spectrum is produced. Support your answer with calculations.

(b) How did de Broglie, and Davisson and Germer contribute to the modification of the Bohr model of the atom? **3**

(q25, 2017 HSC)

Question 28 (7 marks)

(a) Why do stars have different colours? **1**
(2 lines)
(Sample question)

(b) Describe how the distribution of stars on a Hertzsprung–Russell diagram relates to the processes that occur during their evolution. **6**
(12 lines)
(q27, 2016 HSC)

Question 29 (6 marks)

(a) The diagram shows the positions of stars *X* and *Y* on a H–R diagram. **2**

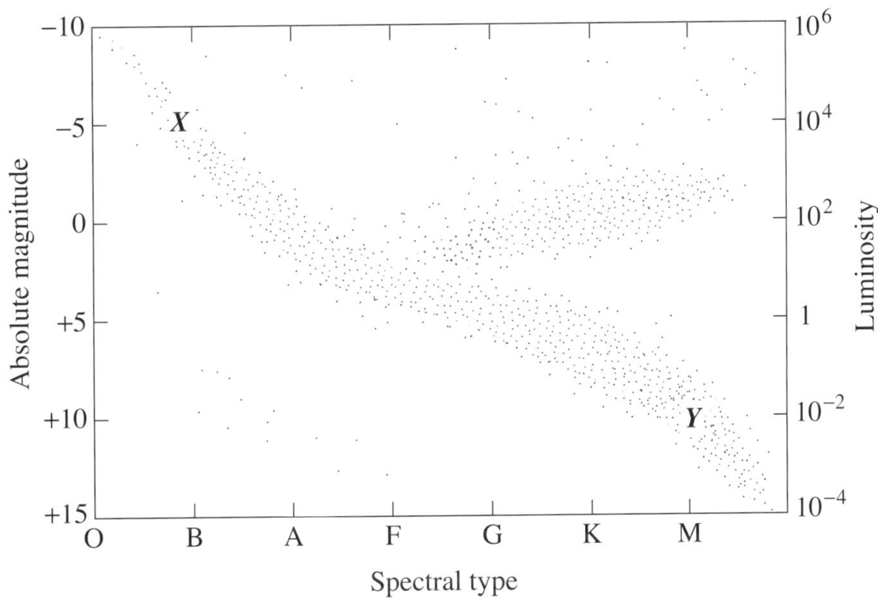

Outline the differences in the spectra of stars *X* and *Y*.

(b) Describe a process which can be used to obtain the spectrum of an individual star. **2**
(q26, 2017 HSC)

(c) Explain why star *X* would remain on the main sequence for a shorter time than star *Y*. **2**
(4 lines)
(Sample question)

Question 30 (5 marks)

(a) State the composition of the He-3 nucleus in terms of fundamental particles. **2**

(b) Outline features of the strong nuclear force. **3**
(q31, 2017 HSC)

Question 31 (6 marks)

Consider the nuclear reaction shown below. You may assume that the uranium nucleus was initially at rest.

$$^{238}_{92}\text{U} \rightarrow {}^{234}_{90}\text{Th} + {}^{4}_{2}\text{He}$$

Nucleus	*Mass (u)*
$^{238}_{92}\text{U}$	238.050 78
$^{234}_{90}\text{Th}$	234.043 59
$^{4}_{2}\text{He}$	4.002 60

(a) Explain why alpha particles can only travel a few centimetres through air. **1**

(b) Qualitatively compare the velocity of the two daughter products after the reaction. Justify your answer. **2**

(c) Determine the energy in joules released in this reaction. **3**

(Sample question)

Question 32 (6 marks)

(a) Define the term 'binding energy'. **1**

(b) Explain why the daughter products of nuclear fission are highly radioactive. **2**

(Sample question)

(c) The diagram shows some components of a nuclear reactor. **3**

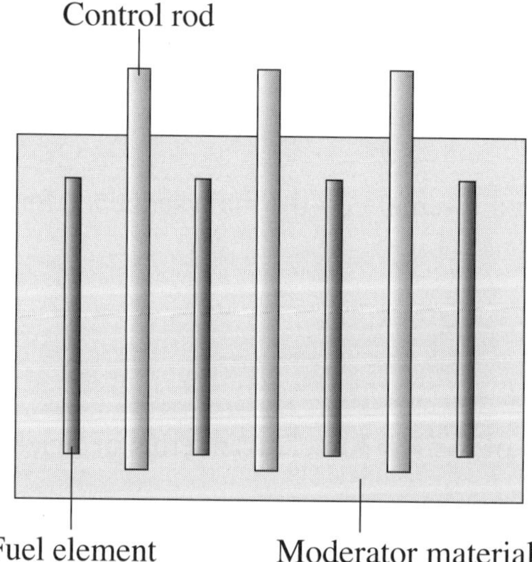

Explain how the labelled components work together to produce a controlled nuclear reaction.

(q32, 2017 HSC)

Question 33 (8 marks)

In the 1920s, Edwin Hubble made a series of telescopic observations that enabled him to plot the velocity of 46 galaxies as a function of their distance from the Earth. Hubble's results are shown below.

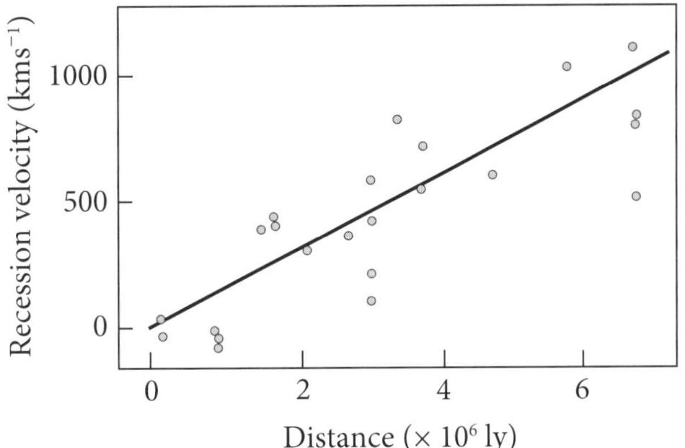

(a) Outline the significance of this graph to our understanding of the universe. **2**

(b) Explain how Hubble used telescopic observations to:

(i) find the distance to each galaxy. **3**

(ii) determine the velocity of each galaxy. **3**

(Sample question)

Question 34 (5 marks)

(a) A star X has a surface temperature of 30 000 K. Determine the wavelength at which star X would emit the most intense electromagnetic radiation. **1**

(2 lines)

(Sample question)

(b) A H–R diagram is shown. Star X is a main sequence star of 12 solar masses. **4**

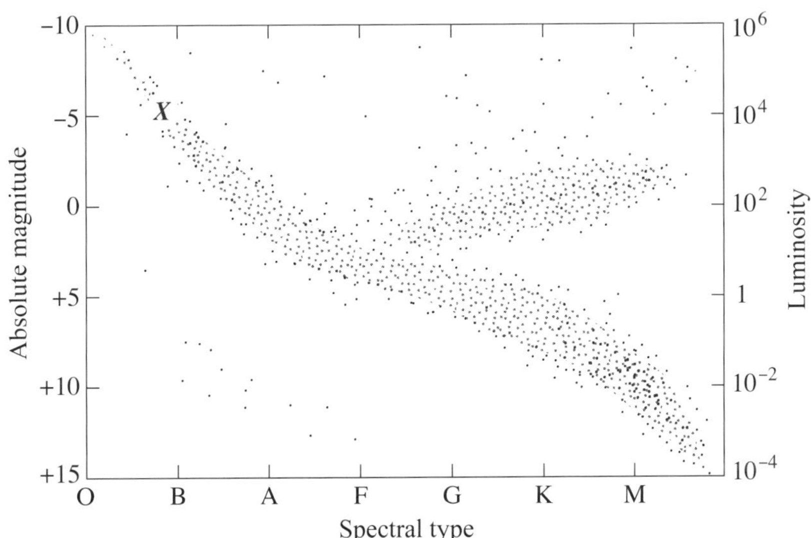

Describe how star X will change physically and chemically as it continues to evolve.

(12 lines)

(q23, 2018 HSC)

Question 35 (6 marks)

(a) What is the missing isotope marked $^{?}_{?}X$ in the reaction shown below? **1**

$$^{234}_{90}Th \rightarrow \, ^{?}_{?}X + \, ^{0}_{-1}\beta + \, ^{0}_{0}\overline{\nu}$$

(2 lines)

(b) The diagram shows apparatus used to investigate subatomic particles. **2**

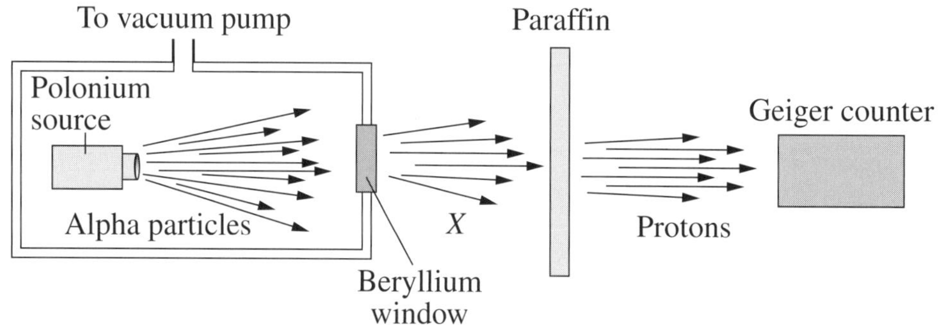

How did Chadwick use a law of physics to identify a property of *X*?

(6 lines)

(c) The following is a nuclear reaction that produces a neutron. **3**

$$^{4}_{2}\alpha + \, ^{9}_{4}Be \longrightarrow \, ^{12}_{6}C + \, ^{1}_{0}n$$

The table shows the masses of the particles in the reaction.

Particle	*Mass (u)*
$^{4}_{2}\alpha$	4.0012
$^{9}_{4}Be$	9.0122
$^{12}_{6}C$	12.0000
$^{1}_{0}n$	1.0087

Using the data from the table, calculate the energy released in this reaction. State your answer in joules.

(6 lines) (q25, 2018 HSC)

Question 36 (8 marks)

(a) Compare quarks and leptons and explain where each type of particle could be found in an atom. **4**

(10 lines)

(b) Compare a controlled and uncontrolled nuclear fission chain reaction and account for the energy released in such reactions. **4**

(12 lines) (q29, 2018 HSC)

Question 37 (7 marks)

In 1910 Robert Millikan conducted a famous experiment in which he used tiny oil droplets to find the charge on an electron. A schematic diagram of this experiment is shown below.

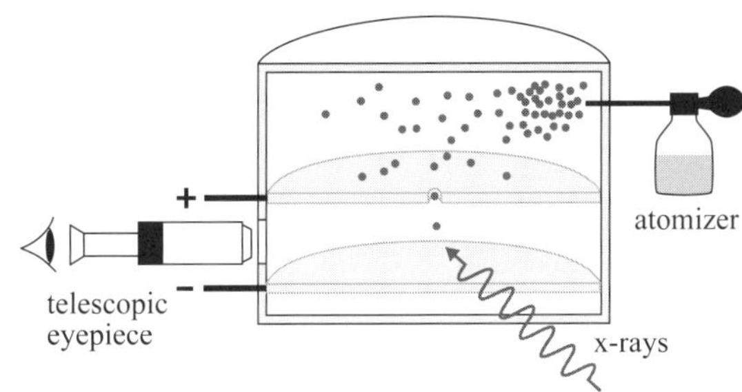

Consider an experiment in which the horizontally charged plates are 2 cm apart. An oil droplet of mass 5 mg is suspended between the plates when a potential difference of 250 V is placed across the charged plates. Note that the top plate is at the higher potential. The following questions refer to this experiment.

(a) What is the purpose of the X-rays in this experiment? **1**
(3 lines)

(b) Find the electric field between the charged horizontal plates. **2**
(3 lines)

(c) Calculate the charge on the suspended droplet. **2**
(3 lines)

(d) How was Millikan able to find the charge on a single electron when each oil droplet was multiply charged? **2**
(5 lines) (Sample question)

Question 38 (5 marks)

Bismuth-210 undergoes alpha decay and has a half-life of five days. **2**

(a) Write a nuclear decay equation for $^{210}_{83}\text{Bi}$.
(2 lines)

(b) What percentage of bismuth-210 would have undergone radioactive decay after three days? **3**
(6 lines) (Sample question)

Question 39 (5 marks)

Outline de Broglie's hypothesis and explain how it was experimentally verified. **5**
(14 lines) (Sample question)

Module 8

From the Universe to the Atom

Sample answers

Part A Objective-response questions

1 D A Hertzsprung–Russell diagram plots luminosity against spectral class (decreasing temperature).

2 C This is how the creation of matter is explained by the Big Bang theory.

3 D The forces created by the fields were in opposite directions but the magnetic field was arranged perpendicular to the electric field because magnetic fields create a force perpendicular to the velocity of the charge.

4 B This was the work of the German scientist Philipp Lenard.

5 A The droplet levitates when the downwards gravitational force is balanced by the upwards electrostatic force or $mg = qE = \frac{qV}{D}$.

Hence $V = \frac{mgd}{q} = \frac{(2 \times 10^{-9} \times 9.8 \times 0.01)}{(2 \times 10^{6} \times 1.6 \times 10^{-19})} = 612$ V (the closest answer is A).

6 B The percentage remaining after six half-lives would be

$100 \times \left(\frac{1}{2}\right)^2 = \frac{100}{64} = 1.56\%$; therefore the amount decayed would be

$100 - 1.56 = 98.4\%$.

7 A The kinetic energy is proportional to the square of the velocity. Doubling the kinetic energy would occur when the velocity was increased by a factor of $\sqrt{2}$. Now as the de Broglie wavelength is inversely proportional to the velocity

($\lambda = \frac{h}{mv}$), the new wavelength will be $\frac{7}{\sqrt{2}} = 5$ nm.

8 C Beta-minus decay occurs when a neutron changes into a proton in the nucleus and an electron (beta particle) is ejected. Hence beta-minus decay always produces a daughter product with one extra proton and the same number of nucleons.

9 C Because the energy gap between levels approaches zero as n approaches infinity, the gap between the $n = 2$ and very large values of n will approach a specific limiting value.

10 C Cadmium or boron control rods are moved in and out of the reactor core to absorb neutrons to control the rate at which fission occurs.

11 B A proton has two up-quarks and one down-quark, and a neutron has one up-quark and two down-quarks.

12 C The standard model only covers three of the four fundamental forces of nature.

13 C This is the correct order for the operation of the forces within the nucleus.

14 C A hadron is a particle composed of quarks and the neutron contains two down quarks (each with a charge of $-\frac{1}{3}$) and one up quark (charge $+\frac{2}{3}$) to ensure the net charge on the neutron is zero.

15 B Binding energy per nucleon is a measure of stability (and mass defect per nucleon). If the binding energy per nucleon increases in a reaction, mass is lost and hence energy released.

16 B Millikan equated the electric and magnetic forces on the droplet and hence $q\left(\frac{V}{d}\right) = mg$ or $q = \frac{mgd}{V}$. Millikan therefore had to measure m, d and V.

17 D The binding energy per nucleon is a measure of the stability of the nucleus. Because each radioactive decay produces a more stable nucleus, each decay must produce a daughter-product with a higher binding energy per nucleon. The final product must therefore be the most stable nucleus that will have the highest-binding energy per nucleon.

18 B Main sequence stars all fuse hydrogen to helium in the core and as this is a red dwarf star it is a very low mass star and hydrogen fusion will occur via the proton–proton (P–P) chain reaction. This is because the P–P chain requires a slightly lower ignition temperature than the CNO cycle.

19 C The lines in star Y's spectrum are blue shifted with respect to star X. If star X was moving away from the Earth more slowly than star Y, the spectra from star X would exhibit a greater red shift and hence the spectra of star X would be shifted to longer wavelengths than the spectra of star Y. The spectral lines of rapidly rotating stars are broadened due to the simultaneous red and blue shift produced by the motion of each side of the star. The broadened spectral lines in star Y's spectrum could therefore be due to the star rotating rapidly.

20 C Bohr suggested that the longest wavelength of the Balmer series would be produced when an electron moved from the $n = 3$ orbital to the $n = 2$ orbital and, hence, by applying Bohr's equation,

$$\frac{1}{\lambda} = R\left[\frac{1}{n_f^2} - \frac{1}{n_i^2}\right]$$

$$\frac{1}{\lambda_\alpha} = R\left(\frac{1}{2^2} - \frac{1}{3^2}\right)$$

$$= \frac{5R}{36}$$

and hence,

$$R = \frac{36}{(5\lambda_\alpha)}$$

Part B Short-answer questions

Question 21 (Total 6 marks)

(a)

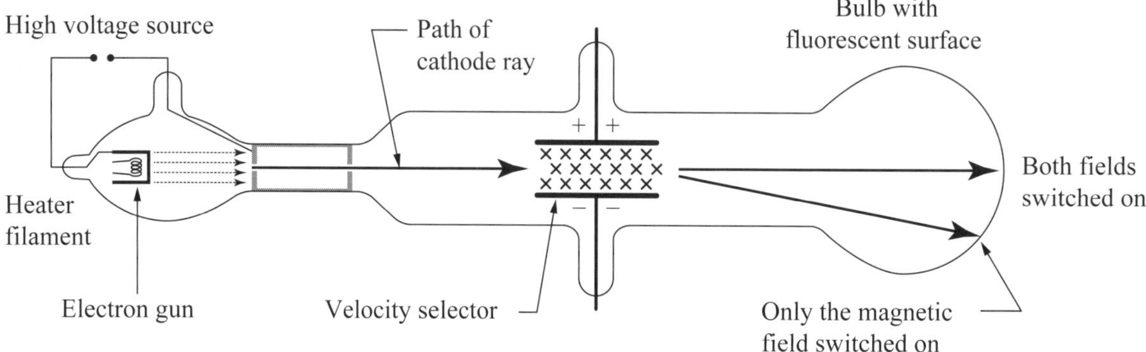

Thomson's cathode-ray tube

(3 marks)

(b) The centripetal force acting on the electron is the magnetic force, so FC ≡ FB.

$$\therefore \frac{mv^2}{r} = Bqv \sin \theta$$

$$\theta = 90° \text{ so } \frac{mv}{r} = Bq$$

$$\therefore r = \frac{mv}{Bq} = \frac{(9.109 \times 10^{-31})(1 \times 10^{7})}{(9 \times 10^{-4})(1.602 \times 10^{-19})} = 6.32 \times 10^{-2} \text{ m}$$ *(3 marks)*

Question 22 (Total 4 marks)

(a) Mass when unbound of individual components in carbon-12 atom, i.e.:

Mass of 6 protons = $6 \times 1.673 \times 10^{-27} = 1.0038 \times 10^{-26}$ kg.

Mass of 6 neutrons = $6 \times 1.675 \times 10^{-27} = 1.0050 \times 10^{-26}$ kg.

Mass of 6 electrons = $6 \times 9.109 \times 10^{-31}$ kg = 5.4654×10^{-30} kg.

Total mass = 2.0093×10^{-26} kg. Converting to amu,

$$\frac{2.0093 \times 10^{-26}}{1.661 \times 10^{-27}} = 12.0972 \text{ amu.}$$

The mass of C-12 is 12.000 amu, showing that a total mass defect of 0.097 amu occurs when the particles become bound together in the atom. *(3 marks)*

(b) 0.097 amu = $1.611\,17 \times 10^{-28}$ kg.

Now $E = mc^2 = 1.6112 \times 10^{-28} \times 9 \times 10^{16} = 1.45 \times 10^{-11}$ J. *(1 mark)*

Question 23 (Total 5 marks)

(a) $\frac{1}{\lambda} = R\left(\frac{1}{n_f^2} - \frac{1}{n_i^2}\right)$

$= 1.097 \times 10^7 \left(1 - \frac{1}{9}\right)$

$= 9\,751\,111$

Therefore $\lambda = 1.02 \times 10^{-7}$ m *(2 marks)*

(b) De Broglie's hypothesis suggested that particles display wave-like behaviour, which can be characterised by $\lambda = \frac{h}{mv}$. He applied this when describing stable electron states using the idea of standing wave patterns. In this model, there needs to be whole numbers of electron wavelengths in each stable state/energy level/circular orbit. This is shown in the equation $n\lambda = 2\pi r$. So, for an electron to absorb a photon, the photon must have an exact amount of energy, which if absorbed will allow the electron standing wave pattern to gain a whole number (n) of wavelengths to fit into a new stable orbit. Since the energy of the photon is E 5 hf, the photon will not be absorbed if it does not have an appropriate frequency. *(3 marks)*

Question 24 (Total 6 marks)

A number of answers are possible. The following is an example.

Rutherford's alpha scattering experiments identified that atoms had a nucleus that was positively charged, contained most of the mass of the atom and was tiny compared with the size of the atom. He postulated that electrostatic forces of attraction between the negative electrons and the positive nucleus provided the centripetal force needed to keep the electrons in orbit. This was a huge change in the understanding of the structure of atoms, previously thought to be solid particles (Dalton and Thompson). Rutherford's model largely persists today.

Experiments, such as those conducted by Thompson and Chadwick, identified that the nuclei of atoms heavier than hydrogen contained multiple protons and neutral neutrons and that atoms of the same element could have nuclei of different mass, i.e. isotopes. This realisation that multiple positive charge particles (protons) were held close together in a tiny nucleus required an explanation as to why the nucleus did not fly apart due to what ought to be enormous forces of electrostatic repulsion. The concept of a force that was much stronger than the electrostatic force, called the strong nuclear force, provided a plausible explanation. This force operated only over very short distances between protons and protons, protons and neutrons, and neutrons and neutrons.

Much more recently, analysis of experiments involving collisions between charged particles accelerated to enormous speeds (energies) in particle accelerators has had a major impact on modern understanding of the atomic nucleus. These experiments have revealed the existence of more than 60 previously unknown subatomic particles and have led to the development of the Standard Model that includes quarks, leptons and bosons. In particular, probing protons with very high energy electrons has shown protons to be made up of three quarks—two up quarks and a down quark. Similarly, neutrons have been found to be made up of one up quark and two down quarks. Up quarks have a charge of +2/3 ($\times$ 1.6 $\times$ 10^{-19} C) and down quarks have a charge of –1/3 (× 1.6 × 10^{-19} C). This explains the net charge value on protons and neutrons. In the Standard Model, the strong force that binds quarks together within the protons and neutrons is due to attraction between quarks with a different colour charge (red, green or blue) and is mediated by bosons called gluons. This strong force is much stronger than the previously named strong nuclear force between nucleons, which is now considered to be only a residual part of the strong force that binds the quarks together inside the protons and neutrons of the nucleus. *(6 marks)*

Question 25 (Total 5 marks)

(a) Proton and neutrons
Protons are positively charged and neutrons have no net charge. (*2 marks*)

(b) The two positive protons experience an electrostatic/*electromagnetic force* of repulsion. This repulsive force is overcome by the much stronger *nuclear force* that exists between all of the nucleons when they are very close together and hence the nucleus holds together. The nuclear force is a residual of the strong nuclear force between the up and down quarks that make up the protons and neutrons. (*3 marks*)

Question 26 (Total 8 marks)

(a) Students could describe the operation of a cyclotron, linear accelerator or synchrotron. For example:
A schematic diagram of a linear particle accelerator is shown in the diagram below.

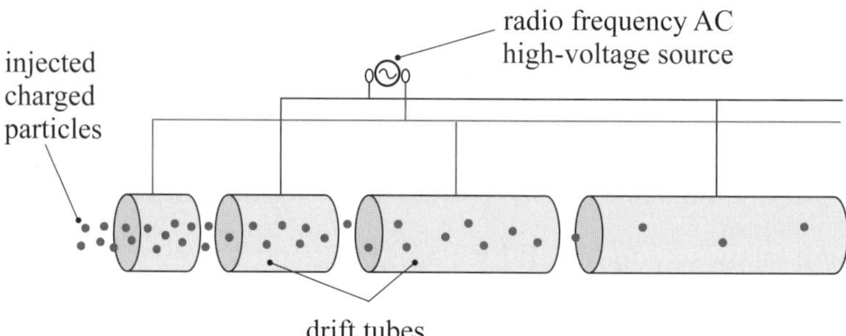

As the name implies, linear accelerators (LINACS), accelerate a particle in a straight line. As shown in the figure above, linear accelerators consist of a large number of evacuated, hollow metal cylinders called drift tubes. The length of the drift tubes increases to ensure the particle will spend the same amount of time in each tube so a constant-frequency AC voltage can be applied to the cylinders.

A high-voltage radio frequency source is connected to each pair of cylinders to accelerate the charged particle as it moves between the cylinders. More energy is imparted to the particle as it moves between each pair of drift tubes. Initially the work done on the charge will accelerate the charge until it approaches the speed of light. As the speed of light is the limiting velocity, the work done on the particle begins to increase the effective mass of the particle rather than its velocity as it approaches the speed of light. This increases the kinetic energy and momentum of the particle without significantly changing the velocity. In this way very high energy particles, travelling very close to the speed of light can be produced by large LINAC particle accelerators. *(4 marks)*

(b) Particle accelerators such as the Large Hadron Collider (LHC), which is fed particles such as protons that have already been accelerated to energies of 450 GeV by a chain of accelerators including linear accelerators and synchrotrons, have been used to gather data about the nature of matter. High energy collisions between these high energy particles and matter, such as lead, in the targets have yielded evidence for a wide range of subatomic particles.

The energies, momentum and charges of these subatomic particles and their decay products have been identified and measured using complex detectors that include tracking devices with strong magnetic fields which measure paths of charged particles; calorimeters which absorb particles and photons and measure their energies; and other special detectors that measure photon energies from Cherenkov radiation. An example of a complex detector is the huge 7000-tonne ATLAS detector in the LHC.

Specific measurements are possible for some longer lived particles such as the Tau lepton. The existence of other particles is inferred through the measurement of their decay products.

Although individual quarks have not been isolated, a wide variety of hadrons (collection of quarks) have been detected and from the properties and decay products of these the six quarks (up, down, strange, charm, top and bottom) have been inferred.

In addition, six leptons (electron, electron neutrino, muon and muon neutrino, tau and tau neutrino) have been identified.

The Standard Model of Matter used evidence of the existence of some particles to predict the existence of many others in a systematic classification. Particle accelerators and detectors have enabled the discovery of many of these particles thus supporting the SMoM.

(4 marks)

Question 27 (Total 7 marks)

(a) Bohr postulated that electrons could orbit the nucleus in specific stationary energy states without emitting radiation. An electron could move from one of these specific states to another only by absorbing or emitting the energy of a photon. The energy difference between the two energy levels is thus $E = hf$ where h is Planck's constant and f is the frequency of the photon. The visible lines in the hydrogen emission spectrum are the Balmer series. For a specific visible wavelength of light to be produced in the hydrogen spectrum, electrons must move from a higher excitation level to level 2, emitting the energy as photons with specific energy equivalent to the energy of the move.

The level from which the electron must move can be determined using the Rydberg equation:

$$\frac{1}{\lambda} = R\left(\frac{1}{n_f^2} - \frac{1}{n_i^2}\right)$$

$$\frac{1}{4.34 \times 10^{-9}} = 1.097 \times 10^7 \left(\frac{1}{2^2} - \frac{1}{n_i^2}\right)$$

$$n_i^2 = \frac{1.097 \times 10^7}{4.39 \times 10^5}$$

$$= 25$$

Therefore $n_i = 5$.

Hence electrons that move from the 5th stationary state to the 2nd stationary state will release the energy as photons with wavelength 434 nm. *(4 marks)*

(b) De Broglie suggested that electrons could behave like waves and have a wavelength λ,

$$\lambda = \frac{h}{p} = \frac{h}{mv}$$

and that when electrons are in the stable states (of the Bohr model), they are in a standing wave situation in which the circumference of the orbitals is exactly equivalent to a whole number of the electron's wavelength. Thus Level 1 was the first stationary state with a circumference of just one wavelength. The second allowable level was when the circumference was exactly equivalent to 2 wavelength, etc.

Davisson and Germer directed beams of electrons at the surface of a nickel crystal and identified that the scattering pattern was equivalent to a diffraction pattern consistent with the electron wavelength predicted by de Broglie.

Thus both the theoretical prediction of de Broglie and the experimental evidence of Davisson and Germer were able to explain the existence of the stable electron states that are part of the Bohr model of the atom and provide justification for the particular energy values associated with each stationary state. *(3 marks)*

Question 28 (Total 7 marks)

(a) Because stars have different surface temperatures they emit light over a different range of wavelengths like black bodies at different temperatures (i.e. $\lambda_{max} = \frac{b}{T}$). Different colours represent different wavelengths of light and hence stars with different surface temperatures will have different colours. *(1 mark)*

(b)

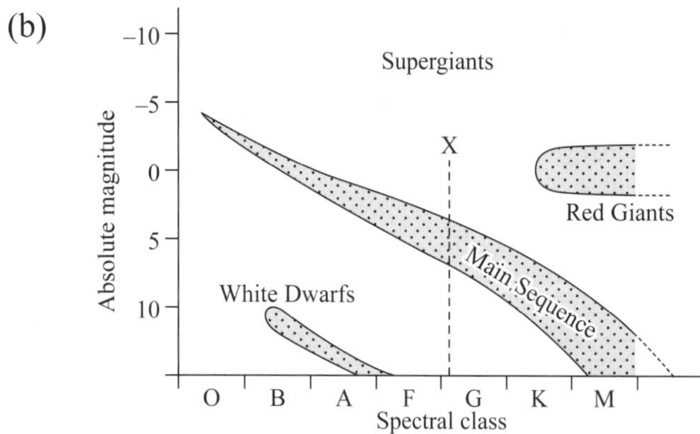

The Hertzsprung-Russell Diagram

[It would be essential to sketch a reasonable H-R diagram, labelling the axes and identifying at least the correct locations of the Main Sequence, Red Giants and White Dwarfs to gain maximum marks for this question.]

When stars evolve sufficiently to begin to emit sufficient light to become part of the H-R diagram, they are located at some point on the Main Sequence, where they will spend most of their 'life' because the rate of fusion in their core is least in this phase. Where that point will be depends entirely on their mass—the greater their mass the greater the gravitational pressure acting on their core, hence the greater the radiation pressure the core must provide to counteract it. Larger stars are hotter, 'bluer' and more luminous. Since a star's mass decreases only slightly during its life, its location barely changes throughout this stage, though its absolute magnitude decreases slightly (the star becomes brighter) so it moves upwards across the Main Sequence. During this phase of its life the star is fusing hydrogen in its core, converting it into helium.

Once the star exhausts the hydrogen fuel in its core it evolves, initially imploding under its own gravitational pressure until the heat released this way causes a new hydrogen shell around the core to fuse, whilst the helium within the core reaches critical temperature and also starts fusing into carbon. The combined fusion activity produces far greater radiation pressure than before so the star expands to around 100 times its original diameter—and its surface area increases 10 000 times. The star now releases more energy than before, but due to the vast increase in surface area the energy released per square metre is far less, so its surface is far cooler; it is now only red-hot. This explains why it is now a Red Giant, spectral class K or M, but thousands of times larger and more luminous than Main Sequence K or M stars. Since the rate of fusion within the core of the star is much greater, this phase is far shorter, so there are far fewer Red Giants than there are Main Sequence stars.

Stars far more massive than the Sun continue the sequence of exhausting their fuel, collapsing, igniting further shells of hydrogen as well as new fusion in their cores, consequently expanding to become supergiants. When these finally have no further fuel and collapse, they supernova, becoming neutron stars or black holes too dim to be included on the H-R diagram any longer.

The extreme rate of fusion of these stars means they exhaust all their fuel yet more rapidly than the Red Giants do, and 'die' more rapidly, so they are comparatively even rarer.

Those, like the Sun, having more moderate masses exhaust their useful energy supply at the end of their Red Giant stage. They implode again under the now irresistible gravitational pressure until their atoms are squeezed so tightly against one another the contraction ceases. They are now similar in size to the Earth. Their core is dead—in fact, the star itself is now a corpse. But the gravitational potential energy released as they collapse is converted to heat, so they are now very hot. They would therefore actually be blue in colour, but the gravitational force at the surface even affects the light they emit, so they are only white-hot and tiny—therefore less luminous—White Dwarfs. Now dead, the heat gradually disperses, and they fade away into obscurity as they become redder. (*6 marks*)

Question 29 (Total 6 marks)

(a) Star *X* is a much hotter main sequence star than star *Y*.

The peak wavelength of the spectra for star *X* will be much shorter than that of star *Y*. The peak for star *X* will be toward the blue end of the visible spectrum, and star *Y* will have a peak wavelength in the red end of the spectrum. The total area under the spectral curve for star *X* will be much greater than star *Y* as it is radiating far more energy.

The absorption spectra for star *X* will show ionised helium lines and moderate hydrogen lines.

Star *Y* will show lines corresponding to sodium, calcium and even molecular species. *(2 marks)*

(b) To obtain the spectra of a star, its light is narrowed to a thin beam by a slit. This narrow beam is then directed to a collimator that creates parallel light rays. These rays are shone onto a diffraction grating which separates the spectrum into its component wavelengths. The diffracted light is then focused onto a CCD which records the intensity of light for each part of the spectrum. *(2 marks)*

(c) Star *X* is much more massive than star *Y* and hence *X* must fuse hydrogen to helium in the core at a much faster rate to balance the inwards pressure of gravity. This increase in the rate of fusion will cause it to form an inert helium core much faster than *Y* and hence *X* will move off the main sequence faster than *Y*. *(2 marks)*

Question 30 (Total 5 marks)

(a) The He-3 nucleus consists of two protons on one neutron. Each proton is made up of two up quarks and one down quark. The neutron consists of one up quark and two down quarks. Hence, in terms of fundamental particles, the He-3 nucleus consists of five up quarks and four down quarks. (*2 marks*)

(b) The strong nuclear force:

- acts between nucleons; i.e. between protons and protons, protons and neutrons; and neutrons and neutrons when nucleons are very close together (approximately 10^{-15} m) and accounts for the stability of the stable nuclei
- is stronger than the repulsive electrostatic forces between protons
- is repulsive when nucleons are closer than 10^{-15} m and thus the nuclear force limits the size of the nucleus
- is not significant at distances outside the nucleus. (*3 marks*)

Question 31 (Total 6 marks)

(a) Alpha radiation is highly ionising (as ${}^{4}_{2}\text{He}$ is a large, doubly charged nucleus) and, consequently, as it passes through the air it rapidly loses kinetic energy by ionising air molecules. This slows the alpha particles and prevents it from travelling more than a few centimetres through the air. *(1 mark)*

(b) As momentum is conserved and the initial momentum was zero, the momentum of the two products must be equal and opposite. As the mass of the alpha particle is much smaller than the mass of the thorium nucleus, the alpha particle must be travelling much faster than the thorium nucleus after the collision. *(2 marks)*

Alternatively, students may answer this question mathematically, as follows.

As momentum is conserved and the initial momentum was zero, the momentum of the two products must be equal and opposite. Calling the mass and velocity of the thorium nucleus *M* and *V*, respectively, and the mass and velocity of the alpha particle *m* and *v*, respectively, we can write

$MV = mv$ or $v = \left(\frac{M}{m}\right)V$.

Hence, as $M >> m$, the alpha particle will have a much higher velocity (v) than the thorium nucleus (V).

(c) We first find the mass defect:

Δm = mass of products – mass of reactants

$= (234.043\,59 + 4.002\,60) - 238.050\,78 = -0.004\,59$ u

The negative sign tells us that mass is lost (i.e. energy is released) in this reaction.

Now the energy released in MeV is given by:

$E = 0.004\,59 \times 931.5 = 4.2756$ MeV

The energy is joules will be given by:

$E = (4.2756 \times 10^6) \times 1.602 \times 10^{-19} = 6.85 \times 10^{-13}$ J *(3 marks)*

Alternatively, students may choose the convert the masses to kilograms and use $E = mc^2$ to find the energy in joules.

Question 32 (Total 6 marks)

(a) Binding energy is the energy required to separate a nucleus into its constituent nucleons. *(1 mark)*

(b) Fission only occurs in some very large nuclei. Large nuclei require a higher neutron to proton ratio for stability than smaller nuclei. Hence when a large nucleus absorbs a neutron and fissions, it splits into two smaller nuclei that will have neutron to proton ratios that are too high for nuclei of this size. Because these smaller nuclei have too many neutrons to be stable they will be highly radioactive and will undergo a series of radioactive decays until their neutron to proton ratio has been reduced to the value that will ensure the nuclei are stable. *(2 marks)*

(c) In a nuclear reactor the fuel elements contain radioactive material, such as uranium, which undergoes nuclear fission—usually with a very long half-life. During the fission reaction fast neutrons are released that can be captured by the nuclei of other atoms in the fuel cells, causing them to undergo fission and release more neutrons.

The rate at which the atoms in the fuel elements can capture the fast neutrons is greatly enhanced if the neutron speed is reduced. This is the role of the moderator material (such as D_2O). Thus the moderator is used to greatly enhance the number of fission reactions and enhance the neutron density inside the reactor. This produces a chain reaction that, if left unchecked, would cause a catastrophic build-up of heat.

The control rods absorb neutrons and are raised or lowered to manage the neutron density, thus controlling the rate of the fission reactions inside the reactor and preventing the chain reaction. Therefore the control rods are used to manage the heat produced by the reactor so that it is at a sustainable level. *(3 marks)*

Question 33 (Total 8 marks)

(a) Before Hubble made these observations, the universe was thought to be static. Hubble's graph shows that the further a galaxy is away from the observer the faster it is moving away from the observer. This would only occur if the universe was expanding, rather than remaining static. Note that the gradient of the line of best fit is called Hubble's constant and is a measure of the rate at which the universe is expanding. *(2 marks)*

(b) (i) Hubble observed a specific type of star called a Cepheid variable star in each galaxy. Cepheid variable stars change intensity with a period that is related to their luminosity. Thus by measuring the period of variability of the Cepheid, Hubble could estimate its luminosity. By assuming the intensity of the light from the star decreased as $\frac{1}{r^2}$ and using the apparent brightness of the star when viewed from earth and its luminosity, Hubble was then able to estimate the distance to the galaxy (r). *(3 marks)*

(ii) Hubble compared the absorption lines in the spectrum from each galaxy to the known absorption spectra from elements on Earth. He found that all the absorption lines in the spectrum from the galaxies were shifted to longer wavelengths. This redshift is produced when the source of the light is moving away from the observer, and the size of the redshift can be used to calculate the velocity of the source of the light (i.e. the recession velocity of the galaxy). *(3 marks)*

Question 34 (Total 5 marks)

(a) Applying Wien's displacement law,

$$\lambda_{max} = \frac{b}{T} = \frac{2.898 \times 10^{-3}}{3 \times 10^{4}} = 9.66 \times 10^{-8}\ \text{m} = 97\ \text{nm}$$ *(1 mark)*

(b) From its position on the H–R diagram (and its mass) we see that star X is a hot, blue–white, B-class giant main sequence star fusing hydrogen to helium in its core via the CNO reaction. Star X will have a comparatively short life span on the main sequence and when it develops a large enough helium core it will collapse until the helium starts to fuse to carbon and the increased core temperature also starts hydrogen fusion in a shell around the core. This will cause the star to greatly expand and the outer layers to cool, forming a red supergiant star.

The star will progressively fuse heavier elements in the core and in shells around the core until an iron core forms. The star slowly loses mass while it is fusing elements because some mass is converting into energy in accordance with Einstein's mass–energy relationship $E = mc^2$. As the fusion of iron consumes rather than releases energy, the star will collapse and then explode in a supernova and spread much of its mass outwards into the surrounding space. If the remaining core is large enough it may collapse into a neutron star with a diameter of about 20 km.

Large stars have short life spans and hence must be second or higher generation stars. Chemically such stars are initially made up mostly of hydrogen with some helium and trace amounts of heavier elements.

As the star passes through its life span, fusing heavier and heavier elements, the percentage of heavier elements (up to iron) gradually increases. When it explodes in a supernova some elements heavier than iron will be produced by the huge amount of energy released. The core may then collapse to a neutron star, which is essentially nuclear material rather than atomic material. *(4 marks)*

Question 35 (Total 6 marks)

(a) By balancing the charge and number of nucleons on each side of the equation we see that the missing isotope is $^{234}_{91}Pa$. *(1 mark)*

(b) Chadwick used kinematics to determine the speed of the particles labelled *X* and the recoil velocity of the protons and nitrogen nuclei after they had been hit by the particles labelled *X* in the diagram.

By using conservation of momentum and energy he was then able to calculate the mass of the particles labelled *X*.

He found that these neutral particles were slightly heavier than a proton and hence discovered that the unknown particles labelled *X* were the long sought-after 'neutrons'. *(2 marks)*

(c) The mass defect will be given by

$$\begin{aligned}\Delta m &= \textit{mass of products} - \textit{mass of reactants} \\ &= 12.0000 + 1.0087 - (4.0012 + 9.0122) \\ &= -0.0047 \text{ u and hence } 0.0047 \text{ u must be converted into energy in this reaction.}\end{aligned}$$

$$\begin{aligned}\text{Energy released} &= 0.0047 \times 931.5 = 4.378 \text{ Mev} \\ &= 4.378 \times 10^{6} \times 1.602 \times 10^{-19} \\ &= 7.014 \times 10^{-13} \text{ J}\end{aligned}$$

OR students may convert the mass defect to kilograms and then use $E = mc^2$ as follows,

$$\begin{aligned}\Delta m &= 0.0047 \times 1.661 \times 10^{-27} \\ &= 7.8067 \times 10^{-30} \text{ kg}\end{aligned}$$

$$E = mc^2 = 7.8067 \times 10^{-30} \times (3 \times 10^8)^2 = 7.026 \times 10^{-13} \text{ J}$$

(Note that the answer differs slightly from the one above because the speed of light is $2.997\,924\,58 \times 10^8$ rather than 3×10^8 as given in the data sheet.) *(3 marks)*

Question 36 (Total 8 marks)

(a) Quarks are comparatively massive fundamental particles with fractional charges.

Quarks are affected by all the fundamental forces and obey the Pauli exclusion principle. Quarks are bound tightly together by gluons to form mesons and baryons. Mesons consist of two quarks, while baryons are made up of three quarks. Protons and neutrons are baryons. Protons are made up of two up quarks each carrying a charge of $+\frac{2}{3}$ and one down quark which carries a charge of $-\frac{1}{3}$, giving a net charge of +1. Neutrons are made up of two down quarks and one up quark giving a net charge of zero. Thus quarks are found in the atomic nucleus.

Leptons are comparatively light fundamental particles that obey the Pauli exclusion principle but which are not affected by the strong (colour) force.

Leptons can be charged particles like the electron or uncharged particles like the electron neutrino. Electrons (leptons) surround the nucleus of the atom. *(4 marks)*

(b) Fission reactions are artificial nuclear reactions, which involve very large nuclei absorbing a neutron and splitting into two smaller nuclei. Fission of uranium-235 and some other very large isotopes are accompanied by the release of two or more neutrons. As these neutrons can be used to initiate further fission reactions, it is possible to set up a fission chain reaction in a suitable (critical) mass of fissionable material. When a large nucleus undergoes fission, the products are more stable than the reactants and hence some mass is converted into energy in the reaction.

An uncontrolled nuclear chain reaction will occur if the mass of fissionable material is greater than the critical mass for the material. Uncontrolled fission reactions run at an increasing rate because each fission reaction initiates more than one further fission reaction.

A fission bomb is an example of an uncontrolled fission reaction.

A controlled nuclear fission reaction is one that runs at a constant rate. That is, only one of the neutrons released in each fission reaction is used to initiate a further fission reaction.

This type of reaction is used in fission power stations where cadmium control rods are used to absorb the excess neutrons and hence control the rate at which the reaction occurs. *(4 marks)*

Question 37 (Total 7 marks)

(a) The X-rays were used to charge the oil droplets.

(X-ray photons are energetic enough to ionise the air molecules between the charged plates and the free electrons tend to stick to the oil droplets.) *(1 mark)*

(b) $E = \frac{V}{d} = \frac{250}{0.2} = 12\,500 \text{ Vm}^{-1}$ downwards *(2 marks)*

(c) First note that 5 mg = 5×10^{-6} kg

As the charge remains stationary, the forces must sum to zero and hence,

$mg = qE$, or $q = \frac{mg}{E} = \frac{5 \times 10^{-6} \times 9.8}{12\,500} = -3.92 \times 10^{-9}$ C

As the electric field is directed downwards the charge must be negative to ensure the electric force opposes the gravitational force on the droplet. *(2 marks)*

(d) Millikan reasoned that each droplet would have a whole number of additional electrons and hence each droplet would have a charge that was a multiple of the electron charge.

To find the charge on the electron he looked for a common denominator that would divide evenly into each value of charge.

This common denominator was the quantum of charge; the charge of the electron. *(2 marks)*

Question 38 (Total 5 marks)

(a) $^{210}_{83}\text{Bi} \rightarrow {}^{206}_{81}\text{Tl} + {}^{4}_{2}\text{He}$ *(2 marks)*

(b) We first find the decay constant, $\lambda = \frac{ln2}{t_{\frac{1}{2}}} = \frac{ln2}{5} = 0.1386$.

The amount remaining will then be given by
$N_t = N_0 e^{-\lambda t} = N_0 e^{-0.1386 \times 3} = 0.66 N_0$

Hence 66% of the initial isotopes will remain after three days and 34% would have decayed.

Alternatively, students could use:

$N_t = N_0(\frac{1}{2})^n$, where n is the number of half-lives that have elapsed.

$n = \frac{3}{5} = 0.6$ half-lives.

Hence, $N_t = N_0(\frac{1}{2})^n = N_0(\frac{1}{2})^{0.6} = 0.66 N_0$ and again we get 66% remaining and so 34% of the bismuth-210 would have undergone radioactive decay. *(3 marks*

Question 39 (Total 5 marks)

De Broglie suggested that particles, such as electrons, had an associated wavelength that was related to the momentum (mv) of the particle by $\lambda = \frac{h}{mv}$.

He also proposed that particles would exhibit both particle and wave properties in different circumstances.

De Broglie's hypothesis was first verified by a series of experiments conducted by Davisson and Germer. They showed that when electrons were scattered by a pure nickel crystal the electrons were not scattered randomly, as they expected for a particle, but instead were scattered predominantly in specific directions.

Previous studies with X-rays had shown that the X-rays scattered by atoms in a crystal interfered with one another to produce maxima and minima of interference in the scattered X-rays. For X-rays with the same wavelength as the electrons, the angle of the X-ray maxima corresponded to the angle at which most electrons were scattered. Thus the scattered electrons in Davisson and Germer's experiment exhibited the wave property of interference and hence behaved like waves rather than particles when they were scattered by the crystal.

In addition when Davisson and Germer changed the velocity (i.e. momentum and hence wavelength) of the incident electrons they found that the angle at which most electrons were scattered changed in the same way that the maximum of interference would change if X-rays with these wavelengths had been used. The Davisson and Germer experiment therefore verified de Broglie's hypothesis that electrons would exhibit wave characteristics and that the wavelength associated with the electron was given by $\lambda = \frac{h}{mv}$. *(5 marks)*

CHAPTER 5

NSW Education Standards Authority

2019 HIGHER SCHOOL CERTIFICATE EXAMINATION

Physics

General Instructions

- Reading time – 5 minutes
- Working time – 3 hours
- Write using black pen
- Draw diagrams using pencil
- Calculators approved by NESA may be used
- A data sheet, formulae sheet and Periodic Table are provided at the back of this paper

Total marks: 100

Section I – 20 marks

- Attempt Questions 1–20
- Allow about 35 minutes for this section

Section II – 80 marks

- Attempt Questions 21–36
- Allow about 2 hours and 25 minutes for this section

Section I

20 marks
Attempt Questions 1–20
Allow about 35 minutes for this section

Use the multiple-choice answer sheet for Questions 1–20.

---

1 A projectile is launched by a cannon as shown.

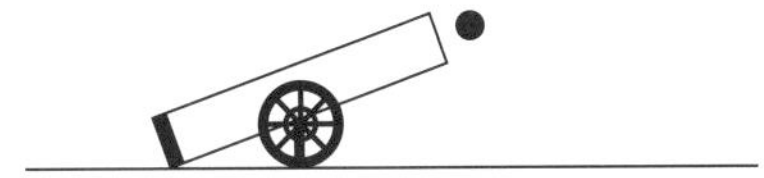

Which arrow represents the velocity of the projectile at its maximum height?

A. ↑

B. ↓

C. ↘

D. →

2 Two stars were observed from Earth. Their spectra are shown with the wavelength in nanometres.

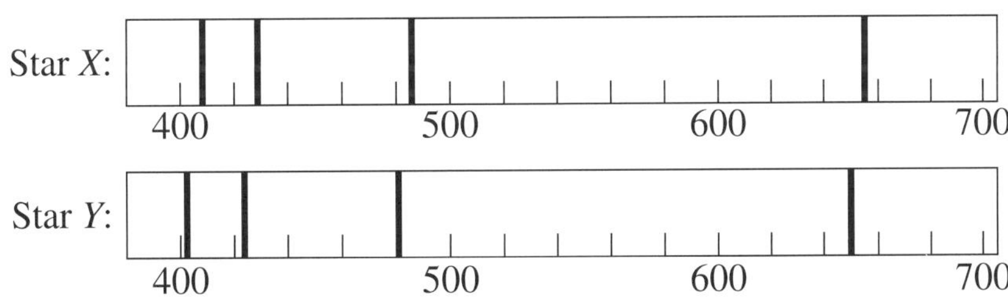

Using these spectra, what can be concluded about the motion of the stars relative to Earth and their chemical compositions?

	Motion relative to Earth	*Chemical composition*
A.	The same	The same
B.	Different	The same
C.	The same	Different
D.	Different	Different

3 Geiger and Marsden carried out an experiment to investigate the structure of the atom.

Which diagram identifies the particles they used and the result that they INITIALLY expected?

A. Alpha particles

B. Protons

C. Alpha particles

D.

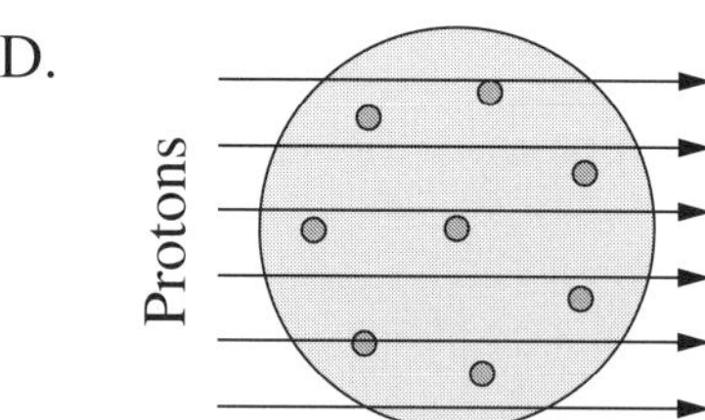

4 Four stars, *P*, *Q*, *R* and *S*, are labelled on the Hertzsprung–Russell diagram.

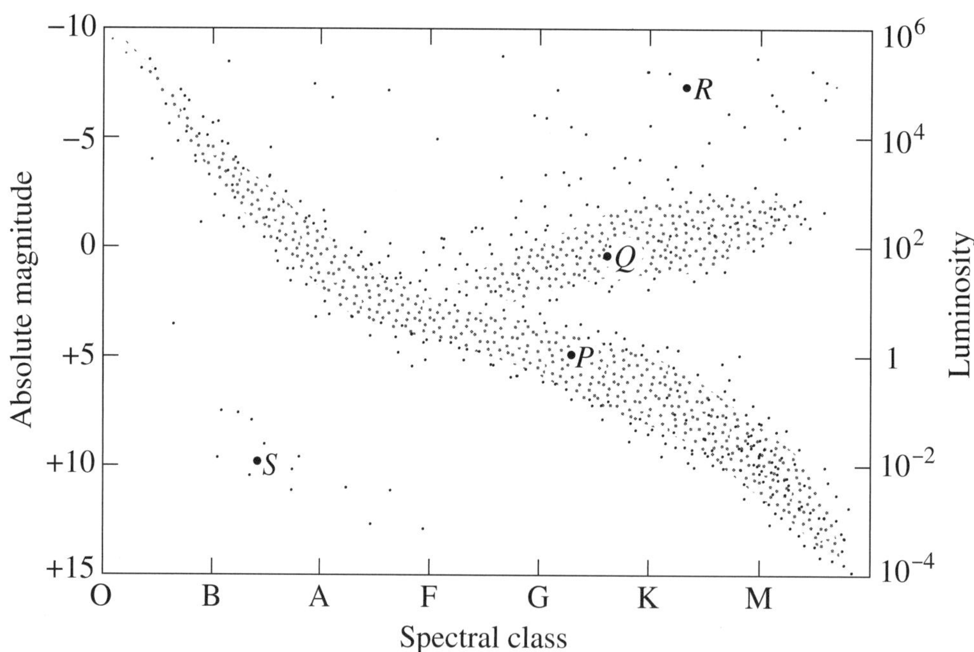

Which statement is correct?

A. *S* has a greater luminosity than *Q*.

B. *R* is a blue star whereas *S* is a red star.

C. *S* has a higher surface temperature than *R*.

D. *P* is at a more advanced stage of its evolution than *R*.

5 The diagram shows two coils wound around a solid iron rod. Initially the switch is closed.

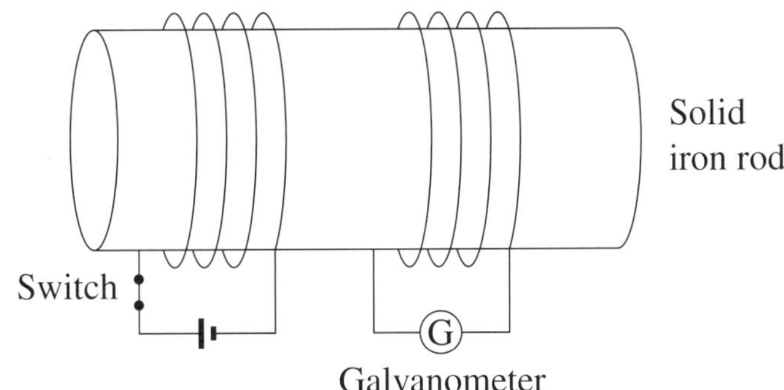

Opening the switch will cause the galvanometer pointer to

A. remain at a constant reading.

B. move from a non-zero reading to a zero reading.

C. move from a zero reading to a non-zero reading, where it remains.

D. move from a zero reading to a non-zero reading, then back to zero.

6 Which graph correctly shows the relationship between the surface temperature of a black body (T) and the wavelength (λ) at which the maximum intensity of light is emitted?

A.

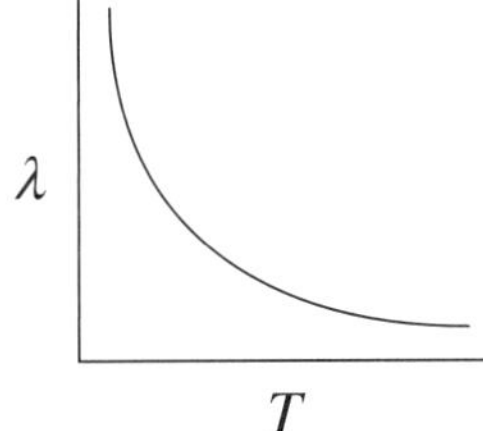

B.

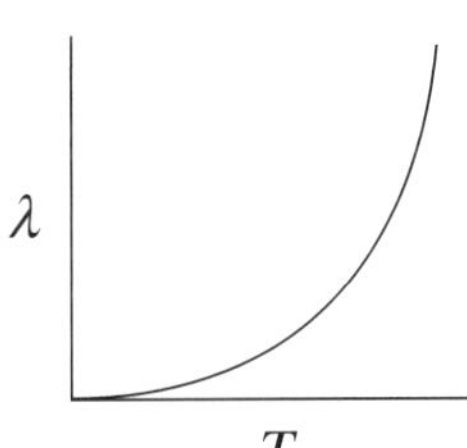

C.

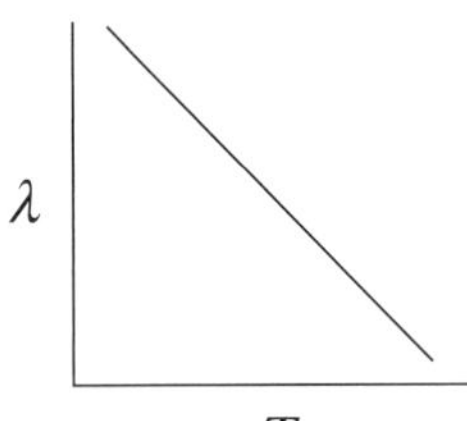

D.

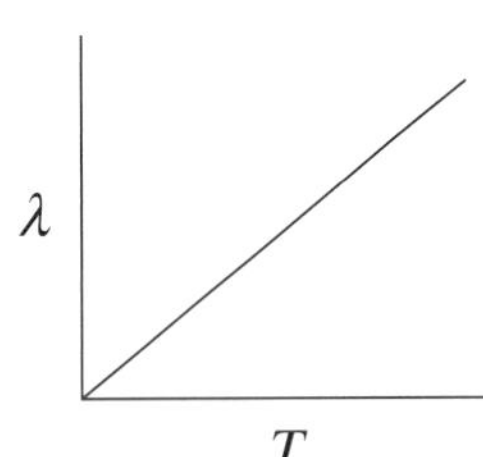

7 A bar magnet is moved away from a stationary coil.

Which diagram correctly shows the direction of the induced current in the coil and the resulting magnetic polarity of the coil?

A.

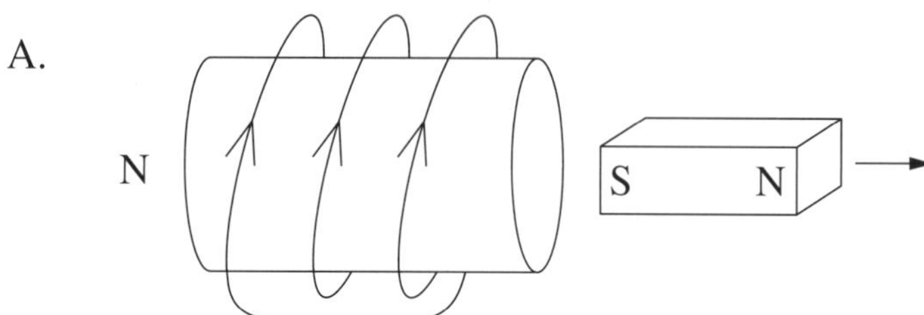

B.

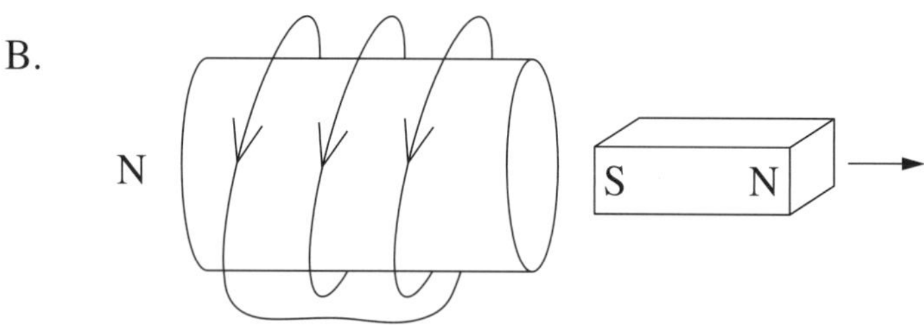

C.

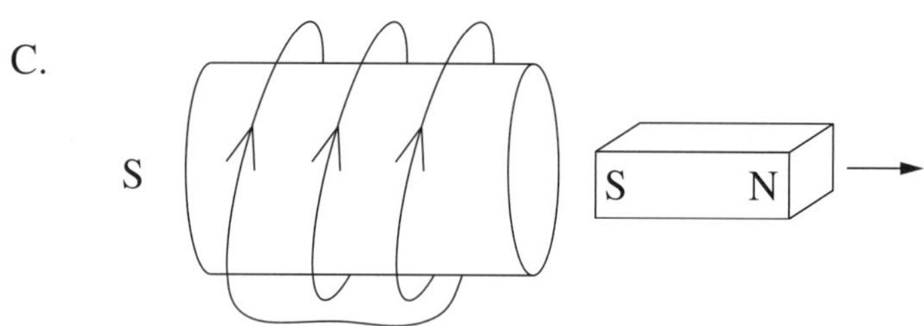

D.

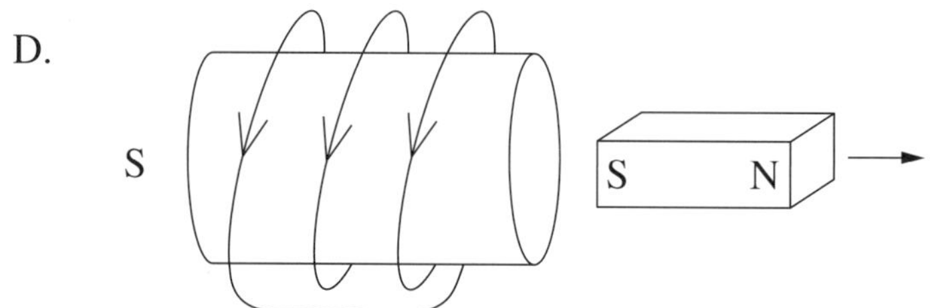

8 A typical galaxy has a diameter of 100 000 light years (~30 000 pc).

Which graph is consistent with Hubble's measurements of the recessional velocity of galaxies?

A.

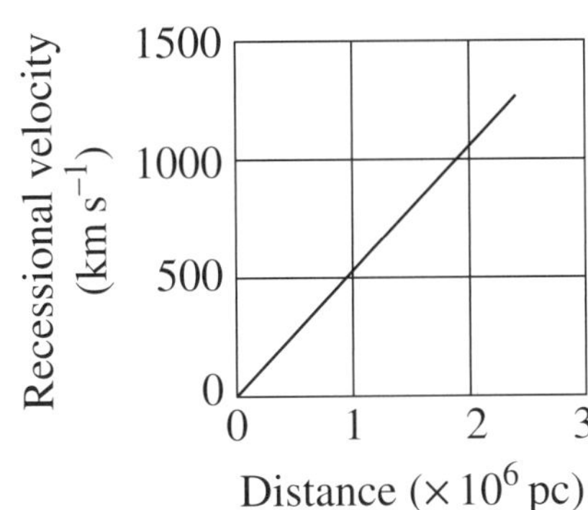

B.

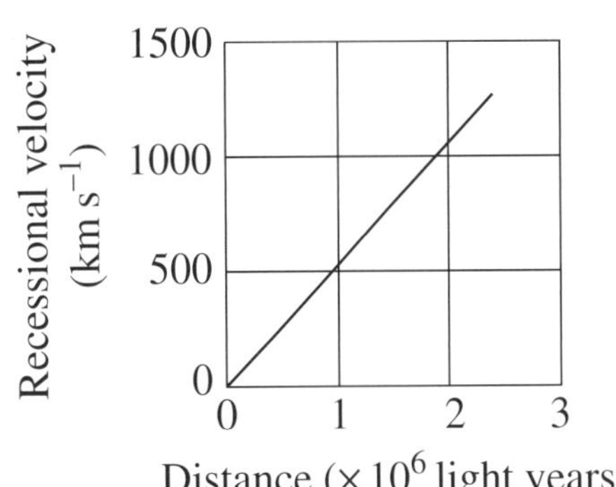

C.

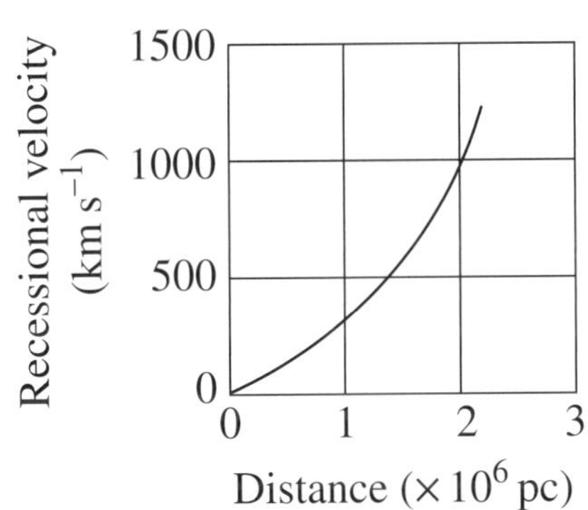

D.

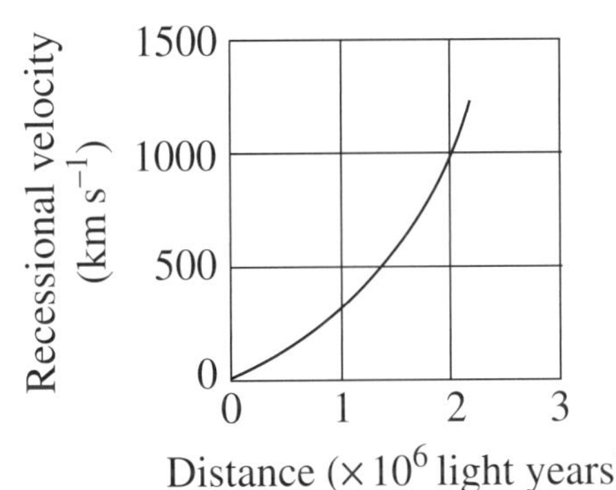

9 Two satellites have the same mass. One (LEO) is in low-Earth orbit and the other (GEO) is in a geostationary orbit.

The total energy of a satellite is half its gravitational potential energy.

Which row of the table correctly identifies the satellite with the greater orbital period and the satellite with the greater total energy?

	Greater orbital period	*Greater total energy*
A.	LEO	LEO
B.	LEO	GEO
C.	GEO	LEO
D.	GEO	GEO

10 A beam of light passes through two polarisers. The second polariser has a transmission axis at an angle of 30° to that of the first polariser. The intensity of the light beam before and after the second polariser is I_0 and I_B respectively.

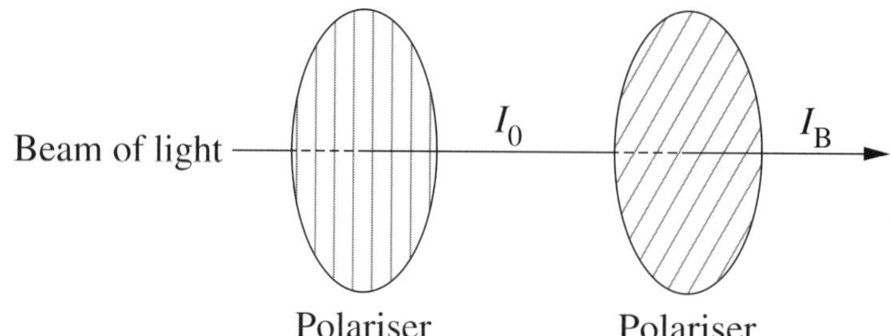

Which row of the table correctly identifies the value of $\frac{I_B}{I_0}$, and the model of light demonstrated by this investigation?

	Value of $\frac{I_B}{I_0}$	*Model of light demonstrated*
A.	0.750	Wave model
B.	0.750	Particle model
C.	0.866	Wave model
D.	0.866	Particle model

11 A dwarf planet orbits the sun with a period of 40 000 years.

The average distance from the sun to Earth is one astronomical unit.

What is the average distance between this dwarf planet and the sun in astronomical units?

A. 34

B. 200

C. 1170

D. 8×10^6

12 The table shows two types of quarks and their respective charges.

Quark	*Symbol*	*Charge*
Up	u	$+\frac{2}{3}$
Down	d	$-\frac{1}{3}$

In a particular nuclear transformation, a particle having a quark composition *udd* is transformed into a particle having a quark composition *uud*.

What is another product of this transformation?

A. Electron

B. Neutron

C. Positron

D. Proton

13 A laser has a power output of 30 mW and emits light with a wavelength of 650 nm.

How many photons does this laser emit per second?

A. 4.6×10^{14}

B. 9.8×10^{16}

C. 3.1×10^{19}

D. 9.3×10^{21}

14 A satellite in circular orbit at a distance r from the centre of Earth has an orbital velocity v.

If the distance was increased to $2r$, what would be the satellite's orbital velocity?

A. $\frac{v}{2}$

B. $0.7v$

C. $1.4v$

D. $2v$

15 Monochromatic light passes through two slits 1 μm apart. The resulting diffraction pattern is measured at a distance of 0.3 m.

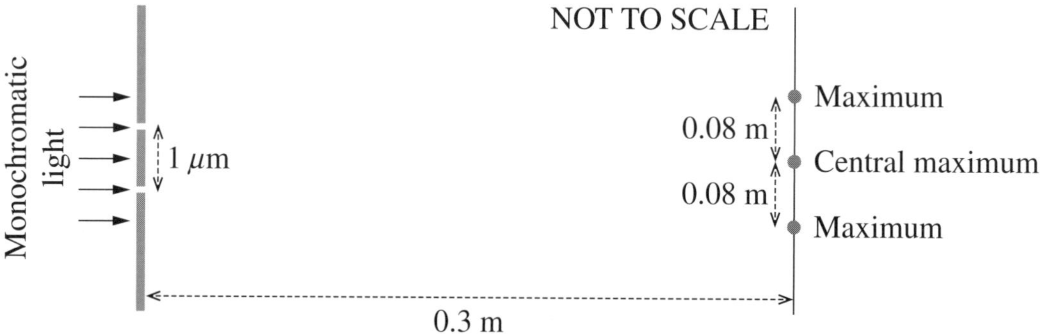

This diffraction pattern can be analysed using the equation $d \sin\theta = \lambda$.

What values of d and θ should be used in the equation?

	d	θ
A.	0.3 m	$\tan^{-1}\left(\frac{0.08}{0.3}\right)$
B.	0.3 m	$\sin^{-1}\left(\frac{0.08}{0.3}\right)$
C.	1 μm	$\tan^{-1}\left(\frac{0.08}{0.3}\right)$
D.	1 μm	$\sin^{-1}\left(\frac{0.08}{0.3}\right)$

16 The diagram shows the trajectory of a particle with charge q and mass m when fired horizontally into a vacuum chamber, where it falls under the influence of gravity.

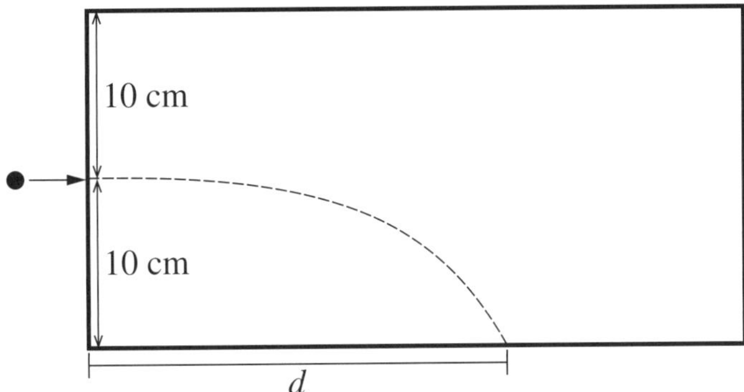

The horizontal distance, d, travelled by the particle is recorded.

The experiment is repeated with a uniform vertical electric field applied such that the particle travels the same horizontal distance, d, but strikes the upper surface of the chamber.

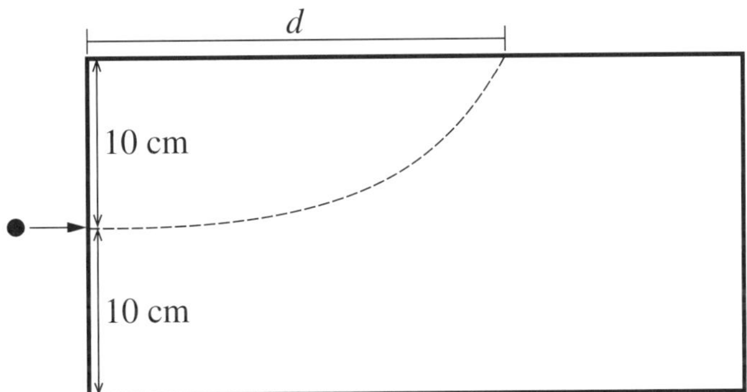

What is the magnitude of the electric field?

A. mgq

B. $2mgq$

C. $\dfrac{mg}{q}$

D. $\dfrac{2mg}{q}$

17 A straight current-carrying conductor, *QR*, is connected to a battery and a variable resistor. *QR* is enclosed in an evacuated chamber with a fluorescent screen at one end.

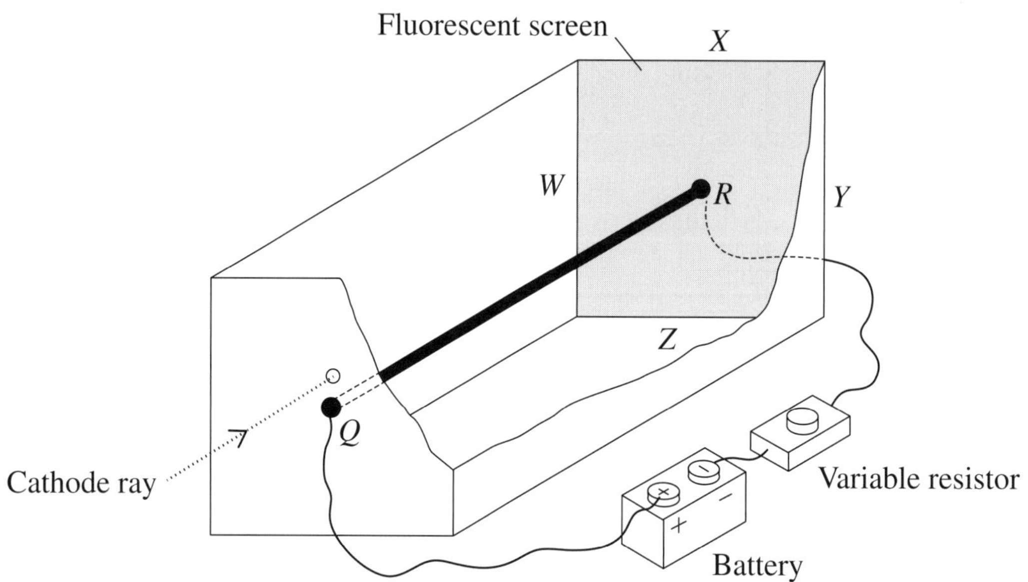

A cathode ray enters the chamber directly above *Q*, initially travelling parallel to *QR*. It passes through the chamber and strikes the fluorescent screen causing a bright spot.

Which direction will this spot move towards if the resistance is increased?

A. *W*

B. *X*

C. *Y*

D. *Z*

18 A circular loop of wire is connected to a battery and a lamp. The apparatus is moved from P to Q along the path shown at a constant velocity through a region containing a uniform magnetic field.

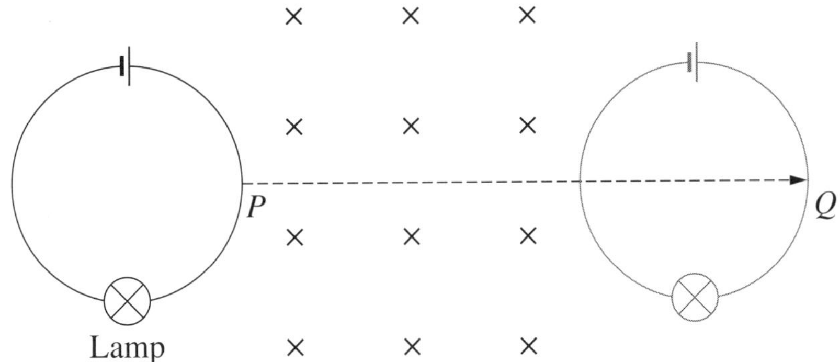

Which graph shows the brightness of the lamp as the apparatus moves between P and Q?

A. Brightness / Time

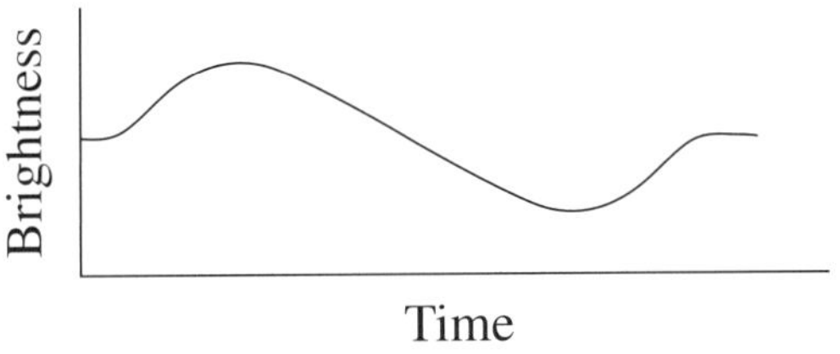

B. Brightness / Time

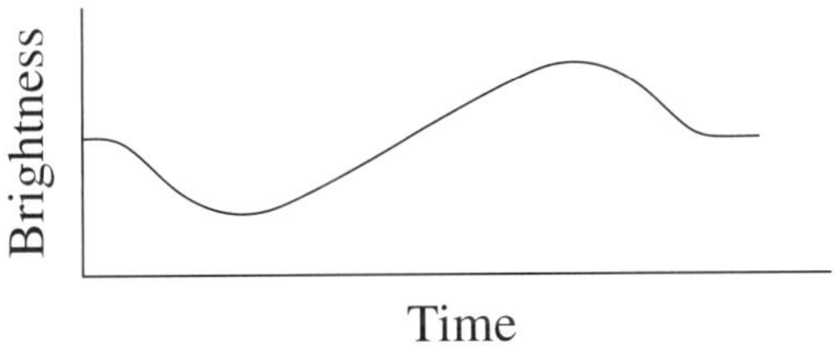

C. Brightness / Time

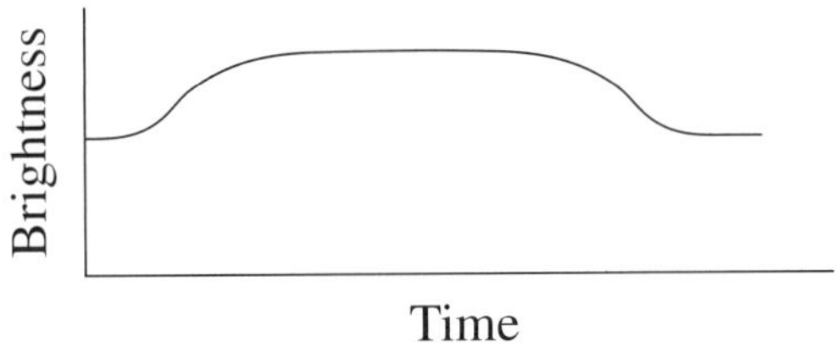

D. Brightness / Time

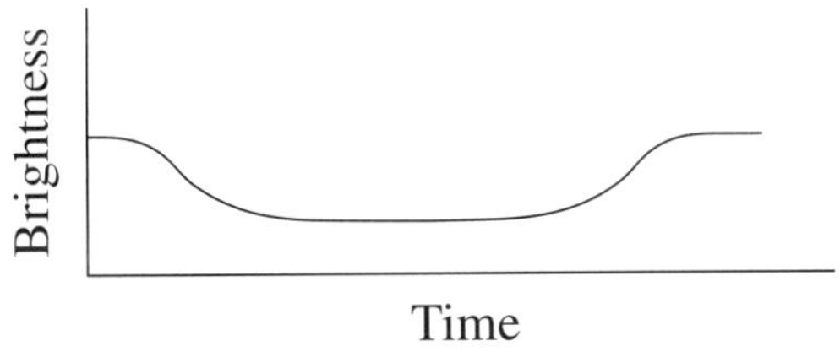

19 Consider the following nuclear reaction.

$$W + X \rightarrow Y + Z$$

Information about W, X and Y is given in the table.

Species	*Mass defect* (u)	*Total binding energy* (MeV)	*Binding energy per nucleon* (MeV)
W	0.00238817	2.224566	1.112283
X	0.00910558	8.481798	2.827266
Y	0.03037664	28.29566	7.073915

Which of the following is a correct statement about energy in this reaction?

A. The reaction gives out energy because the mass defect of Y is greater than that of either W or X.

B. It cannot be deduced whether the reaction releases energy because the properties of Z are not known.

C. The reaction requires an input of energy because the mass defect of the products is greater than the sum of the mass defects of the reactants.

D. Energy is released by the reaction because the binding energy of the products is greater than the sum of the binding energies of the reactants.

20 In the apparatus shown, a backboard is connected by a rod to a shaft. The shaft is spun by an electric motor causing the backboard to rotate in the horizontal plane around the axis X–X′.

A cube is suspended by a string so that it touches the surface of the backboard.

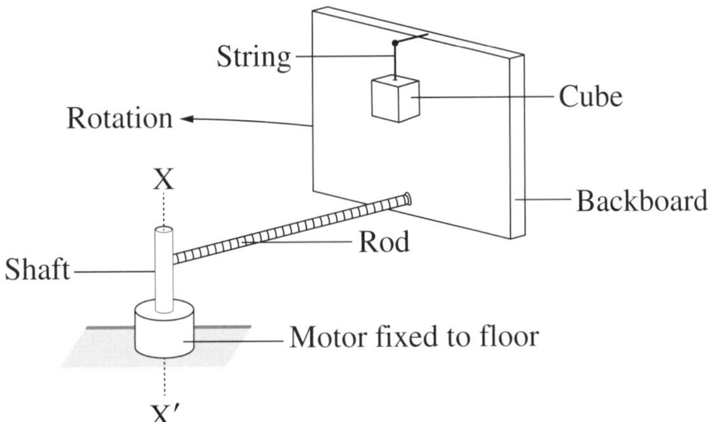

When the angular velocity of the motor is great enough, the string is cut and the position of the cube does not change relative to the backboard.

Which statement correctly describes the forces after the string is cut?

A. The sum of the forces on the cube is zero.

B. The horizontal force of the backboard on the cube is equal in magnitude to the horizontal force of the cube on the backboard.

C. The horizontal force of the backboard on the cube is greater than the horizontal force of the cube on the backboard, resulting in a net centripetal force.

D. The force of friction between the cube and the backboard is independent of the force of the backboard on the cube because these forces are perpendicular to each other.

2019 HIGHER SCHOOL CERTIFICATE EXAMINATION

Centre Number

Student Number

Physics

Section II Answer Booklet

80 marks
Attempt Questions 21–36
Allow about 2 hours and 25 minutes for this section

Instructions

- Write your Centre Number and Student Number at the top of this page.
- Answer the questions in the spaces provided. These spaces provide guidance for the expected length of response.
- Show all relevant working in questions involving calculations.

Please turn over

Question 21 (2 marks)

Outline de Broglie's contribution to quantum mechanics. Support your answer with a relevant equation. 2

....

....

....

....

Question 22 (3 marks)

Spectra can be used to determine the chemical composition and surface temperature of stars. 3

Describe how spectra provide information about OTHER features of stars.

....

....

....

....

....

....

....

....

Question 23 (3 marks)

A student investigated the photoelectric effect. The frequency of light incident on a metal surface was varied and the corresponding maximum kinetic energy of the photoelectrons was measured. **3**

The following results were obtained.

Frequency ($\times 10^{14}$ Hz)	11.2	13.5	15.2	18.6	20.0
Maximum kinetic energy (eV)	0.6	1.3	2.3	3.3	4.2

Plot the results on the axes below and hence determine the work function of the metal in electron volts.

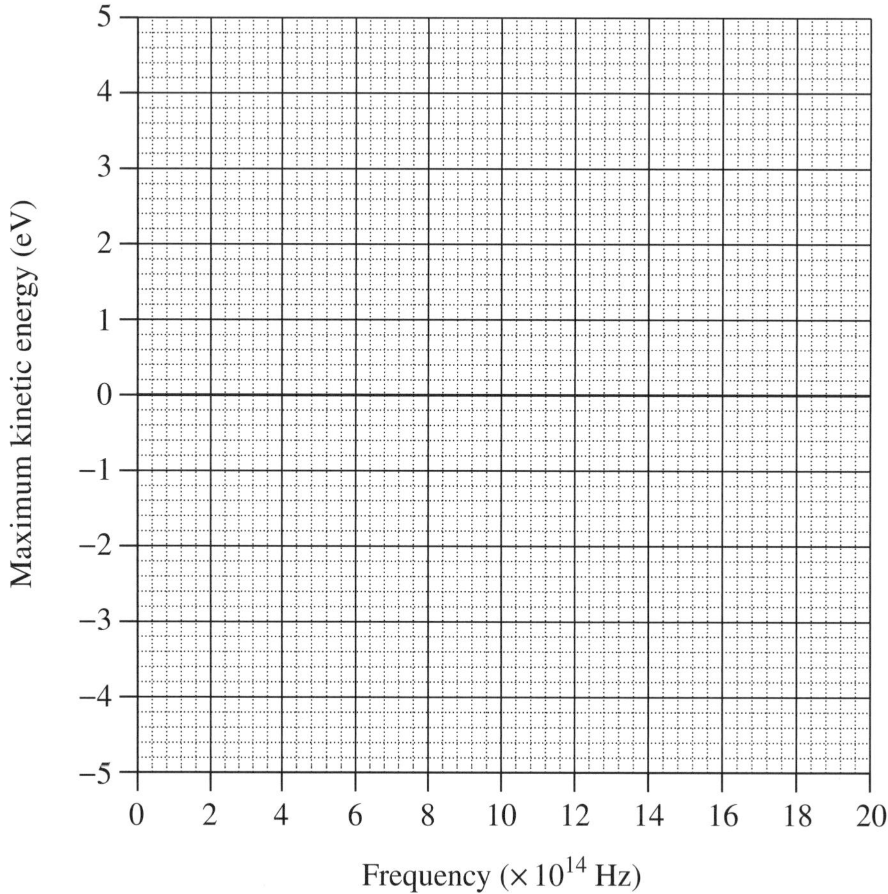

....

....

....

Question 24 (7 marks)

A step-up transformer is constructed using a solid iron core. The coils are made using copper wires of different thicknesses as shown.

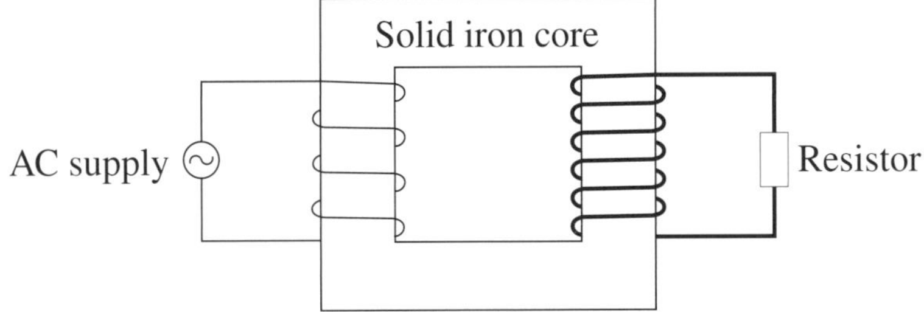

The table shows electrical data for this transformer.

V_s	I_s	$V_p I_p$
50 V	9 A	500 J s^{-1}

(a) Explain how the operation of this transformer remains consistent with the law of conservation of energy. Include a relevant calculation in your answer. **3**

....

....

....

....

....

....

(b) Explain how TWO modifications to this transformer would improve its efficiency. **4**

....

....

....

....

....

....

....

....

Question 25 (4 marks)

The diagram shows a model of electromagnetic waves. **4**

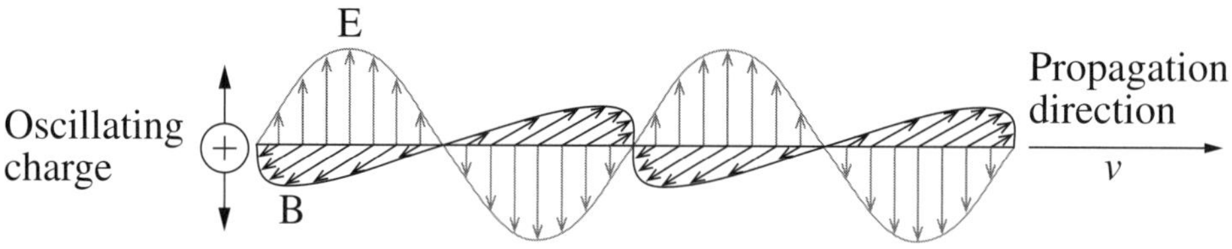

Relate this model to predictions made by Maxwell.

Please turn over

Question 26 (6 marks)

A student carried out an experiment to investigate the relationship between the torque produced by a force and the angle at which the force is applied. A 400 N force was applied to the same position on the handle of a spanner at different angles, as shown.

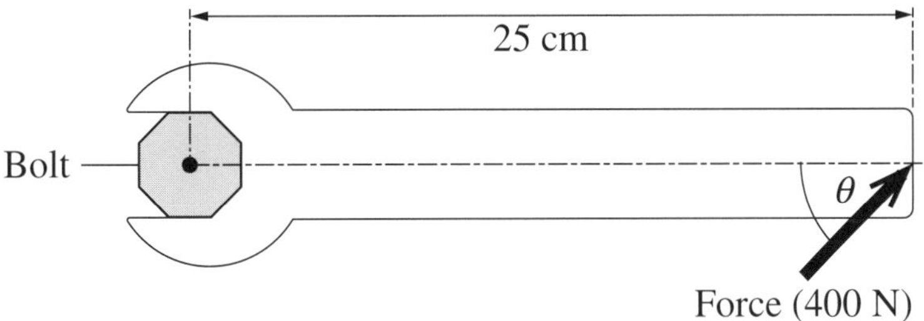

A high-precision device measured the torque applied to the bolt.

The data from the experiment is graphed below.

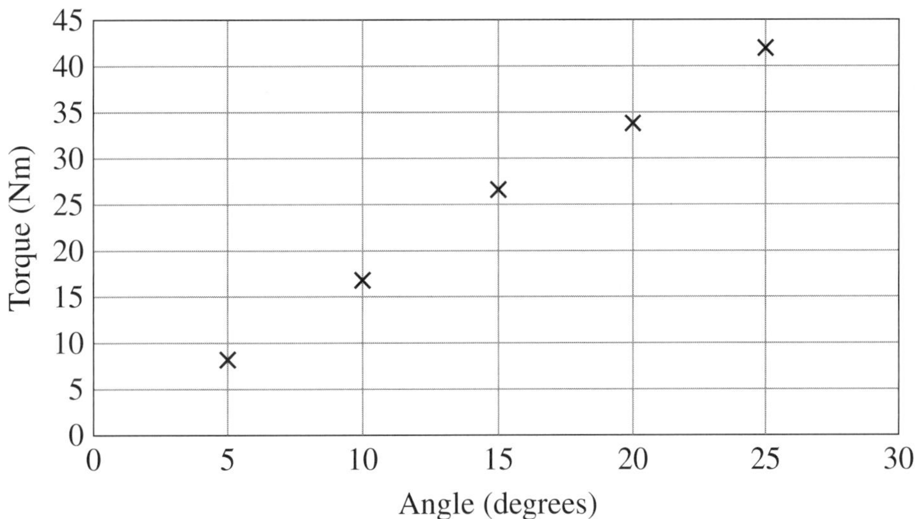

Question 26 continues on the following page

Question 26 (continued)

The student concluded that the torque (τ) was proportional to the angle (θ) and proposed the model

$$\tau = k\theta$$

where $k = 1.7$ Nm/degree.

(a) Justify the validity of the student's model using information from the graph. **3**

....

....

....

....

....

....

(b) What happens to the accuracy of this model's predictions as the angle increases beyond 25°? Justify your answer with reference to a different model. **3**

....

....

....

....

....

....

End of Question 26

Question 27 (6 marks)

(a) Outline a thought experiment that relates to the prediction of time dilation. **3**

....

....

....

....

....

....

(b) Outline experimental evidence that validated the prediction of time dilation. **3**

....

....

....

....

....

....

Question 28 (3 marks)

A metal loop, *WXYZ* is connected to a battery and placed in a uniform magnetic field. A current flows through the loop in the direction shown. **3**

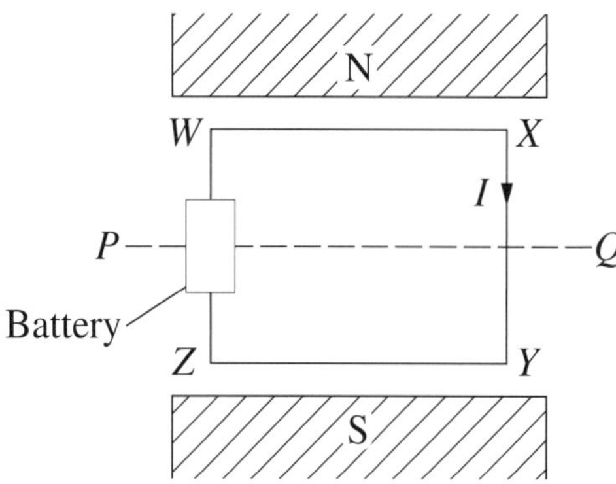

The loop is then allowed to rotate by 90° about the axis *PQ*.

Compare the forces acting on *WX* and *XY* before and after this rotation.

....

....

....

....

....

....

Question 29 (3 marks)

A particle having mass *m* and charge *q* is accelerated from rest through a potential difference *V*. Assume that the only force acting on the particle is due to the electric field associated with this potential difference. **3**

Show that the final velocity of the particle is given by $v = \sqrt{\dfrac{2qV}{m}}$.

....

....

....

....

....

Question 30 (6 marks)

A ball, initially at rest in position P, travels along a frictionless track to point Q and then falls to strike the floor below.

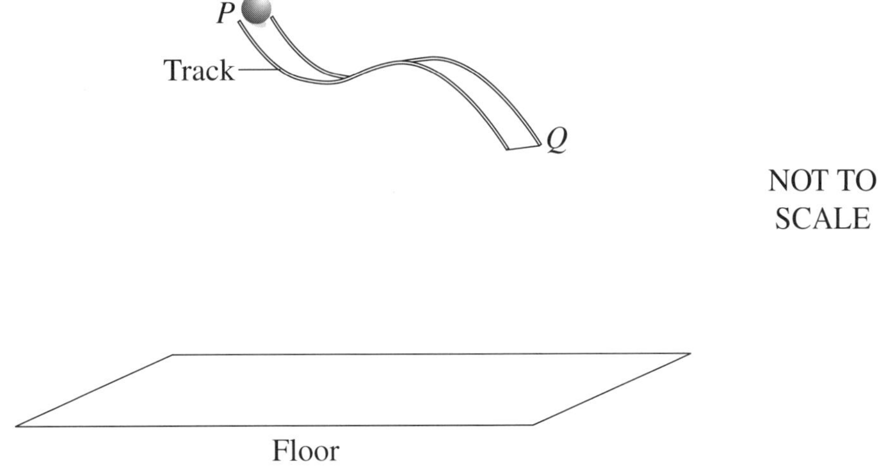

NOT TO SCALE

At the instant the ball leaves the track at Q it has a velocity of 1.5 m s^{-1} at an angle of 50° to the horizontal.

(a) Calculate the difference in height between P and Q. **3**

....

....

....

....

....

....

(b) The ball takes 0.5 s to reach the floor after leaving the track at$Q$. **3**

Calculate the height of Q above the floor.

....

....

....

....

....

....

Question 31 (8 marks)

A student suspends an electric ceiling fan from a spring balance.

The fan is switched on, reaching a maximum rotational velocity after ten seconds.

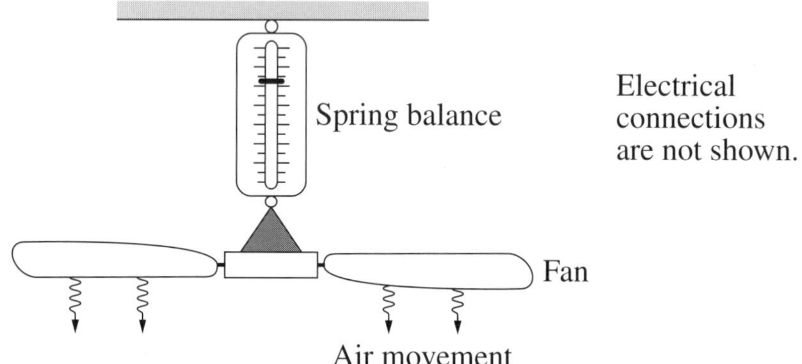

(a) Explain the changes that would be observed on the spring balance in the first 15 seconds after the fan is switched on. **4**

....

....

....

....

....

....

....

....

Question 31 continues on the following page

Question 31 (continued)

(b) The student predicted that the current through the fan's motor would vary as shown on the graph. **4**

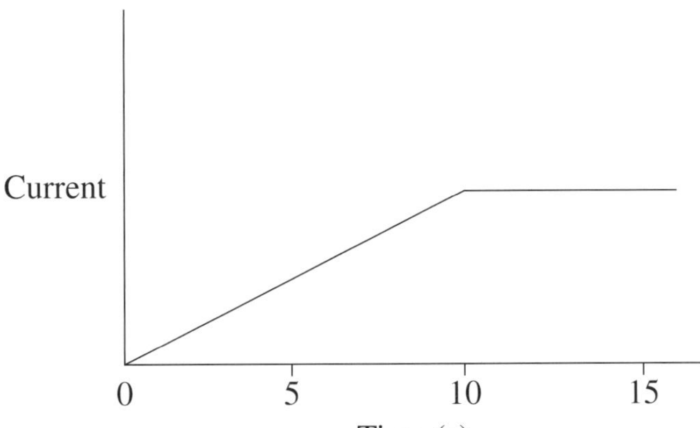

Assess the accuracy of the student's prediction.

....

....

....

....

....

....

....

....

End of Question 31

Question 32 (5 marks)

Describe how specific experiments have contributed to our understanding of the electron and ONE other fundamental particle. **5**

....

....

....

....

....

....

....

....

....

....

....

....

....

....

Question 33 (4 marks)

A proton and an alpha particle are fired into a uniform magnetic field with the same speed from opposite sides as shown. Their trajectories are initially perpendicular to the field. 4

	× × × × × ×	
	× × × × × ×	
	× × × × × ×	
Proton ⟶		⟵ Alpha particle
	× × × × × ×	
	× × × × × ×	
	× × × × × ×	

Explain ONE similarity and ONE difference in their trajectories as they move in the magnetic field.

....

....

....

....

....

....

....

....

....

....

Question 34 (9 marks)

Use the following information to answer this question. **9**

Earth's distance from the sun, r	1.5×10^{11} m
Intensity of solar radiation at a distance r from the sun	1360 W m^{-2}
Surface area of a sphere	$4\pi r^2$

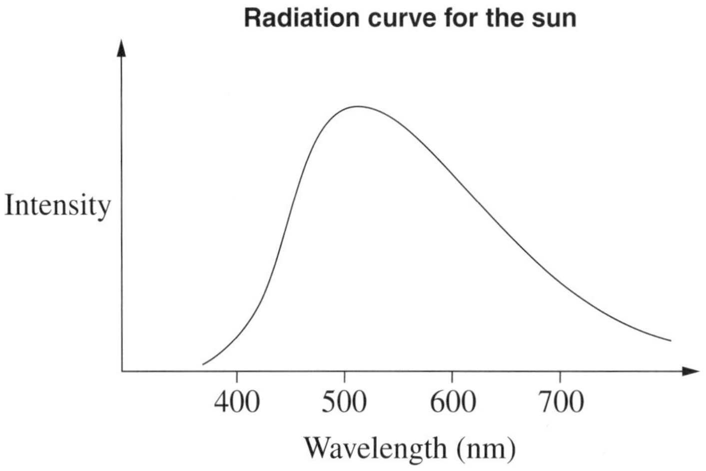

Radiation curve for the sun

Describe both the production and radiation of energy by the sun. In your answer, include a quantitative analysis of both the power output and the surface temperature of the sun.

Question 34 continues on the following page

Question 34 (continued)

....

....

....

....

....

....

....

....

....

....

....

....

....

End of Question 34

Please turn over

Question 35 (4 marks)

The apparatus shown is attached horizontally to the roof inside a stationary car. The plane of the protractor is perpendicular to the sides of the car. **4**

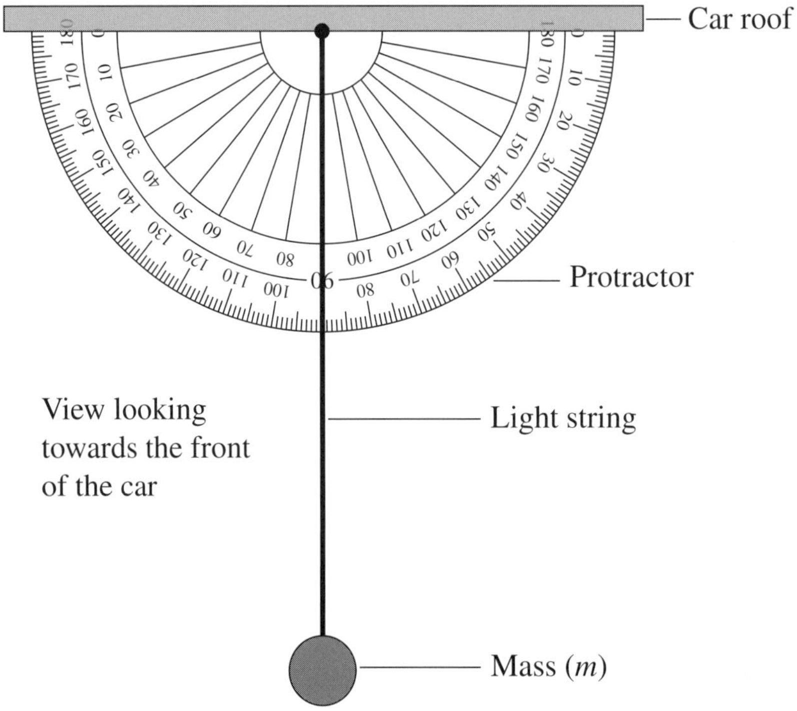

Car roof

Protractor

View looking towards the front of the car

Light string

Mass (m)

The car was then driven at a constant speed (v), on a horizontal surface, causing the string to swing to the right and remain at a constant angle (θ) measured with respect to the vertical.

Describe how the apparatus can be used to determine features of the car's motion. In your answer, derive an expression that relates a feature of the car's motion to the angle θ.

....

....

....

....

....

....

....

....

Question 36 (7 marks)

A radon-198 atom, initially at rest, undergoes alpha decay. The masses of the atoms involved are shown in atomic mass units (u).

radon-198	$\longrightarrow$	polonium-194	+	helium-4
197.999 u		193.988 u		4.00260 u

The kinetic energy of the polonium atom produced is 2.55×10^{-14} J.

By considering mass defect, calculate the kinetic energy of the alpha particle, and explain why it is significantly greater than that of the polonium atom.

....

....

....

....

....

....

....

....

....

....

....

....

....

....

End of paper

2019 HSC Examination paper

Sample answers

Section I (Total 20 marks)

1 D The horizontal velocity remains constant throughout the flight and, at the maximum height, the vertical component of the velocity is zero. Therefore, at the top of the flight, the velocity is equal to the initial horizontal velocity.

2 B The degree to which the spectral pattern is red or blue shifted is determined by the motion of each star relative to the Earth. In this case, the motion of each star relative to the Earth must be different because the pattern of spectral lines from each star has been shifted by a different amount. The chemical composition of each star is the same because both stars have the same spectral lines with the same spacing between the lines indicating the presence of the same element(s).

3 C Because Geiger and Marsden initially relied on Thomson's atomic model they believed that the density of the atom would be so low that high-speed alpha particles would pass straight through the atom with little or no deflection.

4 C The surface temperature of a star is represented by the spectral class of the star. A star with a spectral class *A* or *B* has a much higher surface temperature than a star with a spectral class of *K* or *M*.

5 D By Faraday's law of electromagnetic induction, a voltage will only be induced in the coil connected to the galvanometer when it experiences a changing magnetic flux. When the switch is opened, the current in the coil connected to the battery and the magnetic field produced by this coil will change from its initial value to zero. Only during this period of changing magnetic flux will a voltage be induced on the coil connected to the galvanometer. Hence the galvanometer will go from a zero reading to a non-zero reading when the magnetic flux is changing and then return to a zero reading when the magnetic flux stops changing.

6 A Wien's displacement law for black body radiation states that the wavelength of maximum intensity emitted from a black body is inversely proportional to the absolute surface temperature of the body ($\lambda_{max} = \frac{b}{T}$).

Therefore the relationship between λ_{max} and T is hyperbolic as shown in graph A.

7 D By Lenz's Law the direction of the induced current opposes the change in magnetic flux that produced it. Hence moving the magnet away from the solenoid will induce a current that would produce a north magnetic pole on the right-hand side of the solenoid (and a south magnetic pole on the left-hand side of the solenoid). Thus the induced magnetic field inside the solenoid will be towards the right. To produce a

magnetic field directed to the right inside the solenoid, the right-hand grip rule shows that the induced current must be in the direction shown in answers B and D. However, only answer D is correct as it shows the magnetic polarity of the solenoid due to the induced current on the correct end of the solenoid.

8 A Hubble measured the red shift of a large number of galaxies and their distance from the earth. He believed the results showed a linear relationship between the recession velocity and the distance to the galaxy from the Earth. Thus only graphs A and C reflect Hubble's findings. The correct graph is A because the distances are consistent with Hubble's data and because the gradient of the graph (Hubble's constant) is 500 km s^{-1} per megaparsec which is consistent with Hubble's original analysis. (Note that today the accepted value of Hubble's constant is 73.8 km s^{-1} per megaparsec.)

9 D Kepler's Law of Periods states that the square of the period is proportional to the cube of the orbital radius ($T^2 \alpha r^3$). Hence the geostationary satellite will have the greater period. The gravitational potential energy of a satellite is proportional to the negative inverse of the orbital radius ($U \alpha -\frac{1}{r}$).

As the GEO has the higher gravitational potential energy it will also have the higher total energy. Another way to understand this is to remember that the satellite with the lower orbital radius is more tightly bound to the Earth and hence must have a more negative total energy as it would take more energy to free it from the Earth's gravitational field.

10 A Malus's law states that $I = I_{max}\cos^2\theta$. Hence the transmitted intensity is given by

$I_B = I_0 \cos^2(30^0)$ or $\frac{I_B}{I_0} = \cos^2(30^0) = 0.75$.

Polarisation is historically taken as a demonstration of the wave model of light because it could not be explained by the original particle model of light.

11 C Applying Kepler's Law of Periods ($T^2 \alpha r^3$) with the units of AUs and years, and using ratios for the Earth and the dwarf planet:

$\frac{(40\,000)^2}{r^3} = \frac{1^2}{1^3}$ and hence, $r = \sqrt[3]{(40\,000)^2} = 1170$ years

12 A Charge must be conserved in all reactions. A particle made of *udd* quarks would have zero charge (i.e. a neutron) and a particle with *uud* quarks would have a charge of +1. Thus, to ensure the charge on both sides of the reaction is balanced, a charge of –1 is required (i.e. an electron). We can write this equation as:

neutron → proton + electron

or using fundamental particles and charges in brackets: *udd*(0) →*ddu*(+1) + e(–1)

Students may also recognise this reaction as beta decay in which a neutron decays into a proton and an electron.

13 B Each second the laser beam emits 30 mJ of energy and each photon in the beam has an energy of:

$$E = hf = \frac{hc}{\lambda} = \frac{(6.6626 \times 10^{-34} \times 3 \times 10^{8})}{650 \times 10^{-9}} = 3.058 \times 10^{-19}\text{ J.}$$

The number of photons emitted per second by the laser will be given by:

Number of photons per second = energy emitted per second/energy of each photon.

$$\textit{Number of photons per second} = \frac{30 \times 10^{-3}}{3.058 \times 10^{-19}} = 9.8 \times 10^{16} \text{ photons per second.}$$

14 B The velocity of a satellite can be determined by realising that gravity provides the centripetal force that keeps the satellite in its orbit. This allows us to write:

$F_C = F_G$ or $\frac{mv^2}{r} = \frac{Gmm_s}{r^2}$ and therefore $v = \sqrt{\left(\frac{Gm_s}{r}\right)}$.

Hence $v \propto \frac{1}{\sqrt{r}}$ and if the orbital radius is doubled then the new velocity will become $v_{new} = \frac{v}{\sqrt{2}} = 0.7v$.

15 C The slit separation is represented by the symbol $d = 1\ \mu$m. The angle θ is the angle between the line drawn from the slits to the central maximum and the line drawn from the slits to the first maximum to the side of the central maximum. In this case we can use trigonometry to show the angle would be given by: $\theta = \tan^{-1}\left(\frac{0.08}{3}\right)$.

16 D With no electric field the force on the charge would be the force of gravity ($F = mg$). The force on a charge in an electric field is given by $F = qE$. For the charge to follow the trajectory shown, a net force of $F = mg$ upwards must operate on the charge. To produce this net upwards force, the electric field must apply an upwards force of $2mg$ upwards to the charge. Hence we can write $F_E = qE = 2mg$ or $E = \frac{2mg}{q}$.

17 D The current in the wire moves from *Q* to *R* producing a clockwise magnetic field around the wire. An electron (negative charge) moving above the wire towards the screen *XY* will experience an upwards force in this field (by the right-hand palm rule). If the resistance is increased, the current in the wire will be reduced, which will reduce the upwards force on the electron, causing it to hit the screen at a lower point. Hence the impact point will move towards *Z*.

18 B The battery produces a clockwise current in the loop which causes the globe to glow. When the loop initially moves into the magnetic field, the magnetic flux in the loop will change and this change will induce an anticlockwise current in the loop. This will decrease the net current in the loop and reduce the brightness of the globe. When the loop is completely within the magnetic field, the magnetic flux will not change and the current in the loop will revert to the initial current, increasing the brightness of the lamp. When the loop leaves the field, the magnetic flux again changes in the loop but this time due to Lenz's law the induced current will be clockwise which will increase the current in the loop and the brightness of the lamp. The brightness will return to the initial level when the loop leaves the field as the loop is no longer experiencing a changing magnetic flux.

19 D Because the binding energy of *Y* is significantly greater than the sum of binding energies of *W* and *X*, energy must be released in the reaction. If particle *Z* is a single proton or neutron it will have zero binding energy and if it contains more than one nucleon it would have a binding energy which would require even more energy to be released in the reaction.

20 B By Newton's third law, the horizontal force applied to the cube by the blackboard must be equal to the force applied to the blackboard by the cube. This produces a net force towards the centre of the circle (i.e. a centripetal force) on the cube that causes it to move in a circle.

Section II

Question 21 (Total 2 marks)

De Broglie hypothesised that matter exhibits wavelike behaviour. He showed the wavelength associated with a particle of mass m, travelling with a velocity v, is given by:

$\lambda = \dfrac{h}{mv} = \dfrac{h}{p}$, where the momentum $p = mv$.

De Broglie was able to use his matter-wave hypothesis to explain the existence of the stationary electron states in the hydrogen atomic model proposed by Bohr. *(2 marks)*

Question 22 (Total 3 marks)

Students should describe **three** features of a star that can be deduced from stellar spectra; for example, **any three** of the following:

(i) By measuring the shift of the star's spectrum towards the red or blue end of the spectrum (the optical doppler effect) the velocity of the star relative to the earth can be determined. Stellar spectra shifted towards the red end of the spectrum indicate the star is moving away from the earth.

(ii) If each line in a star's spectrum splits and recombines periodically, it indicates the object is actually two stars in orbit around one another (i.e. a binary star). The period of the orbit can be used to determine the mass of the stars.

(iii) The width of spectral lines can be used to determine the rate at which a star is rotating (rotational broadening).

(iv) The width of spectral lines can also be used to determine the density (pressure broadening) of a star's photosphere. As larger stars have a less dense (lower pressure) photosphere, spectral line width can also give us information on the size of a star.

(v) The magnetic field in a star can be determined by examining the splitting of lines (Zeeman effect) in the star's spectrum. *(3 marks)*

Question 23 (Total 3 marks)

Students should plot the data on the axis provided as shown below:

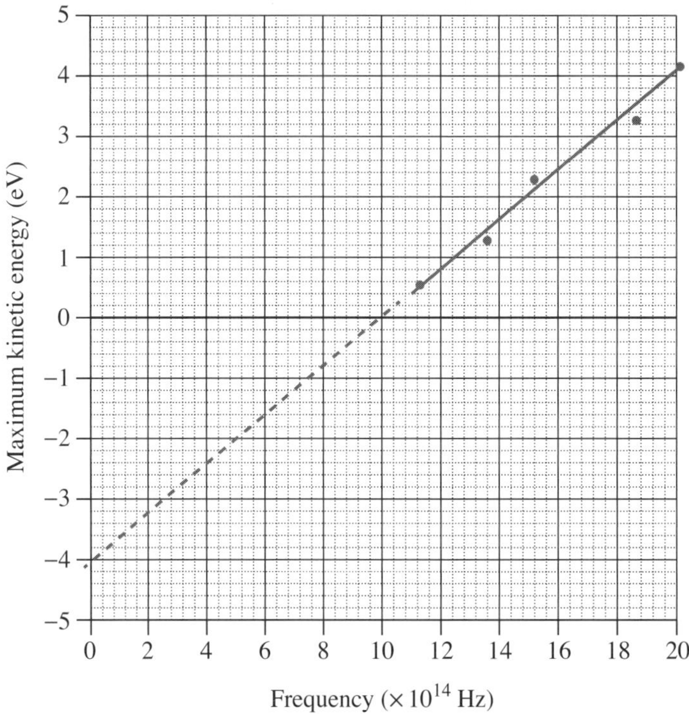

The graph represents the equation, $K_{max} = hf - \phi$. The work function (ϕ) of the metal is therefore given by the intercept of the extrapolated line of best fit with the K_{max} axis, which for this metal is $\phi = 4.0$ eV. *(3 marks)*

Question 24 (Total 7 marks)

(a) The law of conservation of energy states that energy can neither be created nor destroyed; it can only change form. Transformers are not 100% efficient because they lose some energy to heat. Therefore, to be consistent with the law of conservation of energy, the output power ($I_S V_S$) cannot exceed the input power ($I_P V_P$) of the transformer. Using the data for the transformer provided, we see that the input power $I_P V_P = 500$ J s^{-1} is indeed greater than the output power, $I_S V_S = 50 \times 9 = 450$ J s^{-1}. The remaining power (50 J s^{-1}) is lost to heat. *(3 marks)*

(b) The percentage efficiency of the transformer can be defined as:

$$Efficiency = \frac{(useful\ output\ power \times 100)}{(input\ power)} \%.$$

Hence to increase efficiency we must minimise heat losses to maximise the percentage of the input energy that is converted to useful output energy.

Some of the input energy is converted into heat by induced eddy currents, causing resistance heating in the solid iron core. Using a laminated iron core (an iron core cut into thin slices and insulated from one another) rather than a solid iron core, would stop large eddy currents forming and reduce heat loss, resulting in increased efficiency.

Heat loss in the windings is related to the current and resistance of the windings by: $P_{loss} = I^2R$. As the transformer is a step-up transformer, the current is greater in the primary than in the secondary coil. Thus, by using the thicker (lower resistance) wire on the input coil and the thinner wire on the output coil, resistance heating could be reduced. (An alternative answer would be to suggest using thicker wire on both coils as this would also reduce resistance heating in the windings.) *(4 marks)*

Question 25 (Total 4 marks)

Maxwell condensed the laws of magnetism and electricity known at the time into a single theory of electromagnetism. In addition to the existing laws, Maxwell proposed that a changing electric field would produce a changing magnetic field. Faraday had previously shown that a changing magnetic field would produce a changing electromagnetic force (EMF) and hence a changing electric field. Maxwell showed that an oscillating charge would be surrounded by a changing electric field (as the charge is moving) and a changing magnetic field (as the charge is constantly accelerating). He then hypothesised that these changing fields would produce transverse electromagnetic waves made up of mutually perpendicular, changing electric and magnetic fields that would move away from the charge. Finally, by applying his new theory of electromagnetism, he was able to show that these waves would travel at a fixed speed of 3×10^8 m s^{-1} in a vacuum (the speed of light).

The model illustrates many of Maxwell's predictions. It shows a changing electric field producing a perpendicular changing magnetic field which in turn produces a changing perpendicular electric field as Maxwell predicted. It shows an oscillating charge producing a transverse wave made up of mutually perpendicular, changing electric and magnetic fields as Maxwell predicted. The simple model, however, shows waves being produced in one direction only, where Maxwell predicted the waves would actually move away from the oscillating charge in all directions (except parallel to the direction of the charge's oscillation). Maxwell also predicted a specific speed (the speed of light = 3×10^8 m s^{-1}) for the waves, which the model does not show. The model also fails to show the relationship between the frequency and wavelength of the electromagnetic wave and the frequency of oscillation of the charge. *(4 marks)*

Question 26 (Total 6 marks)

(a) The equation proposed by the student, $\tau = k\theta$, is a linear equation and hence a plot of torque against the angle should be a straight line with gradient of $k = 1.7$ Nm/degree. Validation of the model would require experimental data to display this type of linear relationship.

For the small range of angles measured, the data does provide support for the model. The points appear to show a linear relationship and the line of best fit has a gradient of 1.7 Nm/degree, which is in agreement with the student's model. Hence the data does provide some support for the model, provided the model is restricted to this range of angles (i.e. between 0° and 25°). It would not be valid to extrapolate the data to large angles and hence the data shown cannot be used to validate the model for all angles. To validly test the model, the student should have measured the torque over a wider range of angles as the model may be correct for small angles but not for larger angles. *(3 marks)*

(b) The correct model for this situation is $\tau = rF\sin\theta$, because the torque is defined as the product of the applied force and the perpendicular distance between the line of action of the force and the axis of rotation. In terms of the student's graph, instead of a linear relationship, the actual relationship is a sine curve (as $\tau \alpha \sin\theta$). Because $\sin\theta \cong \theta$ for small values of θ, the relationship appears linear for small angles. The sine curve would, however, get further and further from the linear graph as the angle was increased. Thus the student's model would become less accurate as the angle was increased beyond 30°. *(3 marks)*

Question 27 (Total 6 marks)

(a) A thought experiment that illustrates time dilation is shown in the diagram below. Consider a light pulse that travels from the bottom to the top of a train carriage and back. An observer in the carriage (Figure 1A) observes the pulse to go vertically up and down but an observer outside the carriage sees the beam travel much further because of the train's motion (Figure 1B). Both observers agree that the pulse hits the floor at the same time and that the speed of light is constant. The only way to explain that light travels at a different distance at the same speed in each frame and still reaches the floor at the same time is to assume that time passes at a different rate for each observer. Indeed, each will see the other's clock running more slowly than their own.

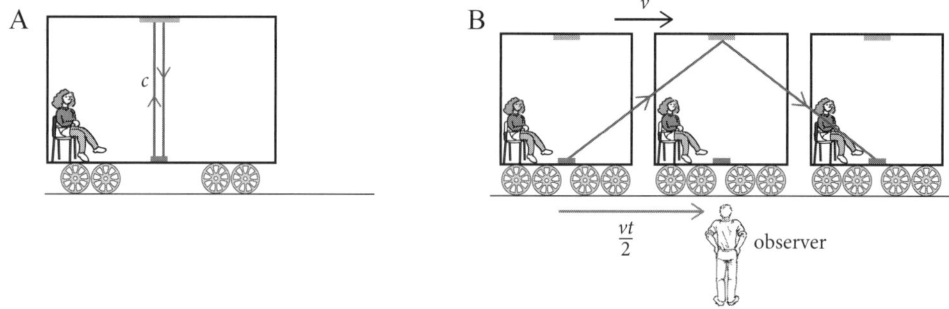

Figure 1 A thought experiment to demonstrate time dilation

(3 marks)

(b) Measurements of the relationship between the half-lives of subatomic particles and their velocity was used to experimentally confirm the theory of time dilation. Muons travelling at speeds close to the speed of light are created in the upper atmosphere by cosmic rays. At rest, muons have a half-life of 2.2 μs and even if they were travelling at the speed of light should decay before reaching the earth's surface. However, observations show most muons reach the earth's surface. This is because the earthbound observer sees the muon's 'clock' running slow due to time dilation and, therefore, the earthbound observer measures the muon's half-life to be over 20 times greater than the rest half-life.

OR

Students could also use the Hafele–Keating experiment, where synchronised atomic clocks were used to confirm the theory of relativity. In this experiment one clock was left in Washington and others were flown around the world in an easterly and westerly direction respectively. After the flights, the time differences between the clocks that had flown around the world and those that had not were found to be consistent with Einstein's relativity theories, including the theory of time dilation. *(3 marks)*

Question 28 (Total 3 marks)

Applying the right-hand grip rule and the relationship $F = ILB\sin\theta$:

The force on WX remains constant and directed into the page, before and after the 90° rotation. This is because WX remains perpendicular to the direction of the magnetic field throughout the rotation (i.e. $F = ILB$).

Before the rotation there is no force on XY as the current is moving parallel to the magnetic field. After rotating 90° the force on XY ($F = ILB$) will be directed towards the right (assuming the rotation is in the direction of the initial torque on the coil). *(3 marks)*

Question 29 (Total 3 marks)

We can use conservation of energy to find the final velocity of the charged particle. When the charge q is accelerated through a potential difference of V, the work done on the charge by the field is equal to the change in the potential energy and is given by:

$$\Delta U = qV$$

Conservation of energy in this situation results in the potential energy being converted into kinetic energy as the charge is accelerated by the field. As the initial kinetic energy is zero, the change in potential energy will be equal to the final kinetic energy:

$\Delta U = qV = \frac{1}{2}mv^2$ and hence $v = \sqrt{\frac{2qV}{m}}$ as required. *(3 marks)*

Question 30 (Total 6 marks)

(a) As the track is frictionless, we can apply conservation of energy to find the height difference between P and Q. The loss of gravitational potential energy will equal the gain in kinetic energy and hence we can write:

$mgh = \frac{1}{2}mv^2$ and hence $h = \frac{v^2}{2g} = \frac{(1.5)^2}{2(9.8)} = 0.115$ m *(3 marks)*

(b) Consider the vertical component of the motion. The vertical component of the initial velocity will be given by: $u = 1.5\sin 50°$ downwards.

Therefore, taking the downwards direction to be positive, the height (s) of Q from the floor will be given by:

$s = ut + \frac{1}{2}gt^2 = (1.5sin50°)(0.5) + \frac{1}{2}(9.8)(0.5)^2 = 1.8$ m

Thus the point Q is 1.8 m above the floor. *(3 marks)*

Question 31 (Total 8 marks)

(a) Before the fan is switched on, the spring balance will read the weight of the fan, *weight* = mg. When the fan is switched on and begins to rotate it will apply a downwards force to the air and the air will apply an equal upwards force F_{up} to the fan (in accordance with Newton's third law). This force will increase as the rotation speed of the fan increases until it reaches a maximum value when the fan reaches its maximum rotational speed at $t = 10$ seconds.

The force measured by the spring balance will be given by: $F = mg - F_{up}$

Thus the force on the spring balance is initially mg but will decrease continually (as the fan speed and as F_{up} increase) until the fan reaches maximum rotational speed at $t = 10$ seconds. From $t = 10$ seconds to $t = 15$ seconds the upwards force will remain constant and, hence, the force on the spring balance will be constant. (This analysis assumes that the upwards force on the fan does not exceed the weight force of the fan.) *(4 marks)*

(b) This question is best answered by assuming that the ceiling fan has a DC motor (despite connections to mains power).

The student's prediction is not an accurate description of the way the current in the fan's motor would change. The initial current in the fan will actually rapidly rise to a maximum and then slowly decrease until it reaches a specific minimum value at $t = 10$ seconds and it will then stay at that value for between 10 and 15 seconds (see Figure 2 below).

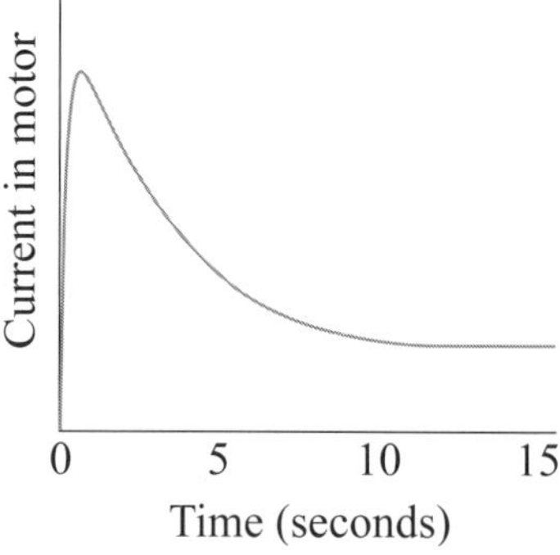

Figure 2 Current in fan motor as a function of time

This is because the back EMF increases as the fan rotates faster and this will decrease the current in the fan motor until the fan reaches its maximum rotational velocity. At the maximum rotational velocity, the current remains constant because the input torque is equal to the load torque at this speed and hence the net torque on the fan is zero. The student's graph also has a discontinuity at $t = 10$ seconds, when in fact the curve would be smooth. (Note the maximum current does not occur instantaneously because some back EMF is produced when the magnetic fields are initially created in the motor coils.) *(4 marks)*

Question 32 (Total 5 marks)

JJ Thomson and Millikan made significant contributions to our understanding of the electron by conducting experiments that enabled the charge and mass of the electron to be determined. Scattering experiments in particle accelerators have also made a significant contribution to our understanding of several fundamental particles, including quarks.

Thomson fired electrons through crossed electric and magnetic fields to find their velocity ($v = \frac{E}{B}$). He then determined the charge-to-mass ratio of the electron ratio by measuring the radius of curvature of their path when they were passed at right angles into a magnetic field. He did this by equating the magnetic force and the centripetal force, $qvB = \frac{mv^2}{r}$ and, hence, $\frac{q}{m} = \frac{v}{Br}$.

Millikan's old drop experiment was used to find the charge on the electron. His experiment used the vertical electric field between a pair of parallel charged plates to balance the force of gravity on charged oil droplets. When the droplets were stationary in the field, he knew that the net force on the droplet was zero and could write $mg = qE = \frac{qV}{d}$, which can be rearranged to give the charge on the droplet, $q = \frac{mgd}{V}$. Millikan found the mass of the droplets by measuring their terminal velocity in a gravitational field. Finding a common denominator for all the droplet charges he measured enabled him to find the electron charge. He was then able to calculate the mass of the electron using the charge-to-mass ratio that had been measured by Thomson.

(Alternatively, students could describe the experiments conducted by Davisson and Germer that involved firing electrons at a Nickel crystal and examining the diffraction pattern produced by the electrons. The scattered electrons exhibited diffraction and interference patterns that confirmed de Broglie's theory of the wave nature of electrons.)

Deep inelastic scattering experiments conducted in particle accelerators were used to prove the existence of quarks and to measure their charge and mass. Initial particle accelerator experiments of this type in 1968, which collided high-speed electrons into protons, showed that protons were not fundamental particles because they contained smaller point-like scattering centres. Later, deep inelastic scattering experiments examined the muons produced in these collisions, and were used to show that these 'scattering points' were the fundamental particles called quarks that had been theoretically predicted earlier. Further scattering experiments proved that six types of quarks (and their antiparticles) existed.

(Alternatively, students could describe the experiments conducted using the high neutrino flux from a nuclear reactor that were used to discover the neutrino or the Large Hadron Collider experiments that led to the discovery of the Higgs boson.) (*5 marks*)

Question 33 (Total 4 marks)

Similarity: As both charges enter the magnetic field perpendicularly to the field, they will both move with uniform circular motion while in the field. This is because each particle will experience a force that will remain perpendicular to the charge's velocity. A force that remains perpendicular to the velocity of a body is called a centripetal force and causes the body to move with a circular trajectory.

Difference: Because an alpha particle has twice the charge and approximately four times the mass of a proton, the circular trajectory taken by the alpha particle will have a radius (r_α) that is twice as great as the radius (r_P) of the circular trajectory taken by the proton.

We can prove this as follows. The magnetic force on the proton is $F_P = qvB$ and because the magnetic force provides the centripetal force we can write for the proton:

$$\frac{mv^2}{r} = qvB \text{ or } r = \frac{mv}{qB}$$

As the charge on the alpha particle is $2q$ and its mass is $4m$, the radius of the path taken by the alpha particle would be $r_\alpha = \frac{(4m)v}{(2q)B} = \frac{2(mv)}{qB} = 2r_P$. *(4 marks)*

Question 34 (Total 9 marks)

Power is produced in the sun by the fusion of hydrogen to helium in the core of the sun. Gravity exerts such a huge pressure on the hydrogen gas in the sun that the core is heated to over 10^7 K. At this temperature some of the collisions between hydrogen nuclei are so energetic that they overcome the coulomb repulsion between the nuclei, enabling them to fuse to helium. Approximately 90% of fusion reactions in the core of the sun occur via the p-p chain (the other 10% of fusion reactions occur via the CNO cycle). The p-p chain and the CNO cycle involve multiple nuclear reactions but the end result is that four hydrogen nuclei (protons) are converted into a helium-4 nucleus. In this process some mass is converted into energy in accordance with Einstein's equation $E = mc^2$. The sun, like all main sequence stars, is in a state of hydrostatic equilibrium with the inwards push of gravity balanced by gas and radiation pressure from the fusion reactions in the core. We can calculate the power emitted from the sun and the rate at which it converts mass to energy by using the data provided.

By multiplying the radiated power per square metre at a distance r by the surface area of a sphere of radius r:

Power emitted by the sun = radiated power per square metre at distance r × surface area of sphere of radius r

$$= 1360 \times 4 \times \pi \times (1.5 \times 10^{11})^2$$
$$= 3.85 \times 10^{26} \text{ watts}$$

Using Einstein's equation, $E = mc^2$, we can, to use the radiated power, calculate the amount of mass converted into energy in the sun per second: $m = \frac{E}{c^2} = 4.27 \times 10^9 \text{ kg s}^{-1}$.

Fusion in the sun causes the sun's surface (photosphere) to heat up to a very high temperature. Energy is then radiated into space in the form of electromagnetic waves from the surface of the sun. The sun acts like a black body (absorbs all the radiation that falls on it and reflects none), radiating electromagnetic radiation over a wide range of wavelengths as shown in the radiation curve provided. Because the sun acts as a black body we can calculate the surface temperature of the sun using the radiation curve from the sun provided and Wien's displacement law. Wien's law, $\lambda_{max} = \frac{b}{T}$, relates the wavelength of maximum emission from a black body to the surface temperature of the black body. From the radiation graph provided, the wavelength of maximum emission from the sun is approximately 510 nm and, hence, the surface temperature of the sun is:

$$T = \frac{b}{\lambda_{max}} = \frac{2.898 \times 10^{-3}}{510 \times 10^{-9}} = 5.7 \times 10^3 \text{ K}.$$

The sun also emits a constant stream of high velocity–charged particles into space. This solar wind mainly consists of protons and electrons. *(9 marks)*

Question 35 (Total 4 marks)

If the car experienced an acceleration with a component that was perpendicular to the car's direction of travel, the mass would move. Hence, whenever the car changes its direction of travel, the mass on the apparatus would move. The apparatus is therefore an accelerometer that could be used to investigate accelerations perpendicular to the direction of motion, such as centripetal acceleration.

If the car moved around a circular corner with radius r, the centripetal force on the mass would cause it to remain displaced by an angle θ when the car was moving through the corner. We can relate this angle to the centripetal acceleration by showing that the forces on the mass (tension and gravity) produce a centripetal force on the mass as shown in Figure 3.

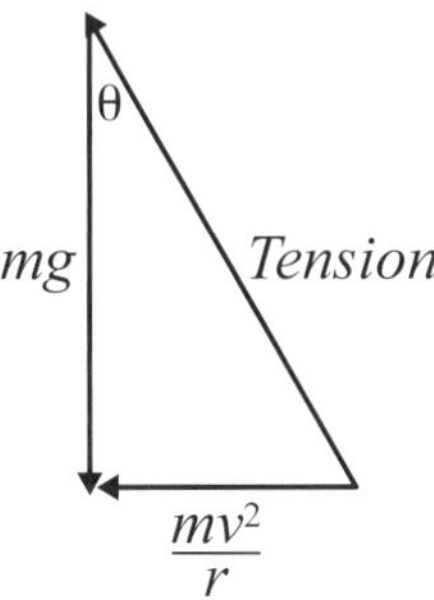

Figure 3 The tension pulling the mass and the weight of the mass sum to produce a net force (centripetal force) on the mass.

From the diagram, $\tan\theta = \frac{\left(\frac{mv^2}{r}\right)}{(mg)} = \frac{v^2}{rg}$ and hence the centripetal acceleration of the car is related to the angle θ by:

$$a_C = \frac{v^2}{r} = g\tan\theta$$

Note that the direction the mass swings will be the opposite direction to that in which the car turns and hence the apparatus can be used to determine the direction in which the car is turning and the centripetal acceleration of the car. *(4 marks)*

Question 36 (Total 7 marks)

We must first determine the mass defect:

Mass defect = mass of reactants – mass of products

$= 197.999 - (193.988 + 4.002\,60) = 0.0084$ u

As the sum of the mass of the products is less than the mass of the reactant, mass must be converted into energy in this reaction. That is, this reaction releases energy.

The total energy released is given by:

$0.0084 \times 931.5 = 7.8246$ MeV, or

$= 7.8246 \times 10^{6} \times 1.602 \times 10^{-19} = 1.254 \times 10^{-12}$ J.

Now by conservation of energy, the alpha particle must have a kinetic energy of:

K = *total energy – kinetic energy of polonium atom*

$= 1.254 \times 10^{-12} - 2.55 \times 10^{-14} = 1.23 \times 10^{-12}$ J.

The alpha particle has a much greater share of the kinetic energy because momentum must also be conserved in the reaction. If we assume the radon atom was initially at rest, the sum of the momentum of the two particles after the collision must also be zero. That is, the particles must have the equal and opposite momentum after the collision. Calling the mass of polonium M and its velocity V after the collision and the mass of the alpha particle m and its velocity after the collision v, we can write:

$$MV = mv \text{ and, hence, } v = \left(\frac{M}{m}\right)V.$$

Thus, as the polonium-194 atom is approximately 48 times larger than the alpha particle, the alpha particle will have 48 times the velocity of the polonium atom and 48 times as much kinetic energy after the reaction.

To prove this, we substitute the velocity of the alpha particle $\left(v = \frac{MV}{m}\right)$ into the kinetic energy equation for the alpha particle:

$$K = \left(\frac{1}{2}mv^2\right) = \frac{1}{2}m\left(\frac{MV}{m}\right)^2 = \frac{M}{m}\left(\frac{1}{2}MV^2\right).$$

Therefore the alpha particle will have $\frac{M}{m}$ times more kinetic energy than the polonium atom after the reaction, which is consistent with the kinetic energy values calculated above. *(7 marks)*

CHAPTER 6

NSW Education Standards Authority

2020 HIGHER SCHOOL CERTIFICATE EXAMINATION

Physics

General Instructions

- Reading time – 5 minutes
- Working time – 3 hours
- Write using black pen
- Draw diagrams using pencil
- Calculators approved by NESA may be used
- A data sheet, formulae sheet and Periodic Table are provided at the back of this paper

Total marks: 100

Section I – 20 marks

- Attempt Questions 1–20
- Allow about 35 minutes for this section

Section II – 80 marks

- Attempt Questions 21–34
- Allow about 2 hours and 25 minutes for this section

Section I

20 marks
Attempt Questions 1–20
Allow about 35 minutes for this section

Use the multiple-choice answer sheet for Questions 1–20.

---

1 The diagram shows a model used to explain the refraction of light passing from medium X into medium Y.

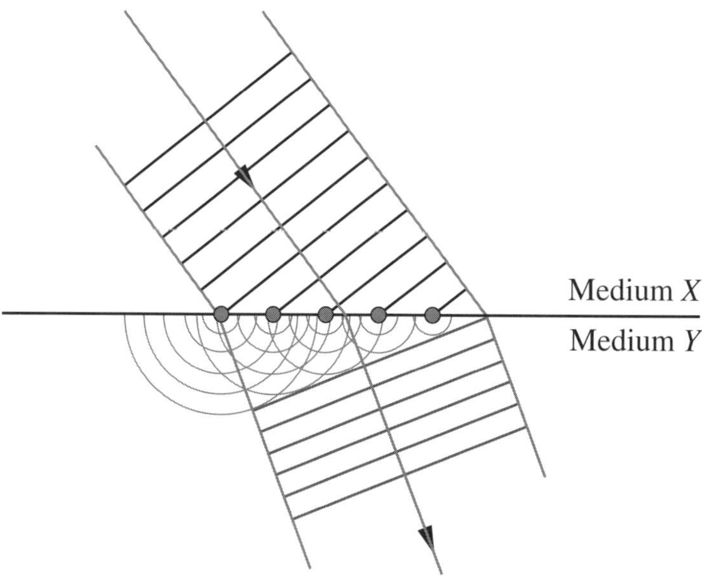

Who proposed this model?

A. Malus

B. Planck

C. Newton

D. Huygens

2 Which of the following is NOT required for the operation of AC induction motors?

A. Brushes

B. Stator winding

C. Magnetic fields

D. Current applied to the rotor

3 What was the basis for Maxwell's prediction of the velocity of electromagnetic waves?

A. Experiments using magnetic fields to accelerate particles

B. Experiments using light and mirrors to establish the finite speed of light

C. Equations showing how oscillating electric and magnetic fields propagate

D. Equations showing how electromagnetic waves are affected by gravitational fields

4 The graph shows the mass of a radioactive isotope as a function of time.

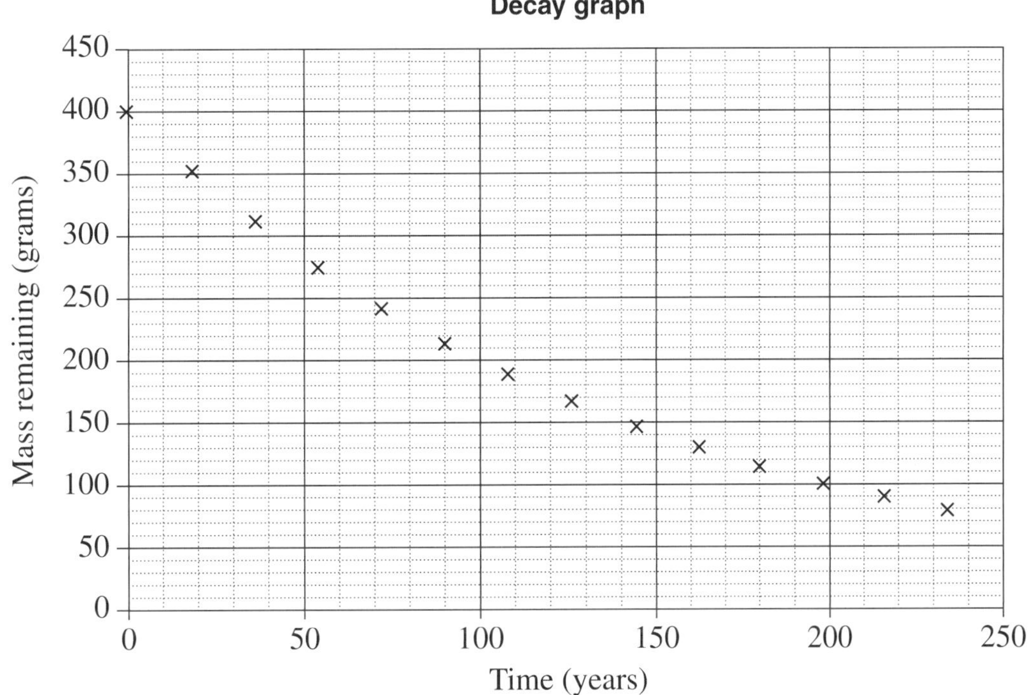

What is the decay constant, in years^{-1}, for this isotope?

A. 0.0030

B. 0.0069

C. 2.0

D. 100

5 A student throws a ball that follows a parabolic trajectory.

What change to the initial velocity would make the ball's time of flight shorter?

A. Increasing only the vertical component

B. Decreasing only the vertical component

C. Increasing only the horizontal component

D. Decreasing only the horizontal component

6 The Hertzsprung–Russell diagram shows characteristics of stars in a globular cluster 100 light years in diameter and 27 000 light years from Earth.

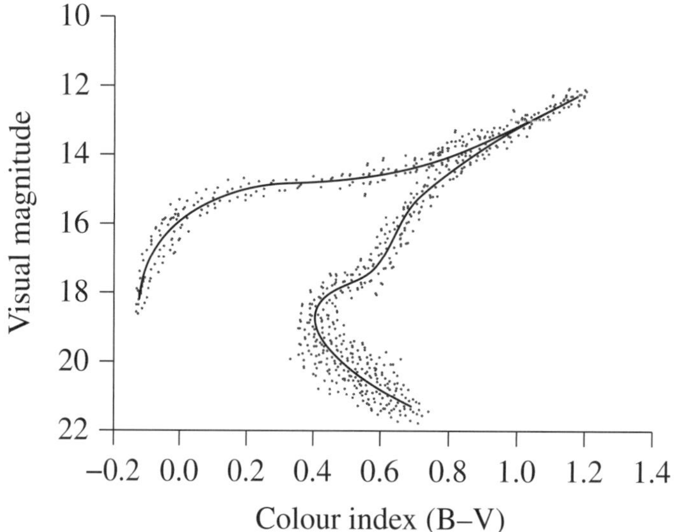

The stars plotted on this Hertzsprung–Russell diagram have approximately the same

A. age.

B. colour.

C. luminosity.

D. mass.

7 The output of a device is shown.

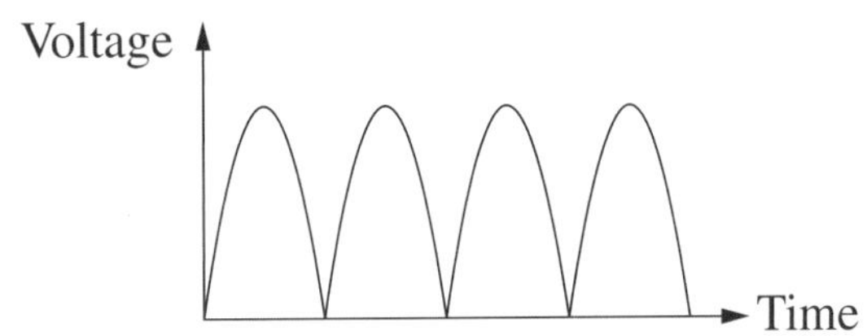

Which diagram represents the device that has the output shown?

A.

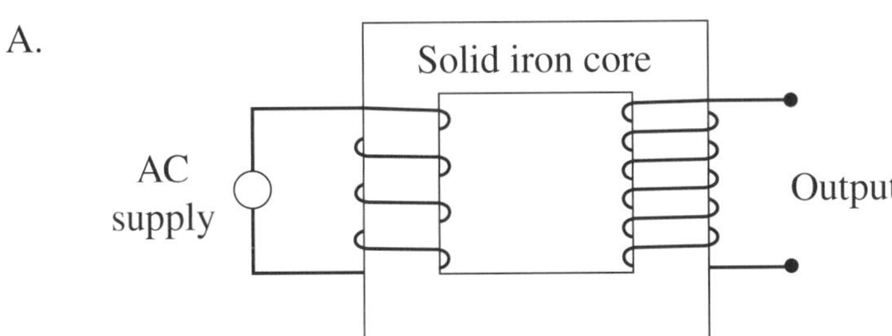

B.

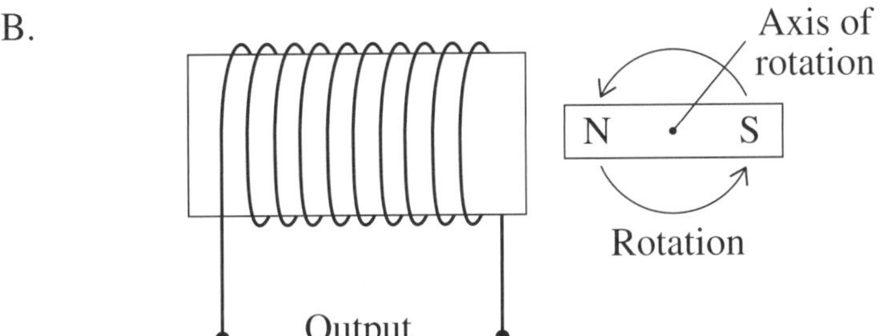

C.

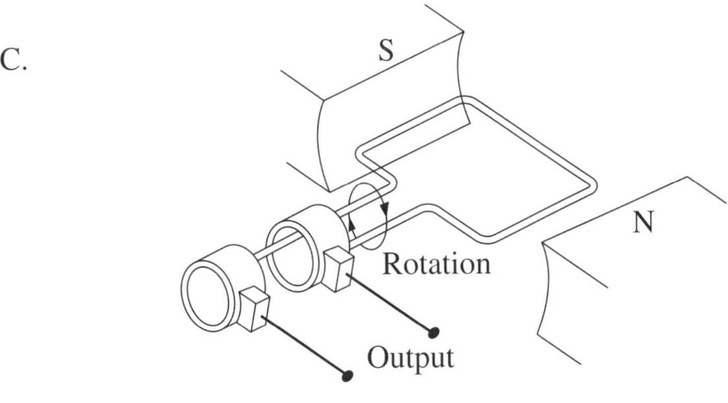

D.

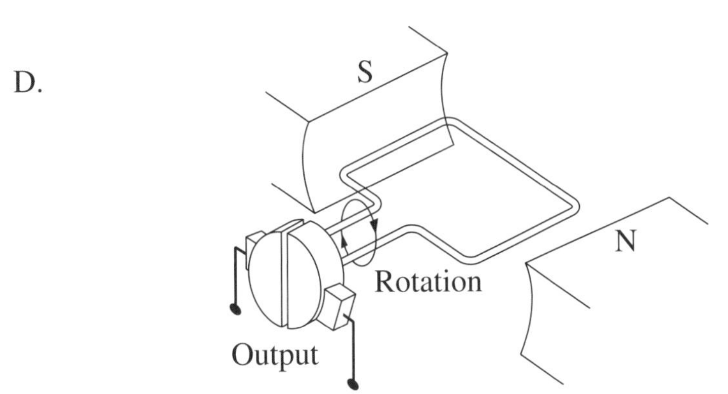

8 A uranium isotope, U, undergoes four successive decays to produce Q.

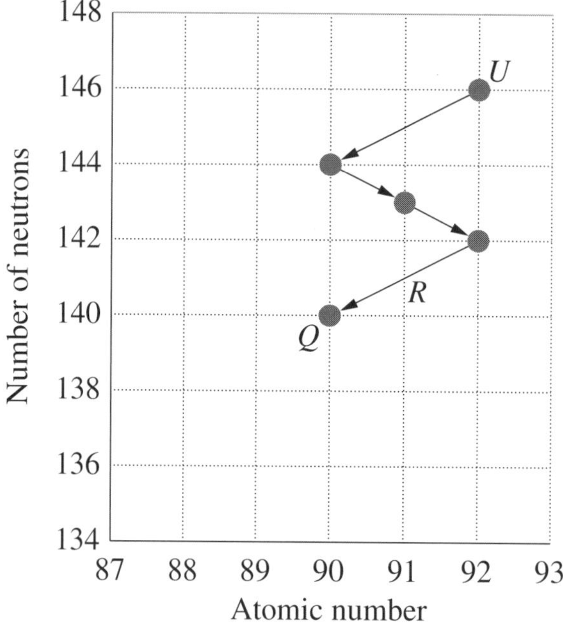

Which row of the table correctly shows the decay process R and product Q?

	Process R	*Product Q*
A.	α	Pa-230
B.	β	Pa-234
C.	α	Th-230
D.	β	Th-234

9 Bohr improved on Rutherford's model of the atom.

Which observation by Bohr provided evidence supporting the improvement?

A. Elements produced unique emission spectra consisting of discrete wavelengths.

B. The collision of an electron and a positron produced two photons that travelled in opposite directions.

C. A small percentage of alpha particles fired at a gold foil target were deflected by angles of more than 90 degrees.

D. A beam of electrons reflected from a nickel crystal produced a pattern of intensity at different angles, consistent with their wave properties.

10 An electron travelling in a straight line with an initial velocity, u, enters a region between two charged plates in which there is an electric field causing it to travel along the path as shown.

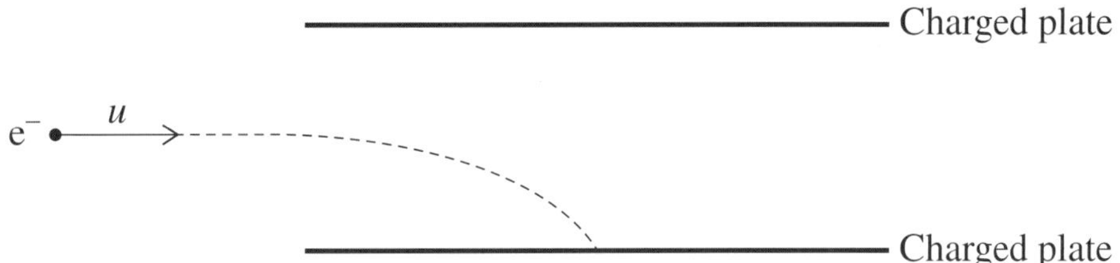

A magnetic field is then applied causing a second electron with the same initial velocity to pass through undeflected.

Which row of the table shows the directions of the electric and magnetic fields when the second electron enters the region between the plates?

	Electric field	*Magnetic field*
A.	Towards top of page	Into page
B.	Towards top of page	Out of page
C.	Towards bottom of page	Into page
D.	Towards bottom of page	Out of page

11 Consider the following nuclear reaction.

$$^{6}_{3}\text{Li} + ^{1}_{0}\text{n} \rightarrow ^{4}_{2}\text{He} + ^{3}_{1}\text{H}$$

The mass of the reactants is 7.023787704 u and the mass of the products is 7.018652532 u.

What type of reaction is this?

A. A fusion reaction in which energy is released

B. A fusion reaction that requires an input of energy

C. A transmutation reaction in which energy is released

D. A transmutation reaction that requires an input of energy

12 In which of the following would the satellite have the greatest escape velocity from Earth?

A.

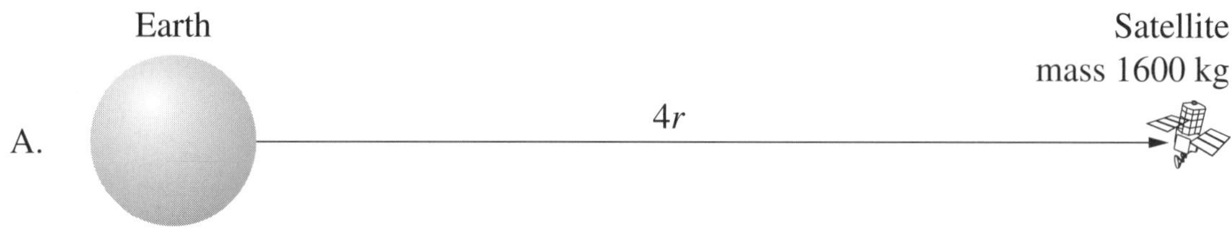

B.

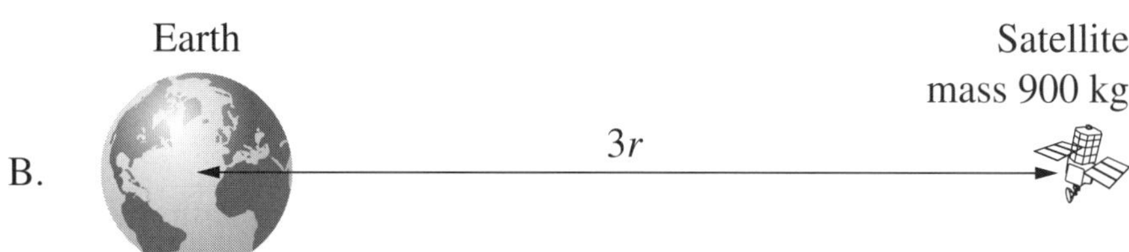

C.

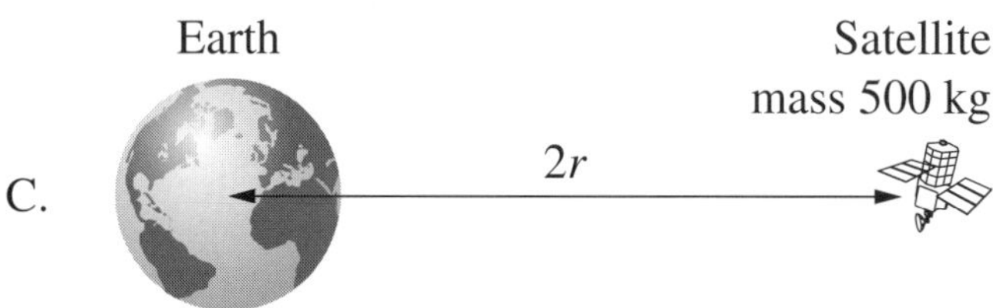

D.

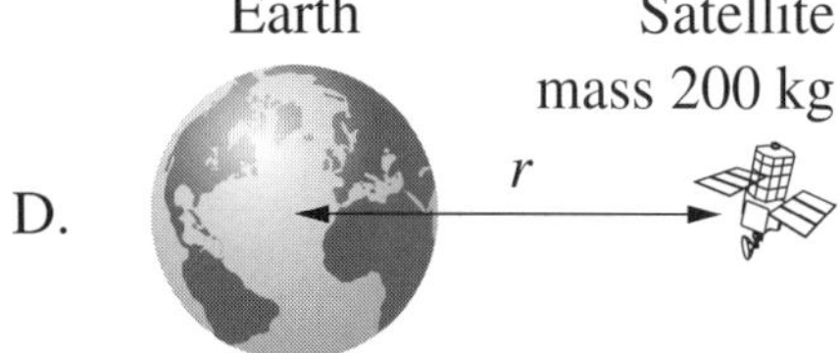

13 The graph shows the relationship between the frequency of light used to irradiate two different metals, and the maximum kinetic energy of photoelectrons emitted.

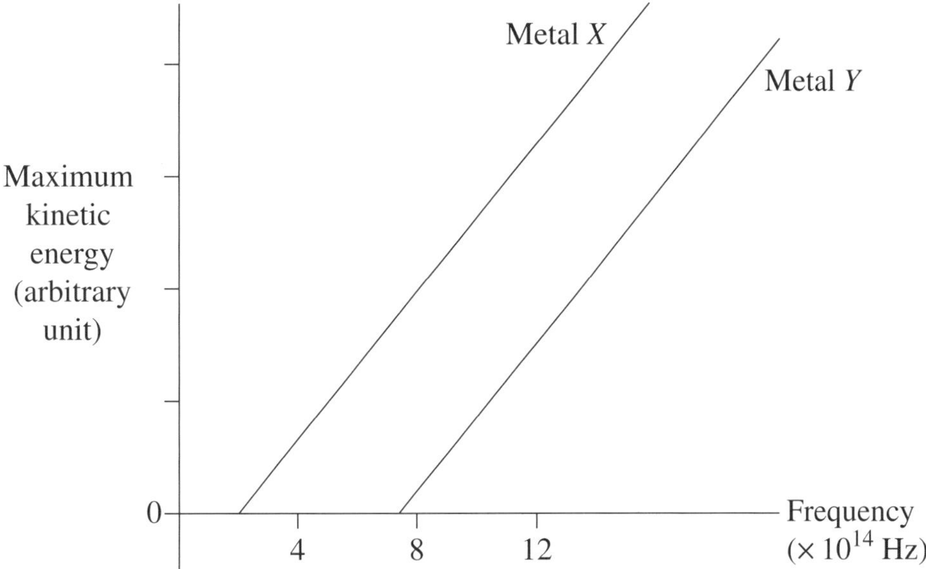

Suppose that light having a frequency of 8×10^{14} Hz is used to irradiate both metals.

Compared to the photoelectrons emitted from metal *X*, photoelectrons emitted from metal *Y* will

A. have a lower maximum velocity.

B. have a higher maximum velocity.

C. take a longer time to gain sufficient energy to be ejected.

D. take a shorter time to gain sufficient energy to be ejected.

14 Two parallel wires, X and Y, each carry a current I.

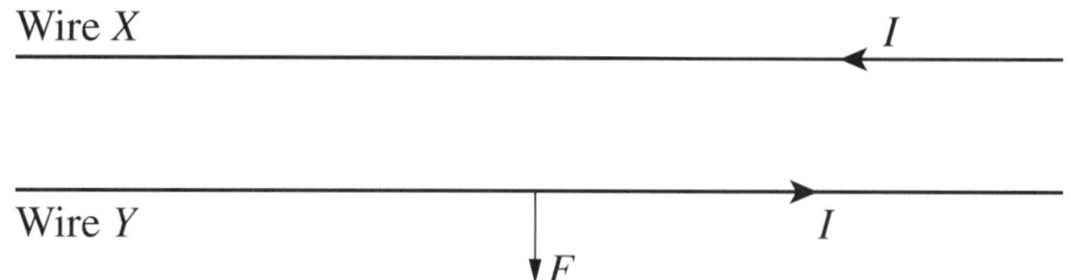

The magnitude and direction of the force on wire Y are represented by the vector F.

The current in wire Y is then doubled and its direction is reversed. The current in wire X remains unchanged.

Which vector arrow represents the force on wire X after the change to the current in wire Y?

A. 

B. 

C. 

D. 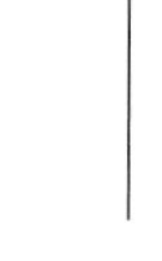

15 A rocket returns to Earth for reuse after launching satellites, using its engines to make a controlled landing.

The rocket having a mass of 7800 kg is on approach to the ground, travelling horizontally at 20 m s^{-1} as shown in the diagram, when the engine thrust is changed to 90 000 newtons.

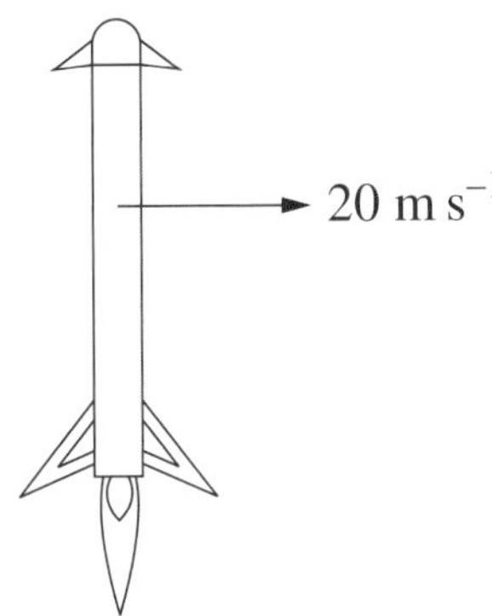

Which diagram shows the trajectory of the rocket following this change of thrust?

A.

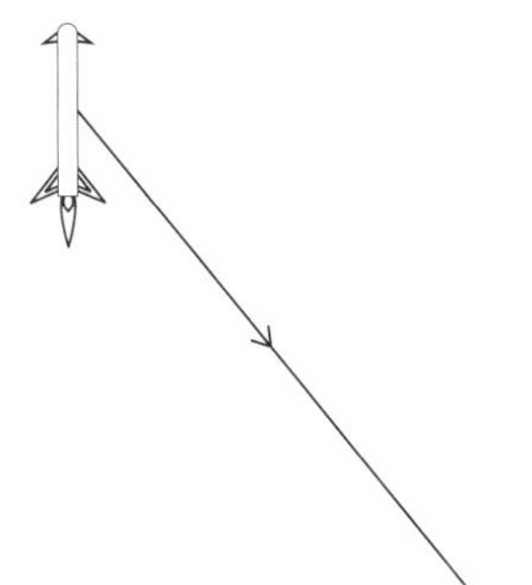

Ground

B.

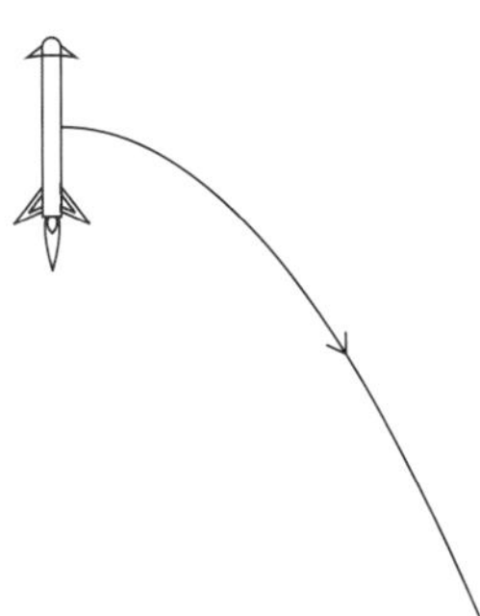

Ground

C.

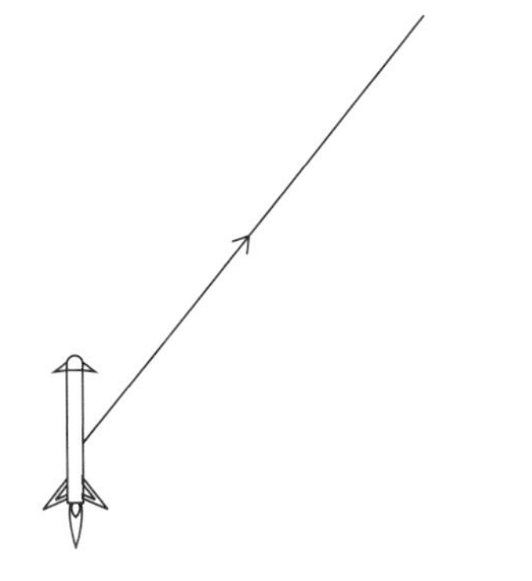

Ground

D.

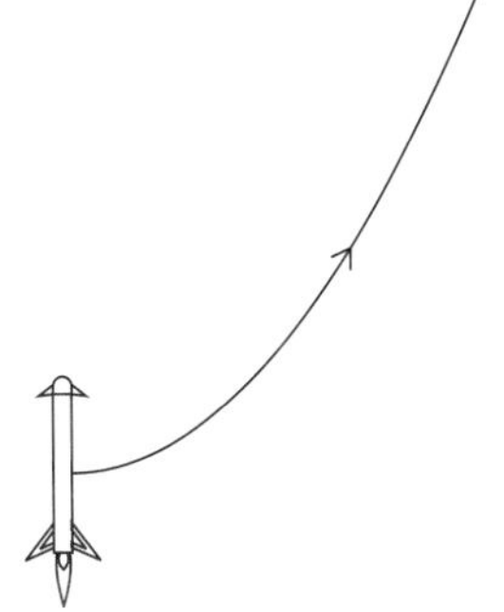

Ground

16 A model of the core of a nuclear fission reactor is shown.

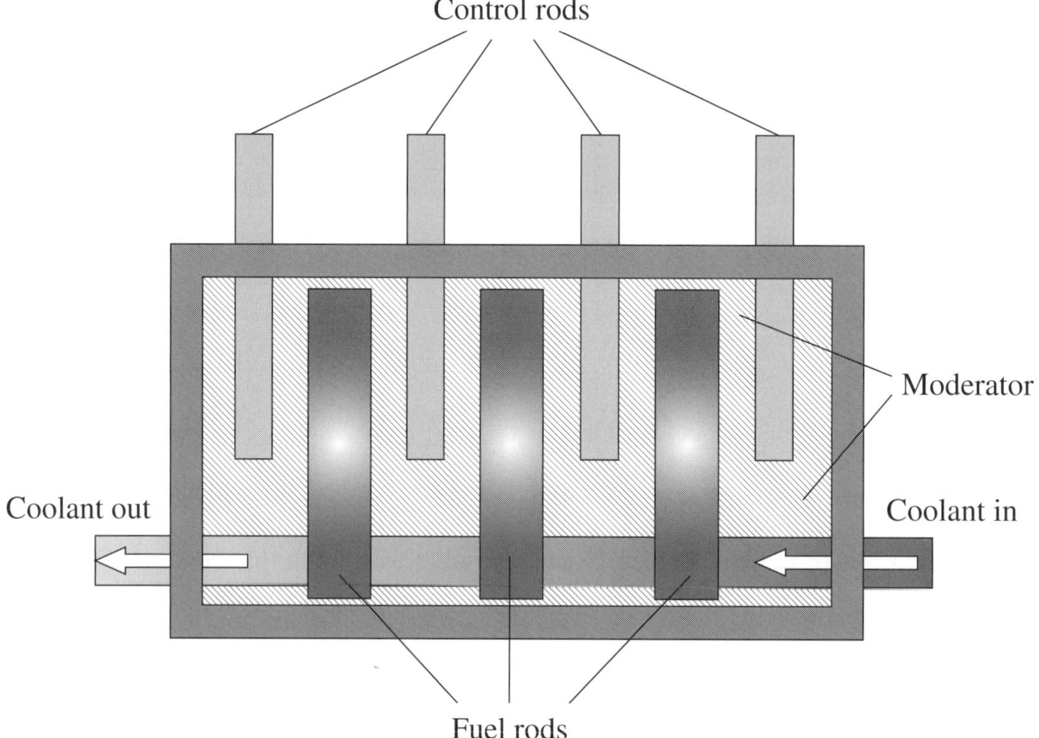

When the reactor is operating normally, the moderator, control rods and coolant work in combination to maintain a controlled nuclear reaction in the fuel rods.

The moderator is a liquid which slows down neutrons to increase the rate of fission. The control rods absorb free neutrons. The coolant reduces the core temperature.

A fault causes some of the moderator to leak out of the core.

Which action would compensate for the effect of the loss of moderator?

A. Withdraw the control rods from the core.

B. Lower the control rods further into the core.

C. Pump the coolant through the core at a faster rate.

D. Reduce the temperature of the coolant before pumping it into the core.

17 In a thought experiment, observer *X* is on a train travelling at a constant velocity of 0.95c relative to the ground. Observer *Y* is standing on the ground outside the train. As observer *X* passes observer *Y*, observer *X* sends a short light pulse towards the sensor.

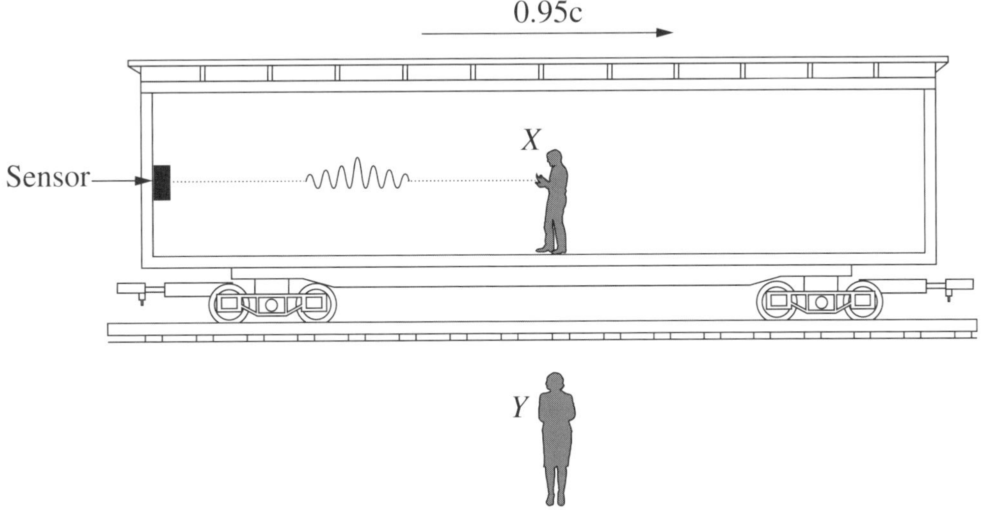

Which statement about the light pulse is correct as observed by *X* or *Y* in their respective frames of reference?

A. Its velocity observed by *Y* is 0.05c.

B. *X* sees it travel a shorter distance to the sensor than *Y*.

C. *X* sees it take a longer time to reach the sensor than *Y*.

D. Both *X* and *Y* see it travel the same distance in the same amount of time.

20 The diagram shows a smooth, semi-circular, vertical wall with radius, r.

A ball is launched from the position shown with a velocity u towards north at an angle to the horizontal.

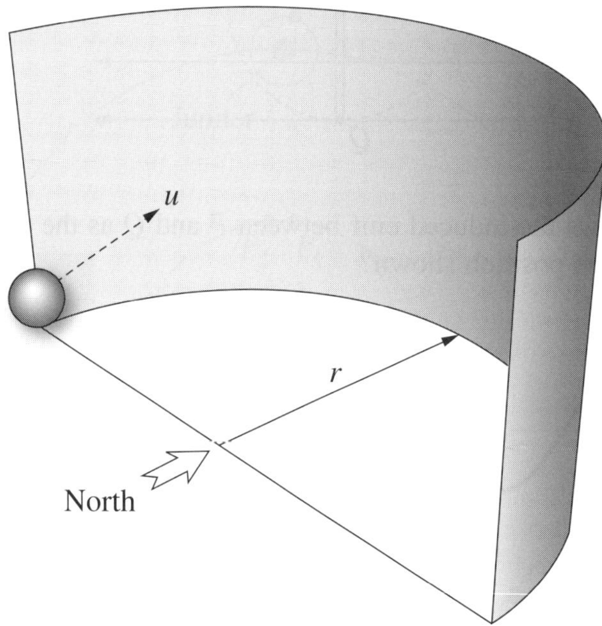

The ball follows a trajectory around the wall before landing on the ground, opposite its starting point. It does not reach the top of the wall.

Assume that there is no friction between the ball and the wall.

Which statement correctly describes the net force acting on the ball during its motion?

A. The magnitude of the net force remains constant.

B. The direction of the net force is vertically downwards.

C. The direction of the net force is perpendicular to the wall.

D. The magnitude of the net force reaches a minimum when the ball is at its highest point.

2020 HIGHER SCHOOL CERTIFICATE EXAMINATION

Centre Number

Student Number

Physics

Section II Answer Booklet

80 marks
Attempt Questions 21–34
Allow about 2 hours and 25 minutes for this section

Instructions

- Write your Centre Number and Student Number at the top of this page.
- Answer the questions in the spaces provided. These spaces provide guidance for the expected length of response.
- Show all relevant working in questions involving calculations.

Please turn over

Question 21 (5 marks)

(a) Calculate the wavelength of light emitted by an electron moving from energy level 3 to 2 in a Bohr model hydrogen atom. **2**

....

....

....

....

....

....

....

(b) Describe the behaviour of electrons in the Bohr model of the atom with reference to the law of conservation of energy. **3**

....

....

....

....

....

....

....

Question 22 (5 marks)

A capsule travelling at 12 900 m s^{-1} enters Earth's atmosphere, causing it to rapidly slow down to 400 m s^{-1}.

(a) During this re-entry, the capsule reaches a temperature of 3200 K. **2**

What is the peak wavelength of the light emitted by the capsule?

....

....

....

....

(b) Outline TWO limitations of applying special relativity to the analysis of the motion of the capsule. **3**

....

....

....

....

....

....

Question 23 (3 marks)

The graph shows data for a motor connected to a 240 V power supply. 3

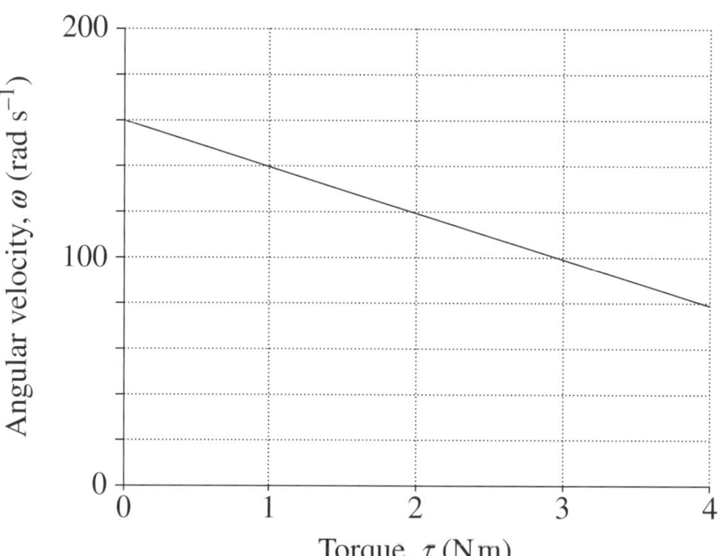

The equation for the torque, τ, produced by the motor is $\tau = \dfrac{VI\eta}{\omega}$

where τ = torque (N m)
V = voltage (V)
I = current (A)
η = efficiency = 0.3
ω = angular velocity (rad s^{-1})

A circuit breaker cuts the current to the motor if the current exceeds 5 A.

Determine what will happen when the motor produces a torque of 2.95 N m. Show relevant calculations.

....

....

....

....

....

....

Question 24 (4 marks)

The graph shows the vertical displacement of a projectile throughout its trajectory. The range of the projectile is 130 m. **4**

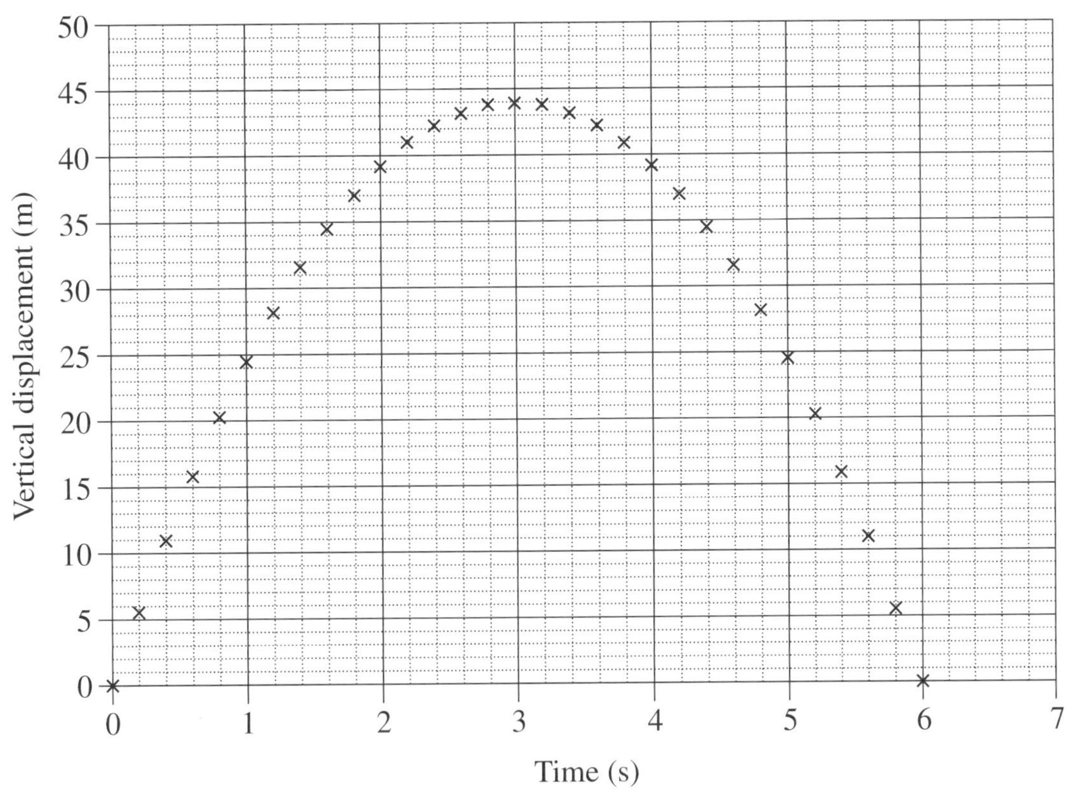

Calculate the initial velocity of the projectile.

....

....

....

....

....

....

....

....

....

....

....

....

Question 25 (4 marks)

Describe the hydrogen atom in terms of the Standard Model of matter. **4**

....

....

....

....

....

....

....

....

....

....

....

....

Question 26 (8 marks)

(a) Describe the difference between the spectra of the light produced by a gas discharge tube and by an incandescent lamp. **2**

....

....

....

....

(b) The graph shows the curves predicted by two different models, *X* and *Y*, for the electromagnetic radiation emitted by an object at a temperature of 5000 K. **2**

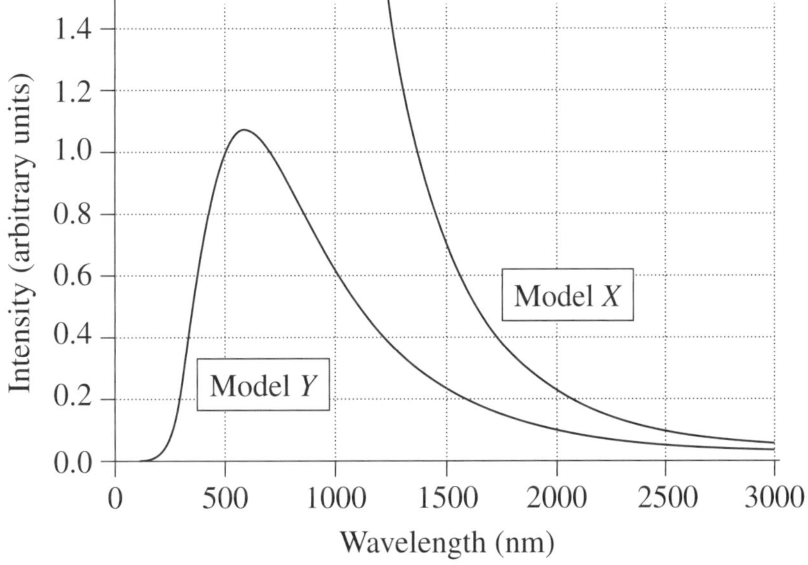

Identify an assumption of EACH model which determines the shape of its curve.

....

....

....

....

Question 26 continues on the following page

Question 26 (continued)

(c) The diagram shows the radiation curve for a black body radiator at a temperature of 5000 K. **4**

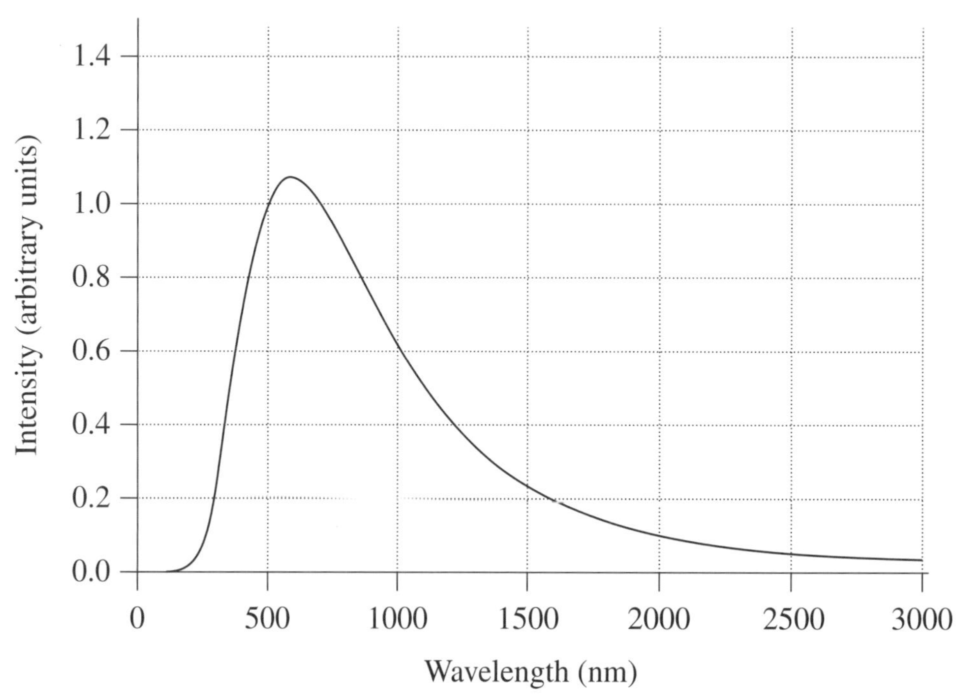

On the same diagram, sketch a curve for a black body radiator at a temperature of 4000 K and explain the differences between the curves.

....

....

....

....

....

....

....

....

....

....

End of Question 26

Question 27 (4 marks)

The following apparatus is used to investigate light interference using a double slit. **4**

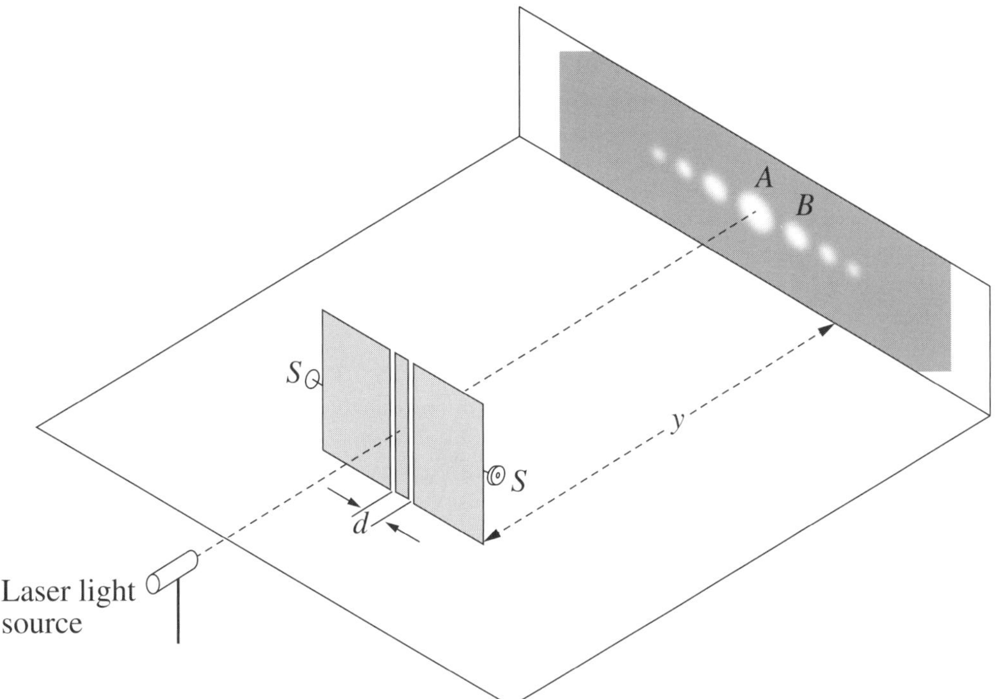

The distance, y, from the slits to the screen can be varied. The adjustment screws (S) vary the distance, d, between the slits. The wavelength of the laser light can be varied across the visible spectrum. The diffraction pattern shown is for a specific wavelength of light.

Explain TWO methods of keeping the distance between the maxima at A and B constant when the wavelength of the laser light is reduced.

....

....

....

....

....

....

....

....

....

....

....

Question 28 (7 marks)

A metal rod sits on a pair of parallel metal rails, 20 cm apart, that are connected by a copper wire. The rails are at 30° to the horizontal.

The apparatus is in a uniform magnetic field of 1 T which is upward, perpendicular to the table.

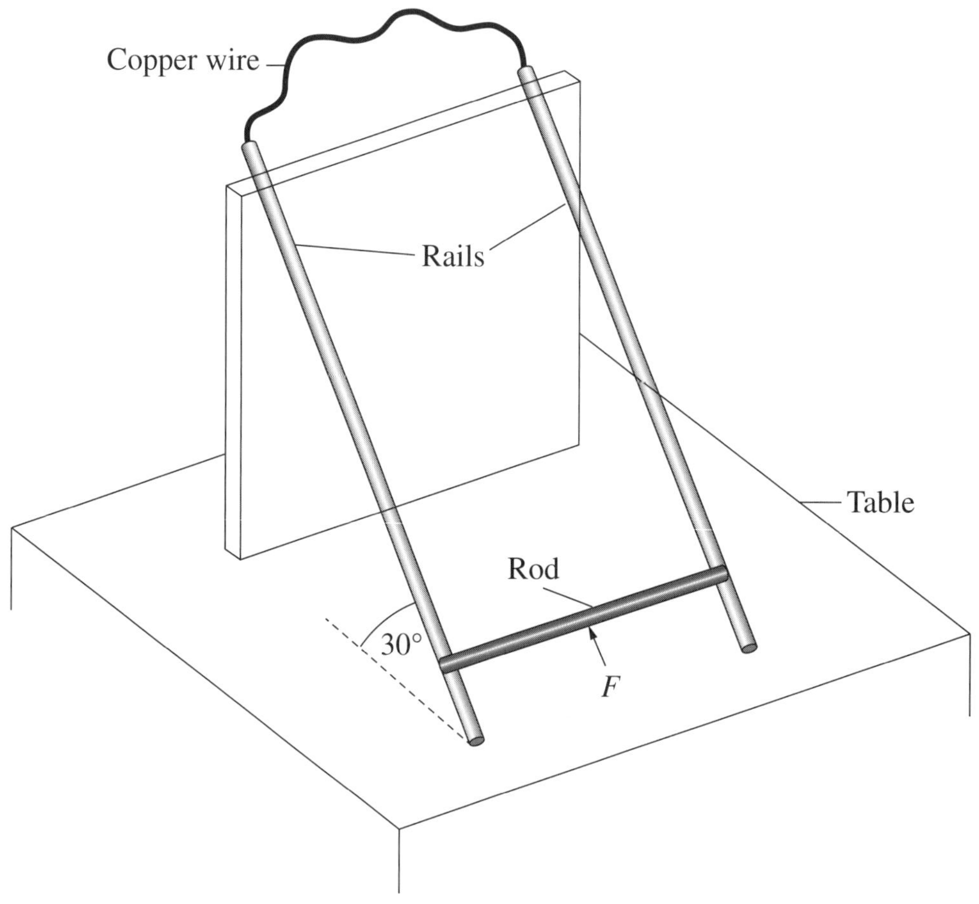

A force, F, is applied parallel to the rails to move the rod at a constant speed along the rails. The rod is moved a distance of 30 cm in 2.5 s.

(a) Show that the change in magnetic flux through the circuit while the rod is moving is approximately 5.2×10^{-2} Wb. **2**

....

....

....

....

Question 28 continues on the following page

Question 28 (continued)

(b) Calculate the emf induced between the ends of the rod while it is moving, and state the direction of flow of the current in the circuit. **2**

....

....

....

....

(c) The experiment is repeated without the magnetic field. **3**

Explain why the force required to move the rod is different without the magnetic field.

....

....

....

....

....

....

End of Question 28

Question 29 (5 marks)

In an experiment, alpha particles were fired into a thin sheet of beryllium. Unknown radiation was detected.

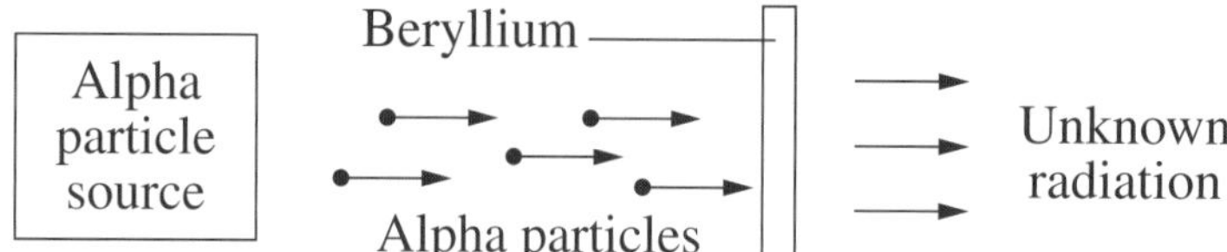

Further experiments were conducted in which it was observed that the unknown radiation:

- was not deflected by an electric field
- caused protons to be ejected from a block of paraffin
- could not produce the photoelectric effect.

Scientists debated the nature of this unknown radiation, hypothesising that it was gamma radiation.

(a) Explain why the hypothesis was proposed and then rejected, with reference to the observations. **3**

....

....

....

....

....

....

....

....

....

....

....

(b) How did these experiments change the model of the atom? **2**

....

....

....

....

Question 30 (7 marks)

(a) Explain, using an example, how a particle accelerator has provided evidence for the Standard Model of matter. **3**

....

....

....

....

....

....

....

(b) (i) Calculate the wavelength of a proton travelling at 0.1c. **2**

....

....

....

....

....

(ii) Explain the relativistic effect on the wavelength of a proton travelling at 0.95c. **2**

....

....

....

....

....

Question 31 (6 marks)

(a) The orbit of a comet is shown. 

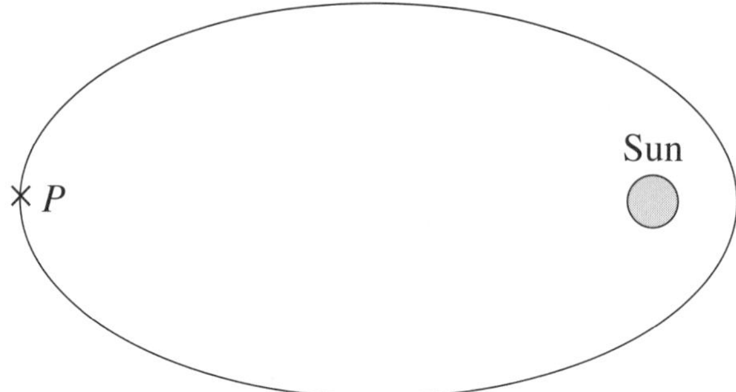

Account for the changes in velocity of the comet as it completes one orbit from position P.

(b) Two stars, A and B, of equal mass m, separated by a distance x, interact gravitationally such that the speed of A is constant. **3**

Derive an expression for the speed of B.

Question 32 (7 marks)

A rope connects a mass on a horizontal surface to a pulley attached to an electric motor as shown. **7**

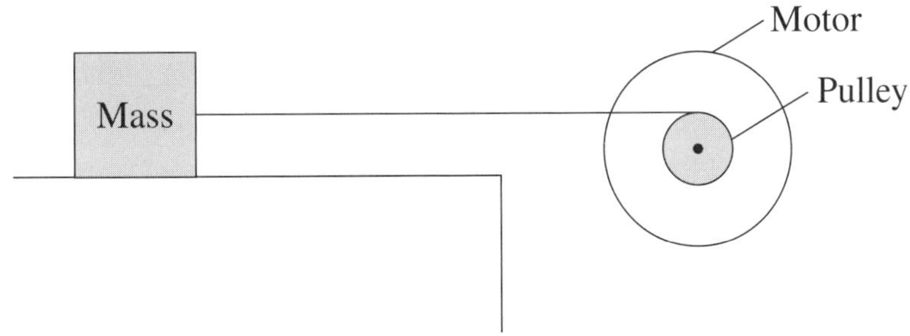

Explain the factors that limit the speed at which the mass can be pulled along the horizontal surface. Use mathematical models to support your answer.

Question 33 (9 marks)

A strong magnet of mass 0.04 kg falls 0.78 m under the action of gravity from position X above a hollow copper cylinder. It then travels a distance of 0.20 m through the cylinder from Y to Z before falling freely again. **9**

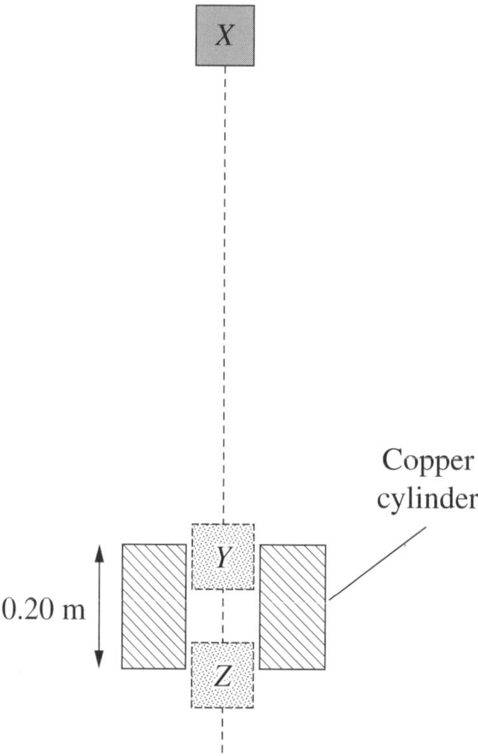

The magnet takes 0.5 seconds to pass through the cylinder. The displacement–time graph of the magnet is shown.

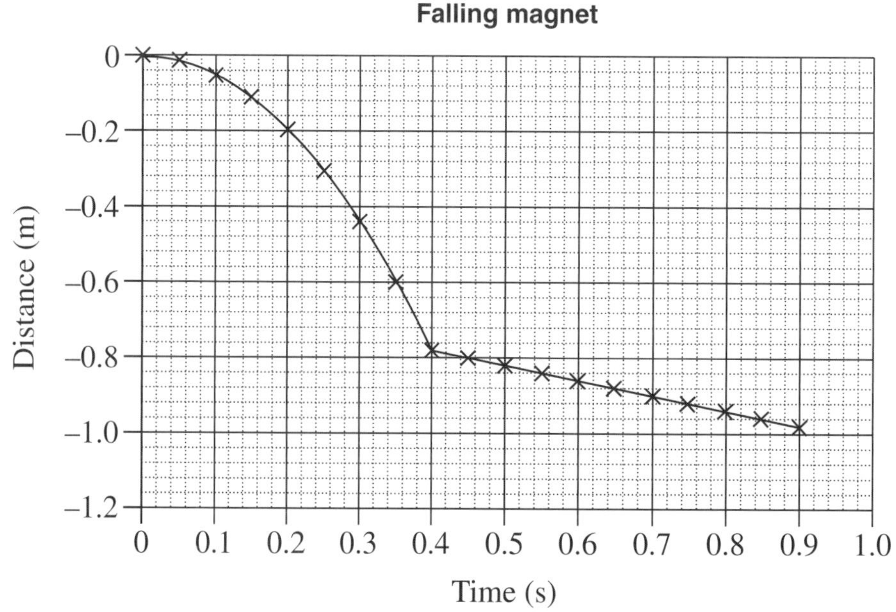

Question 33 continues on the following page

Question 33 (continued)

Analyse the motion of the magnet by applying the law of conservation of energy.

Your analysis should refer to gravity and the copper cylinder, and include both qualitative and quantitative information.

End of Question 33

Question 34 (6 marks)

A charged particle, q_1, is fired midway between oppositely charged plates X and Y, as shown in Figure 1. The voltage between the plates is V volts.

The particle strikes plate Y at point P, a horizontal distance s from the edge of the plate. Ignore the effect of gravity.

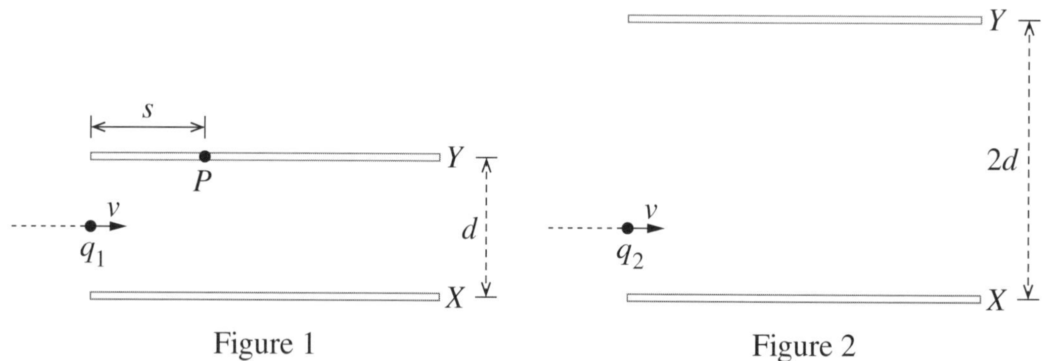

Figure 1

Figure 2

Plate Y is then moved to the position shown in Figure 2, with the voltage between the plates remaining the same.

An identical particle, q_2, is fired into the electric field at the same velocity, entering the field at the same distance from plate X as q_1.

(a) Compare the work done on q_1 and q_2. **3**

....

....

....

....

....

....

....

....

....

....

Question 34 continues on the following page

Question 34 (continued)

(b) Compare the horizontal distances travelled by q_1 and q_2 in the electric field.

....

....

....

....

....

....

....

....

....

....

....

....

End of Question 34

End of paper

2020 HSC Examination paper

Sample answers

Section I (Total 20 marks)

1 D This analysis of refraction assumes light is a wave and uses Huygens' Principle, which states that every point on the wavefront acts as an independent source of secondary waves.

2 A DC motors require brushes to connect the external power source to the rotor coil but AC induction motors do not require brushes. AC induction motors do require stator windings and magnetic fields. Current is applied to the rotor in AC motors by electromagnetic induction in the rotor, produced by the changing magnetic fields generated by the AC stator coils.

3 C Maxwell derived the speed of electromagnetic waves by combining his equations of electromagnetism to produce a wave equation. This electromagnetic wave equation showed that all electromagnetic waves would travel at the same speed (2.99×10^8 ms^{-1}) in a vacuum. Earlier experiments had found that light travelled at this speed.

4 B Using the data plotted on the graph, the time for half of the isotopes in the sample to undergo radioactive decay (i.e., the half-life) is $t_{\frac{1}{2}} = 100$ years.

The half-life ($t_{\frac{1}{2}}$) is related to the decay constant by:

$$\lambda = \frac{ln2}{t_{\frac{1}{2}}} = \frac{ln2}{100} = 0.0069\ yr^{-1}$$

5 B The time of flight $\left(t = \frac{u_y}{g}\right)$ is determined by the vertical component of the initial velocity and the local acceleration due to gravity. Hence decreasing the initial vertical velocity will decrease the time of flight.

6 A The H–R diagram provided shows the evolution of stars that have a similar age but have different colours, masses and luminosities. We can see that the stars are the same age because the stars leave the main sequence at the same point. Because the stars are the same age, the more massive the stars, the further along their evolutionary path they are. We can check this answer by recalling that the horizontal axis of the H–R diagram is related to colour (surface temperature), while the vertical axis is related to luminosity. Because stars have different positions on the diagram, they have different colours and luminosities. The position of main sequence stars on a H–R diagram depends on their mass. Because the main sequence stars in the lower right fall at different places on the main sequence (diagonal line from lower right to upper left of H–R diagram) they must have different masses.

7 D Answer **D** shows a DC electric generator that would produce the voltage signal shown. All the other devices would produce an AC output.

8 C By reading the axes on the diagram, the decay process R involves the loss of two protons and two neutrons and is hence an alpha (α) decay. Again, reading from the axes on the diagram the final product Q has 90 protons and 140 neutrons and hence it has a mass number of 90 + 140 = 230. By referring to the periodic table we see that thorium has a mass number of 90 and hence the product Q must be Th-230.

9 A Bohr proposed that electrons orbiting nuclei in atoms could only exit in specific energy states and that an electron could move to a higher or lower energy level by absorbing or emitting a photon. To conserve energy the photon would have to have an energy equal to the energy difference between these energy states. As photon energy is related to the wavelength of the photon $\left(E = \frac{hc}{\lambda}\right)$ this would mean that elemental atoms when excited would emit specific wavelengths (line spectra) that were characteristic of the energy difference between electron energy states in the atom.

10 B Because an electron is a negative particle it would experience a downwards force (as shown) when an electric field was applied in an upwards direction. The applied magnetic field would have to produce an upwards force on the electron to balance the downwards force produced by the electric field to ensure the electron would move through the plates without being deflected. Using the right-hand palm rule for a negative particle we see a magnetic field directed out of the page would be required to produce an upwards force on the (negatively charged) electron.

11 C Because this reaction does not involve two or more nuclei combining, it is a transmutation reaction rather than a fusion reaction. In addition, as the total mass of the products is less than the total mass of the reactants, energy would be released in the reaction (i.e., to ensure energy is conserved, the lost mass is converted to energy in accordance with $E = \Delta mc^2$).

12 D Escape velocity depends on the mass of the central body (M) but is independent of the mass (m) of the escaping body $\left(v_{esc} = \sqrt{\frac{2GM}{r}}\right)$. The satellite closest to the Earth (smallest value of r) will be the most tightly bound to the Earth and hence the one which will require the greatest escape velocity.

13 A The incident light frequency is above the threshold frequency for both metals and hence photoelectrons will be emitted from both metals. Reading from the graph provided, we see that the maximum kinetic energy of the photoelectrons emitted from metal ***Y*** (and hence the maximum velocity of the photoelectrons) is less than the maximum kinetic energy of the photoelectrons emitted from metal X. Alternatively, you could note that the threshold frequency is lower for metal X than for metal Y and hence the photoelectrons emitted from metal Y will have less maximum kinetic energy than the photoelectrons emitted from metal X.

14 C Reversing the direction of the current in one of the wires will change the repulsive force between the wires to a force of attraction between the wires. This will produce a downwards force on wire ***X***. As the force between the wires is proportional to the product of the current in each wire, doubling the current in one of the wires will double the force between the wires. Hence the changes will result in the magnitude of the force on each of the wires being doubled.

15 D When the thrust is changed, the net force on the rocket will be:

$F = thrust - mg = 90\,000 - 7800 \times 9.8 = 13\,560$ N upwards.

This force will cause the rocket to accelerate upwards at:

$a = \frac{F}{m} = \frac{13\,560}{7800} = 1.74 \text{ ms}^{-2}$ upwards.

Because the rocket has an initial horizontal velocity of 20 ms^{-1}, this acceleration will cause the rocket to follow a parabolic path upwards as shown in answer **D**. (Note that this is similar to a projectile fired horizontally but because the acceleration is upwards rather than downwards the object follows a parabolic path upwards rather than downwards.)

16 A If the amount of moderator in the core is reduced, the rate of reaction will decrease and the core will cool. One way to compensate for the loss of reaction rate and maintain the core temperature would be to increase the rate at which fission reactions occur in the core. Of the options given only withdrawing the control rods **(A)** will increase the rate of reaction and maintain the core temperature. It does this by making more neutrons available to initiate further fission reactions. Pumping coolant through the core at a faster rate or reducing the coolant temperature before it was pumped into the core would only cool the core further.

17 C Both observers will see the light pulse travelling at the same speed but *Y* sees the pulse travel a shorter distance because the sensor has moved to the right as the pulse travelled towards it. Hence observer *X* will measure the time to be longer than observer *Y*. Alternatively, we can apply Einstein's time dilation theory and note that observer *Y* will see observer *X*'s clock (i.e. his time) running slow with respect to *Y*'s clock. Thus, *Y* will see *X* measure a longer time for the pulse to reach the sensor because *X*'s clock is running slower than *Y*'s clock.

18 B Light takes time to travel from Jupiter to the Earth. At position *Q* the Earth moves further from Jupiter while the orbit of Jupiter's moon is being observed. The time that light takes to travel the extra distance moved by the Earth will be added to the orbital period measured by the earth-bound observer. This will cause the orbital period measured at position *Q* to be longer than the orbital periods measured from the other positions shown.

19 C Because the moving rod does not cut any magnetic field lines as it rotates, the rotation will not induce a voltage across the ends of the wire. Alternatively, consider the magnetic force that would be produced on the charges in the wire as it is rotated in the magnetic field. Using the right-hand rule, we see that the charges on the rod would only experience forces in and out of the page when the rod is rotated in the magnetic field. As the charges experience no forces along the rod, no potential will be produced between the ends of the rod as it is rotated.

20 A There are two forces operating on the ball as it moves around the cylindrical wall. Gravity acts downwards and the wall exerts a centripetal force (normal reaction force) perpendicular to the wall, towards the centre of the cylinder. The net force on the ball will be the vector sum of these two perpendicular forces. The magnitude and direction of the gravitational force (mg) remain constant and because the horizontal velocity (v_y) remains constant, the magnitude of the centripetal force ($\frac{mv_y^2}{r}$) will also remain constant. The magnitude of the net force on the ball will therefore remain constant.

Section II

Question 21 (Total 5 marks)

(a) Applying Bohr's equation:

$$\frac{1}{\lambda} = R\left(\frac{1}{n_f^2} - \frac{1}{n_i^2}\right) = 1.097 \times 10^7\left(\frac{1}{2^2} - \frac{1}{3^2}\right)$$

$$\lambda = \frac{1}{\left(1.097 \times 10^7\left(\frac{1}{2^2} - \frac{1}{3^2}\right)\right)} = 656 \times 10^{-9} = 656 \text{ nm}$$

(2 marks)

(b) In the Bohr and Rutherford models, negatively charged electrons orbit a positively charged nucleus. But unlike Rutherford's model, in which electrons can take any orbit, Bohr's model postulated that the electron orbital momentum was quantised and that electrons could therefore only orbit in orbitals with specific energies that he called stationary states. Bohr postulated that these stationary states were stable because electrons would not emit electromagnetic radiation when orbiting in one of these states. Bohr proposed that an electron could move to a more energetic orbital or to a less energetic orbital by absorbing or emitting a photon of light respectively. To ensure energy was conserved when photons are emitted or absorbed, the energy of the emitted or absorbed photon ($E = hf$) must be equal to the energy difference between the two stationary states. For example, if an electron moved from a stationary state with an electron energy E_i to a less energetic level with an energy of E_f a photon of light would be emitted with an energy of $E_{photon} = hf = (E_i - E_f)$. *(3 marks)*

Question 22 (Total 5 marks)

(a) Applying Wein's displacement law,

$$T = \frac{b}{\lambda_{max}} \text{ and hence,}$$

$$\lambda_{max} = \frac{b}{T} = \frac{2.898 \times 10^{-3}}{3200} = 905 \times 10^{-9}\,\text{m} = 906\text{ nm}$$ *(2 marks)*

(b) Special relativity can only be meaningfully applied to situations involving constant velocity (inertial reference frames). It cannot be applied to analyse motion involving acceleration. Because the capsule is experiencing a negative acceleration due to its interaction with the atmosphere it is in a non-inertial frame and special relativity cannot be meaningfully applied to the situation.

Second, there is nothing to be gained in applying the special theory of relativity to objects, such as the capsule, that are moving with velocities that are only a tiny fraction of the speed of light. The effects of special relativity increase as the relative speeds approach the speed of light. At speeds that are low compared to the speed of light the predictions of the special theory of relativity are essentially identical to those of Newtonian mechanics.

Students may also refer to the capsule being in a gravitational field as special relativity does not account for the effects of gravitational fields on space and time. *(3 marks)*

Question 23 (Total 3 marks)

From the graph we see that when the motor produces a torque of $\tau = 2.95$ Nm the angular velocity of the motor will be $\omega = 100$ rad s^{-1}. We can now calculate the current in the motor by rearranging the equation given:

$$\tau = \frac{VI\eta}{\omega}$$

Hence, $I = \frac{\tau\omega}{V\eta} = \frac{(2.95)(100)}{(240)(0.3)} = 4.10$ A

As this is below the circuit breaker current, the motor will continue to operate with an angular velocity of $\omega = 100$ rad s^{-1}. *(3 marks)*

Question 24 (Total 4 marks)

Because the horizontal velocity of a projectile is constant, we can find the initial horizontal velocity using the horizontal component of the motion as follows:

$$u_x = \frac{\textit{range}}{\textit{time of flight}} = \frac{130}{6} = 21.67\text{ ms}^{-1}$$

As the vertical velocity is zero at the top of the flight, we can find the initial vertical velocity of the particle using the time to reach the maximum height OR from the maximum height reached. We will call the upwards direction positive and downwards negative.

As it takes 3 seconds to reach the maximum height and the final vertical velocity is zero at the top of the flight, we can apply the relationship,

$v = u + gt$

And hence, $u_y = -gt = -(-9.8)(3) = 29.4$ ms^{-1}.

OR

As the projectile reaches a height of 44 m and the vertical velocity is zero at the top of the flight, the vertical component of the motion gives:

$v^2 = u^2 + 2as$, and as $v = 0$ when $s = 44$ m,

$u_y = \sqrt{-2as} = \sqrt{-2(-9.8)(44)} = 29.4 \text{ ms}^{-1}$

The initial velocity is determined by adding the vertical and horizontal parts vectorially:

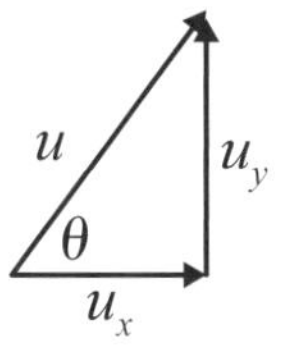

$$u = \sqrt{u_x^2 + u_y^2} = \sqrt{21.67^2 + 29.4^2} = 36.5 \text{ ms}^{-1}$$

$$\tan\theta = \frac{u_y}{u_x} \text{ or } \theta = \tan^{-1}\left(\frac{29.4}{21.67}\right) = 54^{\circ}$$

The initial velocity is therefore 36.5 ms^{-1} at 54^{0} above the horizontal as shown in the vector diagram above. *(4 marks)*

Question 25 (Total 4 marks)

The standard model describes matter in terms of fundamental particles and the forces that operate between them. The hydrogen atom consists of a tiny positively charged central nucleus (a single proton) and a fast-moving negatively charged electron that can only assume specific quantised energies in the region around the nucleus. The standard model suggests the hydrogen atom is constructed from three types of fundamental matter particles called fermions (up quark, down quark, electron) and two types of fundamental force-mediating particles called bosons (gluons and photons). The electron is also called a lepton because it belongs to a subgroup of fermionic particles that interacts with the electromagnetic force but does not interact with the strong force.

The standard model tells us that the proton that forms the hydrogen nucleus is made up of three quarks that are tightly bound together by the strong force, which is mediated by the exchange of gluons between the quarks. Protons contain two up quarks each carrying a charge of $+\frac{2}{3}$ and one down quark, which carries a charge of $-\frac{1}{3}$. This gives the proton a total charge of +1. The electron is a lepton with charge –1 and hence it interacts with the electromagnetic force but does not interact with the strong force. The electromagnetic force that holds the electron in the atom is mediated by the exchange of photons between the negative electron and the positive nucleus. *(4 marks)*

Question 26 (Total 8 marks)

(a) An incandescent lamp emits a continuous spectrum (i.e. a continuous spread of wavelengths). The intensity and wavelengths of peak emission are determined by the temperature of the globe filament as it is essentially a black body spectrum. In contrast, a gas discharge tube produces line spectra. The light in this case is not emitted at all wavelengths but is emitted at a number of specific wavelengths, which are characteristic of the elemental atoms in the discharge tube. *(2 marks)*

(b) Model *X* shows what was known as the ultraviolet catastrophe. It reflects the prediction that is made by classical physics, which assumes atoms (called electromagnetic oscillators at the time) can emit and absorb a continuous spectrum of electromagnetic waves. Using this assumption leads to a prediction that hot bodies should emit an infinite intensity of light at high frequencies. Model *Y* was proposed by Planck and assumes light can only be emitted or absorbed in packets or quanta (later called photons) that have an energy that is a whole number multiple of some minimum amount. Model *Y* assumes only photons with energy $E = n(hf)$ can be emitted or absorbed by the hot body. This assumption leads to a model that limits the intensity of light emitted at high frequencies from a hot body. *(2 marks)*

(c)

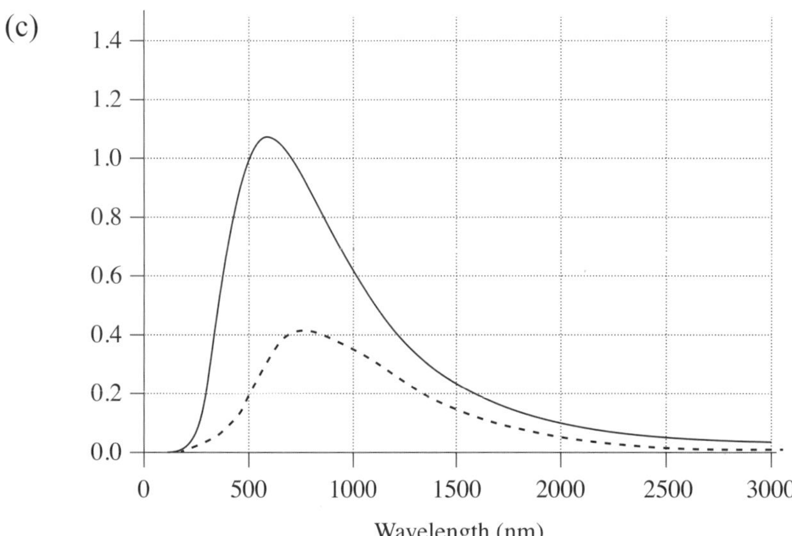

The emission at 4000 k is shown by the dotted line on the graph above. By applying Wein's Displacement Law, $\lambda = \frac{b}{T}$, we see that reducing the temperature of the black body from 5000 K to 4000 K increases the wavelength at which the most intense radiation is emitted from 579 nm to 724 nm. This occurs because the atoms have less average energy at the lower temperature resulting in fewer high energy transitions being excited and (as $\lambda = \frac{hc}{E}$) fewer short wavelength photons being emitted. The intensity emitted at all wavelengths will be lower for a body at 4000 K than for a body at 5000 K, because the body would have less internal energy at the lower temperature and hence would radiate significantly less total power (area under the curve). *(4 marks)*

Question 27 (Total 4 marks)

Young's double slit equation shows the angle (θ) from the central maxima (A) to the beam producing the first ($m = 1$) interference maxima (B) is related to the wavelength of light by $\lambda = d\sin\theta$. This equation is derived by assuming light from each slit is superimposed at each point on the screen, but only where the two light beams are in phase will a maximum be produced.

Decreasing the incident wavelength will decrease θ and hence decrease the separation between *A* and *B*. To maintain the separation, we therefore need to change the experiment in ways that will increase the separation between *A* and *B*.

Moving the screen further from the slits would increase the separation between the maxima A and B on the screen. We could therefore move the screen further from the slits, to maintain the same separation when the incident wavelength was reduced.

We see from Young's equation that reducing the slit separation (d) at a specific wavelength increases the angle to the first maxima. Therefore, reducing the slit separation could be used to maintain the separation between A and B when the incident wavelength was reduced.

(4 marks)

Question 28 (Total 7 marks)

(a) The rod reduces the magnetic flux linking the circuit as it moves up the rails. When the rod moves 30 cm up the rails, the magnetic flux linking the circuit will be reduced by:

$\Delta\Phi = BA\cos\theta = 1 \times (0.30 \times 0.20)\cos 30^{\circ} = 5.2 \times 10^{-2}$ WB *(2 marks)*

(b) Given that the flux change occurred over a time of 2.5 seconds,

$$EMF = \frac{N\Delta\Phi}{\Delta t} = \frac{-5.2 \times 10^{-2}}{2.5} = -0.021V$$

The negative sign reminds us to apply Lenz's law. In this case as the magnetic flux is decreasing, the current flow must be in an anticlockwise direction (viewed from above) around the circuit to oppose the change in magnetic flux.

Thus, the rod will produce an $EMF = -0.021V$ in the circuit and a current that will flow in an anticlockwise direction. *(2 marks)*

(c) Some force is required to overcome the component of the weight of the rod and some force is required to generate the electric current in the rod. We can explain in two ways why the required force would be lower when the magnetic field is removed.

For energy to be conserved, the force required to move the rod must decrease when the magnetic field is switched off. This is because when the magnetic field is switched on, moving the rod generates an EMF, which produces a current in the circuit. This in turn means that electrical power ($P = IV$) is being produced and hence mechanical work is being converted into electrical energy as the rod is moved. Therefore, some of the work ($W = Fs$) done pushing the rod up the rails is converted into electrical energy. When the magnetic field is switched off, no electrical power is produced and hence the force needed to move the rod will be reduced.

OR

When the magnetic field is switched on, moving the rod generates an EMF and a current that passes from left to right through the rod. Because the rod carries current perpendicular to a magnetic field a force ($F = LI_{\perp}B$) will be exerted on the rod. This force will be perpendicular to the magnetic field and the direction of current in the rod. Using the right-hand rule we see that the rod will experience a force parallel to the table top and perpendicular to the rail, in a direction that will have a component that would oppose the motion of the rod. This force opposing the motion would only appear when the magnetic field was on and the rod was being moved. Switching the magnetic force off would remove this opposing force, which would reduce the force required to move the rod up the rails. *(3 marks)*

Question 29 (Total 5 marks)

(a) This experiment was first conducted by Bothe and Becker in 1930. Because they found the unknown radiation was very penetrating and was not deflected by electric fields, they suggested the unknown radiation could be gamma rays. Two years later, Curie and Joliot showed that when paraffin wax was bombarded with the radiation, very energetic protons were emitted. But ejecting protons from paraffin wax would require gamma rays with energies that were higher than any gamma rays that had previously been studied and this raised doubts about the gamma ray hypothesis. The matter was settled in 1932 by Chadwick who measured the range of protons scattered by the unknown radiation and the recoil velocities of various gas atoms after they were irradiated with the unknown radiation. By applying the laws of conservation of energy and momentum Chadwick was able to show that the radiation could not be gamma rays. He then showed his results were consistent with the hypothesis that the unknown radiation was an uncharged particle with about the mass of a proton. The particle was the long sort neutron. *(3 marks)*

(b) These experiments led to the discovery of the neutron. Before the neutron, the atomic model consisted of a nucleus containing protons and orbiting electrons. After the discovery, it was clear that the atomic nucleus contained protons and neutrons. The discovery of the neutron changed the atomic model and enabled the existence of isotopes and the difference between atomic number and mass number to be explained. With the discovery of the neutron, scientists thought they had found the three fundamental building blocks of the atom. *(2 marks)*

Question 30 (Total 7 marks)

(a) The following answer looks at W and Z bosons but students could choose one of the confirmations of a variety of predictions made by the standard model; for example, the use of particle accelerators to confirm the existence of the W and Z bosons, quarks or the Higgs boson.

An important part of the Standard Model was Electroweak theory. This theory was developed in the 1960s and combines electromagnetic force and the weak force into a single theory. The theory predicted that the electroweak force would be mediated by massive fundamental particles called W and Z bosons. Because the particles were very massive (m) and $E = mc^2$, a particle accelerator would require an energy $E \geq mc^2$ to produce such particles. Particle accelerators at the time could not produce such energetic collisions. It was not until 1983 that CERN constructed a synchrotron accelerator with sufficient energy to experimentally confirm the existence of the long-predicted W and Z bosons. This discovery provided strong evidence to support the standard model. *(3 marks)*

(b) (i) As the proton is only moving at $\frac{1}{10}$th the speed of light we can use the non-relativistic relationship to calculate the wavelength:

$$\lambda = \frac{h}{mv} = \frac{6.626 \times 10^{-34}}{1.673 \times 10^{-27} \times 0.1 \times 3 \times 10^{8}} = 1.32 \times 10^{-14} \text{ m}$$

(2 marks)

(ii) As the velocity of the proton approaches the speed of light its relativistic momentum becomes progressively greater than its non-relativistic momentum $\left(p_v = \dfrac{p_0}{\sqrt{1 - \frac{v^2}{c^2}}}\right)$.

Because the wavelength of the proton is inversely proportional to the momentum $(\lambda = \frac{h}{p})$, this increase in momentum means the relativistic wavelength will be shorter than the non-relativistic wavelength.

OR

Students may choose to calculate the actual difference as part of their explanation.

Substituting $v = 0.95c$ into Einstein's equation for relativistic momentum, we get

$$p_v = \frac{p_0}{\sqrt{1 - 0.95^2}} = 3.2p_0$$

and therefore the relativistic wavelength (λ_v) at $0.95c$ is related to the non-relativistic wavelength (λ_0) by $\lambda_v = \dfrac{\lambda_0}{3.2}$

Hence, the relativistic wavelength is significantly shorter than the non-relativistic wavelength.

(2 marks)

Question 31 (Total 6 marks)

(a) Students can answer this question using conservation of energy, dynamics or Kepler's 2nd Law. Answers using energy and dynamics follow.

Because the comet is in free flight in a gravitational field its total energy will be constant at each point in the orbit. At the point P the comet is furthest from the Sun and hence its gravitational energy will be at a maximum $(U = -\frac{Mm}{r})$ and therefore its kinetic energy $(K = \frac{1}{2}mv^2)$ must be at a minimum. Thus, at point P the velocity will be the lowest in the orbit. As the comet moves in its orbit closer to the planet from P, the distance to the planet will decrease, the gravitational energy will decrease (i.e. it becomes a greater negative number) and hence the kinetic energy and velocity will increase. Halfway through the orbit, when the comet is closest to the Sun, its potential energy will be a minimum and hence to conserve energy its kinetic energy (velocity) will be a maximum. As the comet moves back towards the point P it will slow down until its velocity reaches a minimum again at P.

OR

Alternatively, students may choose to answer this question by referring to the direction of force on the comet. When the comet moves closer to the Sun the gravitational force on the comet from the Sun has a component parallel to the comet's velocity which will cause the comet to accelerate $(a = \frac{F}{m})$ in the direction of its motion. This increase in velocity will act from P until the comet is closest to the Sun where its velocity will be a maximum. When the comet moves from this point of closest approach towards P it will experience a gravitational force from the Sun that has a component in the opposite direction to the comet's velocity. This force will cause the comet to slow down until it reaches a minimum velocity at point P. *(3 marks)*

(b)

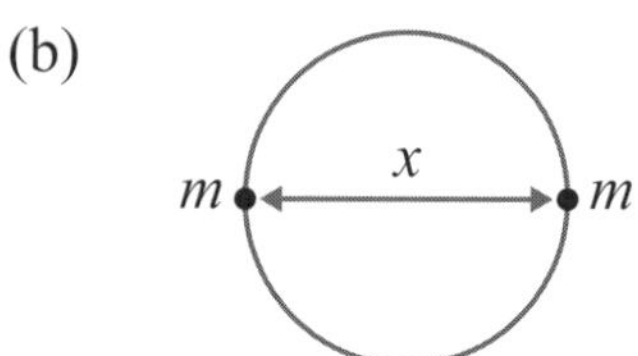

As the velocity of the stars is constant and they have equal mass they must be orbiting one another in a circular orbit. As the distance between the stars is x, the radius of the orbit will be $\frac{x}{2}$ and the gravitational force on each star will be:

$$F = \frac{Gmm}{x^2}$$

The centripetal force will be given by: $F = \frac{mv^2}{\left(\frac{x}{2}\right)} = \frac{2mv^2}{x}$

As the centripetal force is produced by the gravitational attraction between the stars, we can equate the equations above:

$$\frac{Gmm}{x^2} = \frac{2mv^2}{x}$$

Therefore: $v = \sqrt{\frac{Gm}{2x}}$

(3 marks)

Question 32 (Total 7 marks)

To simplify the analysis, we will assume the pulley and string are frictionless, the velocity is low enough for air friction to be ignored and the surface is long enough for the block to reach maximum speed.

Provided the motor torque is great enough to overcome the static friction between the block and the plane (i.e., $\tau > (\mu_s mg)r$), the maximum velocity of the block will be determined by the output power of the motor and the force required to pull the block against dynamic friction.

If we call the co-efficient of dynamic friction μ_v, the force required to move the block at constant velocity will be given by $F = \mu_v mg$ and as $P = Fv$,

the maximum velocity will be related to the output power (P_{out}) of the motor by

$$v = \frac{P_{out}}{F} = \frac{P_{out}}{\mu_v mg}$$

The input power of the motor ($P_{in} = IV$) is related to the output power by the efficiency (ε),

$\varepsilon = \frac{P_{out}}{P_{in}} = \frac{P_{out}}{(IV)}$ and hence, $P_{out} = \varepsilon(IV)$.

This gives a final expression for the velocity: $v = \frac{P_{out}}{\mu_v mg} = \frac{\varepsilon(IV)}{\mu_v mg}$

Hence, the speed of the block is proportional to the electrical power used by the motor (IV) and the efficiency of the motor but inversely proportional to the mass of the block, the local acceleration due to gravity and the friction between the block and the plane. All these factors will limit the maximum velocity of the block.

(Note that the efficiency of an electric motor is determined by the design of the motor and would involve the number of windings on the armature, the area of the armature coils, number of armature coils, strength of magnetic field, friction of the bearings, etc.)

OR

Students may also answer in terms of the torque produced by the motor and the effect of back EMF on the current and torque as the angular velocity of the motor increases, as follows.

The maximum speed of the mass will be determined by the motor operating speed that would produce enough torque to move the mass at constant velocity. When the motor is switched on, the angular velocity of the motor (and the velocity of the block) will increase until the back *EMF* reduces the current in the motor enough for the motor torque to be just sufficient to move the mass at a constant velocity. Therefore, the angular velocity of the motor and the velocity of the block will increase until the output torque of the motor ($\tau = nIA \perp B$) is equal to the torque required to move the block at constant velocity against friction ($\tau = (\mu_s mg)r$). We can write this relationship as $\tau = nIA \perp B = (\mu_v mg)r$, where r is the radius of the pulley.

The speed of the block will therefore be limited by the torque required to move the block and the output power of the electric motor. Increasing the mass of the block, local acceleration due to gravity or the friction coefficient between the block and the plane would reduce the speed because the motor would have to supply a greater torque. Increasing the input power to the motor or its efficiency would increase the speed of the block because the motor would have a higher output power and be able to supply the required torque at a higher rate of rotation. (Note that the efficiency of the motor could be improved by increasing the number of coils and/or turns on the armature, increasing the strength of the magnetic field and/or area of the coils, lowering friction in the bearings, etc.).

OR

The previous paragraph detailing the limitations to the speed could be replaced by a discussion of the power and force:

the power required to pull the block at constant velocity is given by

$P = Fv$ and hence $v = \dfrac{P}{F}$

where P is the power required to move the block at constant velocity and F is the force required to move the block against the force of friction at this velocity. Now as the frictional force on the block is $F_f = \mu_v mg$ we can write:

$$v = \frac{P}{F_f} = \frac{P}{\mu_v mg}.$$

Hence, increasing the output power of the electric motor will increase the speed of the block but increasing the mass of the block, local acceleration due to gravity and the friction coefficient between the block and the plane will decrease the speed of the block. Note that the output power is determined by the electrical input power (IV) and the efficiency of the motor. *(7 marks)*

Question 33 (Total 9 marks)

In the first 0.4 seconds of the flight, the magnet is falling freely under the influence of gravity from position X to Y and consequently it accelerates at 9.8 ms^{-2} downwards during this time interval. Because the magnet is in free flight during this time, conservation of energy requires the sum of the gravitational potential energy and kinetic energy of the magnet to remain constant. The gravitational potential energy of the magnet will therefore be progressively converted to kinetic energy as it falls. The distance moved by the block in the first 0.4 seconds would be:

$$s = ut + \frac{1}{2}at^2 = \frac{1}{2}(9.8)(0.4)^2 = 0.784 \text{ m, which agrees with the graphed data.}$$

We can find the velocity of the magnet after 0.4 s at point Y by using kinematics or by using conservation of energy. To use conservation of energy we equate the loss of gravitational potential energy to the gain in kinetic energy:

$$mg\Delta h = \frac{1}{2}mv^2 \text{ and therefore } v = \sqrt{2gh} = \sqrt{2 \times 9.8 \times 0.78} = 3.9 \text{ ms}^{-1} \text{ downwards.}$$

During the next 0.5 seconds the magnet falls at this constant velocity downwards (indicated by a constant gradient of -3.9 ms^{-1} shown on the displacement–time graph provided) and hence its kinetic energy remains constant as it moves from point Y to point Z. To ensure energy is conserved, the potential energy lost as the magnet moves from Y to Z must have been converted into electrical energy (i.e. heat) in the copper ring rather than into kinetic energy. The average electrical power produced in the ring during this period of the fall would therefore be:

$$P = \frac{\Delta E}{\Delta t} = \frac{mg\Delta h}{\Delta t} = \frac{0.04 \times 9.8 \times 0.2}{0.5} = 0.16 \text{ W}$$

By Faraday's law, the changing magnetic flux in the copper cylinder produced by the falling magnet will induce an EMF and produce a current that will flow around the copper cylinder. This current (by Lenz's law) will be in a direction that will produce magnetic fields to oppose the motion of the magnet. The magnet must have experienced an upwards force equal to its weight to have moved with constant velocity and hence the force on the magnet due to the induced currents in the copper cylinder must have been:

$$F = mg = 0.04 \times 9.8 = 0.392 \text{ N upwards.}$$

As expected, the work done by the block against the magnetic force is equal to the potential energy lost.

$$\Delta U = mg\Delta h = 0.04 \times 9.8 \times 0.2 = 0.078 \text{ J}$$

$$W = Fs = 0.39 \times 0.2 = 0.078 \text{ J}$$

Thus, energy is conserved as the magnet falls through the ring. *(9 marks)*

Question 34 (Total 6 marks)

(a) The electric field between a pair of parallel plates is given by: $E = \frac{V}{d}$

The force on a charged particle between charged parallel plates is given by:

$$F = qE = \frac{qV}{d}$$

And the work done is given by: $W = F_{\parallel}s$

Because the force on each particle is upwards (as it hits the top plate at *P*) and only the displacement in this direction is caused by the force, the work done on particle 1 will be:

$$W = F_{\parallel}s = \left(\frac{q_1V}{d}\right)\left(\frac{d}{2}\right) = \frac{q_1V}{2}$$

The work done on particle 2 will be:

$$W = F_{\parallel}s = \left(\frac{q_2V}{2d}\right)\left(\frac{3d}{2}\right) = \frac{3q_2V}{4}$$

As the charges are identical, the field does $\frac{3}{2}$ times more work on q_2 than it does on q_1.

(3 marks)

(b) Students could answer this question qualitatively or quantitatively as follows:

The vertical force $\left(F = \frac{qV}{d}\right)$ and hence vertical acceleration of q_2 will be half as great as the vertical acceleration of q_1. As the vertical distance to the top plate is three times further for q_2 than for q_1 and as q_2 is accelerating less rapidly, q_2 will be in flight for a longer time than q_1 before it hits the top plate. Because both charges have the same horizontal velocity, q_2 will therefore travel a greater horizontal distance before it hits the top plate than q_1.

OR

The horizontal component of the velocity is equal and constant in both cases. Given the upwards acceleration will be $a = \frac{F}{m} = \frac{\frac{qv}{d}}{m} = \frac{qV}{dm}$ and the initial vertical component of the velocity is zero, the time of flight for the first particle can be determined using

$$s = ut + \frac{1}{2}at^2 \text{ and hence } t = \sqrt{\frac{2s}{a}} = \sqrt{\frac{2\left(\frac{d}{2}\right)dm}{qV}} = \sqrt{\frac{md^2}{qV}}$$

The horizontal distance travelled by q_1 in this time will be: $s = vt = v\sqrt{\frac{md^2}{qV}}$

Similarly, for q_2,

$$a = \frac{F}{m} = \frac{\frac{qv}{2d}}{m} = \frac{qV}{2dm}$$

$$t = \sqrt{\frac{2s}{a}} = \sqrt{\frac{2\left(\frac{3d}{2}\right)2dm}{qV}} = \sqrt{\frac{6md^2}{qV}}$$

$$s = vt = v\sqrt{\frac{6md^2}{qV}}$$

Therefore particle 2 will travel a horizontal distance of $(\sqrt{6})s$, where s is the horizontal distance travelled by q_1.

(3 marks)

CHAPTER 7

NSW Education Standards Authority

2021 HIGHER SCHOOL CERTIFICATE EXAMINATION

Physics

General Instructions

- Reading time – 5 minutes
- Working time – 3 hours
- Write using black pen
- Draw diagrams using pencil
- Calculators approved by NESA may be used
- A data sheet, formulae sheet and Periodic Table are provided at the back of this paper

Total marks: 100

Section I – 20 marks

- Attempt Questions 1–20
- Allow about 35 minutes for this section

Section II – 80 marks

- Attempt Questions 21–35
- Allow about 2 hours and 25 minutes for this section

Section I

20 marks
Attempt Questions 1–20
Allow about 35 minutes for this section

Use the multiple-choice answer sheet for Questions 1–20.

---

1 A marble is rolled off a horizontal bench and falls to the floor.

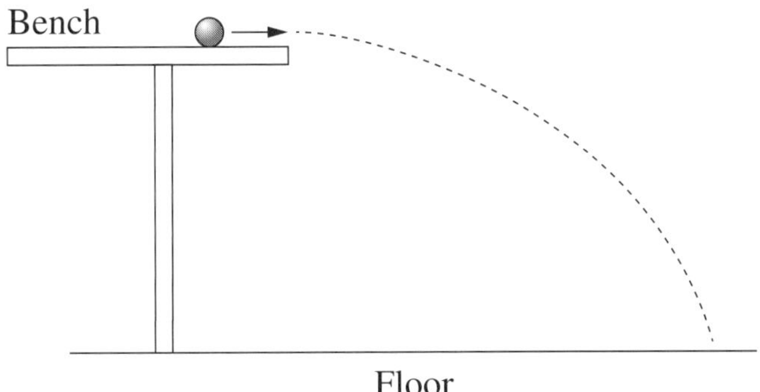

Rolling the marble at a slower speed would

A. increase the range.

B. decrease the range.

C. increase the time of flight.

D. decrease the time of flight.

2 A positively charged particle is moving at velocity, v, in an electric field as shown.

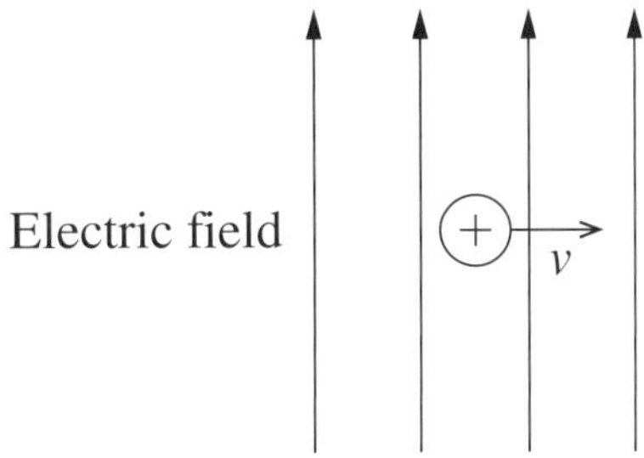

What is the direction of the force acting on the particle due to the electric field?

A. Into the page

B. Out of the page

C. Up the page

D. Down the page

3 Which of the following is NOT a fundamental particle in the Standard Model of matter?

A. Electron

B. Gluon

C. Muon

D. Proton

4 An astronaut is travelling towards Earth in a spaceship at $0.8c$. At regular intervals, a radio pulse is sent from the spaceship to an observer on Earth.

Which quantity would the astronaut and the observer measure to be the same?

A. Length of the spaceship

B. Speed of the radio pulses

C. Momentum of the astronaut

D. Time interval between the radio pulses

5 The spectrum of an object is shown.

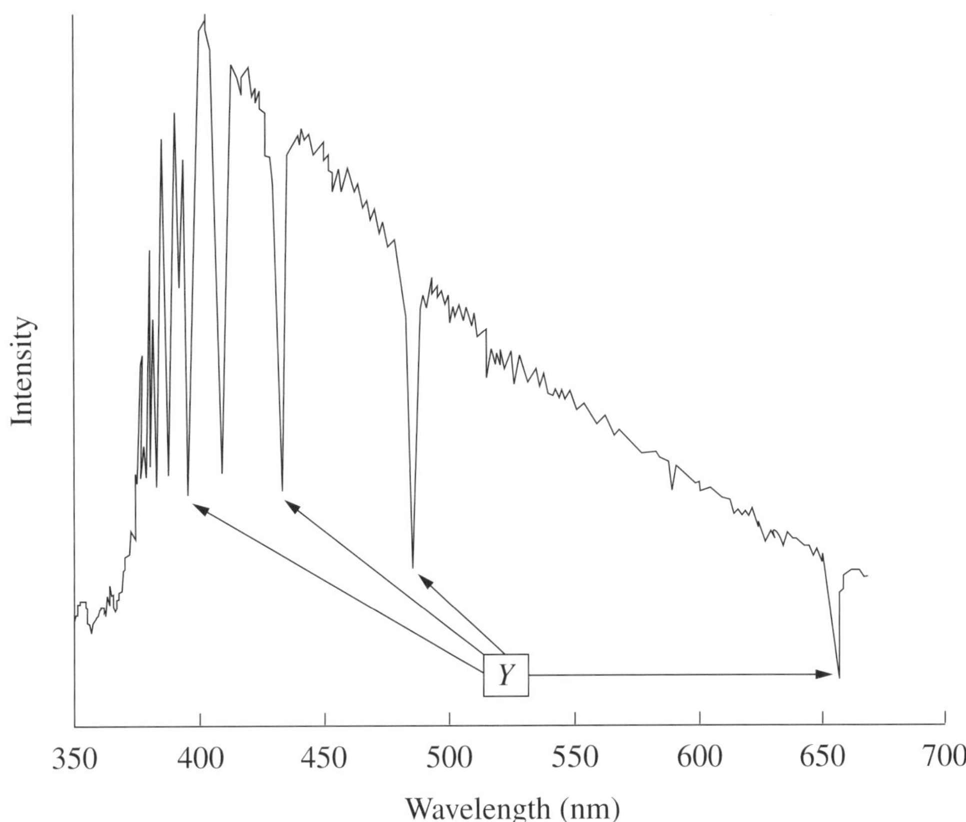

Which row of the table correctly identifies the most likely source of the spectrum and the features labelled *Y*?

	Source of spectrum	*Features labelled Y*
A.	Star	Absorption lines
B.	Discharge tube	Absorption lines
C.	Star	Emission lines
D.	Discharge tube	Emission lines

6 The diagram shows part of a nuclear fusion process that occurs in stars.

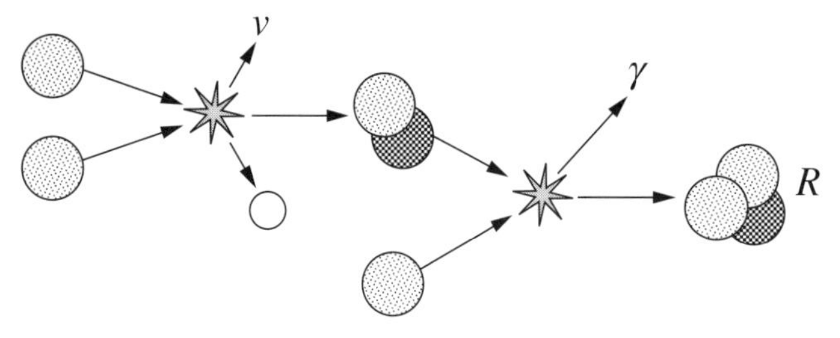

Proton
Neutron
Positron
γ Gamma ray
v Neutrino

What is the isotope labelled *R*?

A. H-2

B. He-2

C. H-3

D. He-3

7 In a certain ideal transformer, the current in the secondary coil is four times as large as the current in the primary coil.

Which row of the table correctly identifies the type of transformer and the ratio of turns?

	Type of transformer	*Ratio of turns in primary coil to turns in secondary coil*
A.	Step up	4 : 1
B.	Step up	1 : 4
C.	Step down	4 : 1
D.	Step down	1 : 4

8 Light from a point source is incident upon a circular metal disc, forming a shadow on a screen as shown. A bright spot is observed in the centre of the shadow.

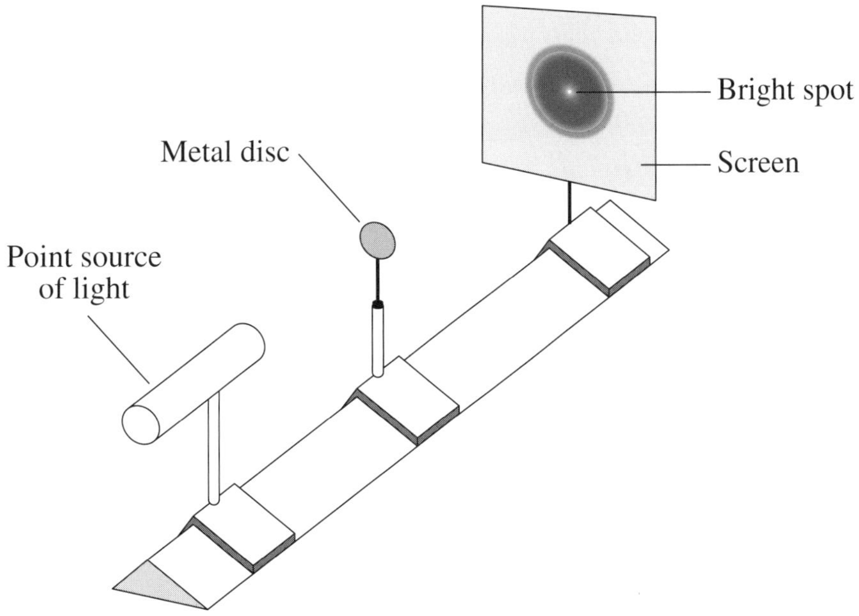

The bright spot is caused by a combination of

A. interference and refraction.

B. refraction and polarisation.

C. polarisation and diffraction.

D. diffraction and interference.

9 A mass, *M*, is positioned at an equal distance from two identical stars as shown.

The mass is then moved to position *X*.

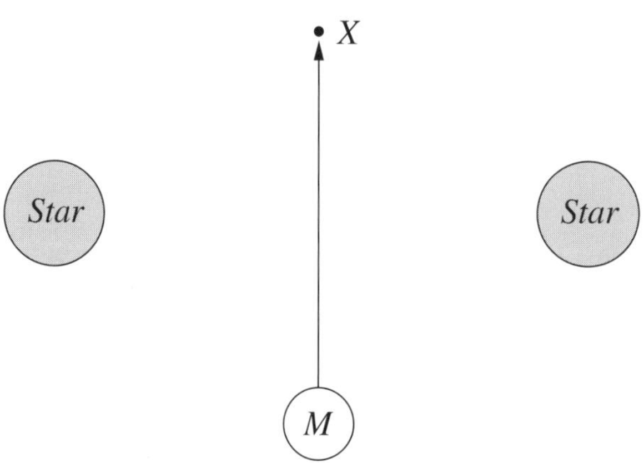

Which graph best represents the gravitational potential energy, *U*, of the mass during this movement?

A.
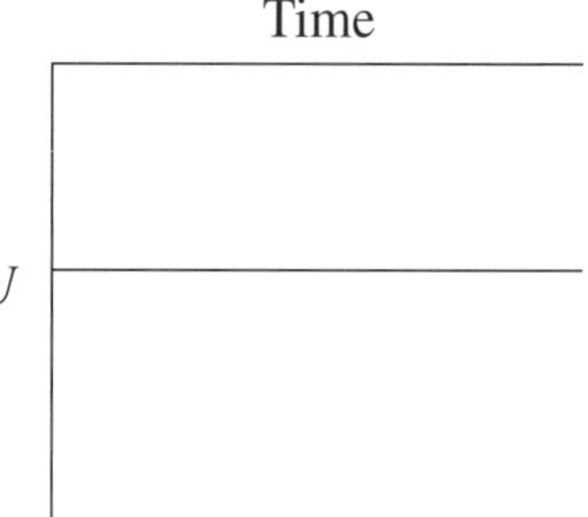

B.
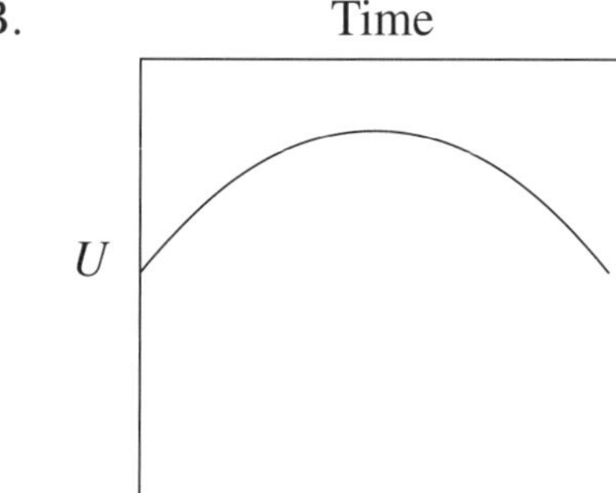

C.
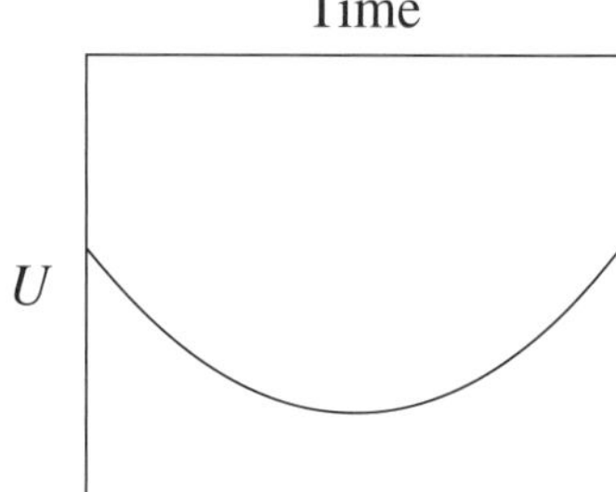

D.
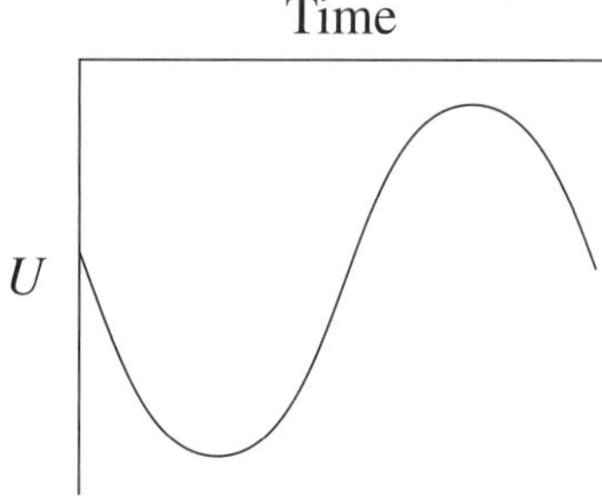

10 A strong magnet is moved past a copper block at a constant speed as shown.

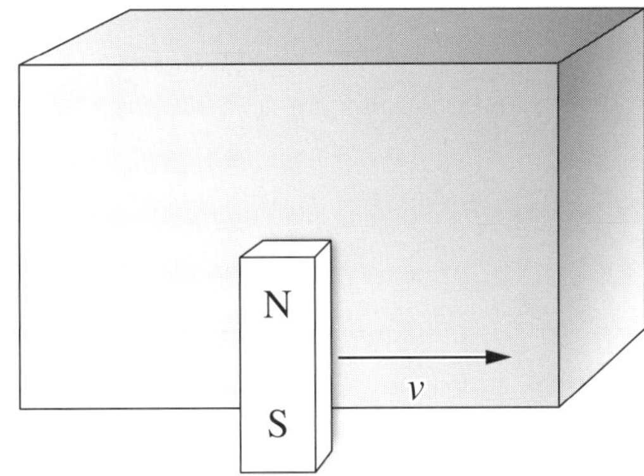

What is the direction of the force acting on the copper block?

A. To the left

B. To the right

C. Into the page

D. Out of the page

11 What is the peak wavelength of electromagnetic radiation emitted by a person with a body temperature of 37°C (310 K)?

A. 9.3×10^{-6} m

B. 7.8×10^{-5} m

C. 9.3×10^{-3} m

D. 7.8×10^{-2} m

12 Which graph shows the magnitude of back emf induced in a DC motor rotating continuously at different angular velocities?

A.

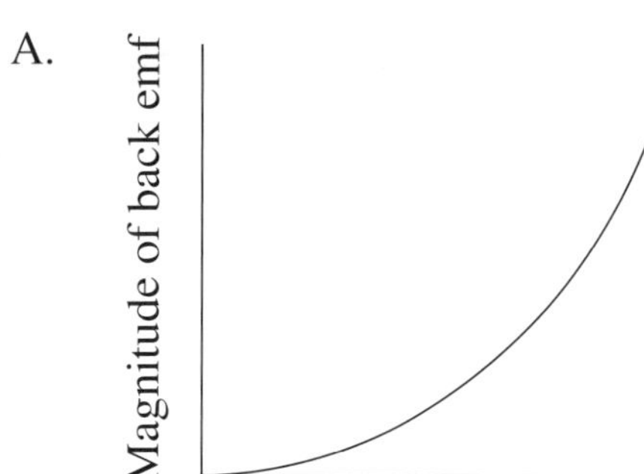

B.

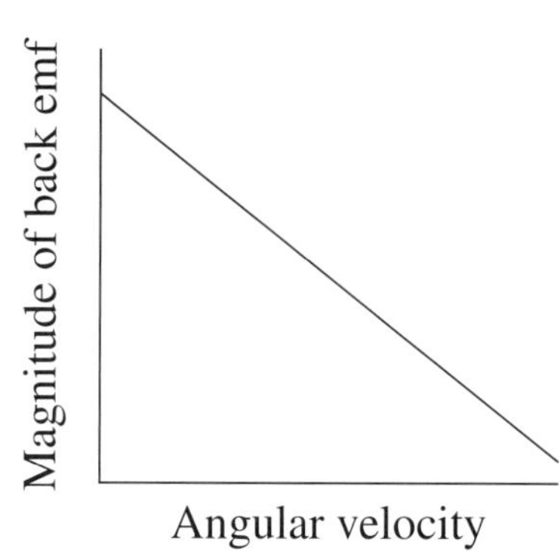

C.

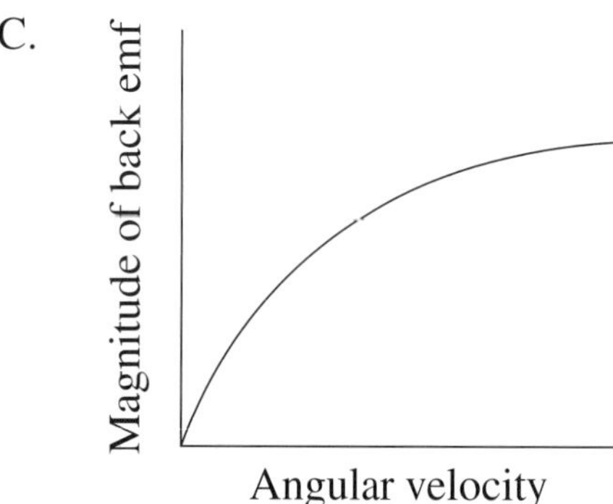

D.

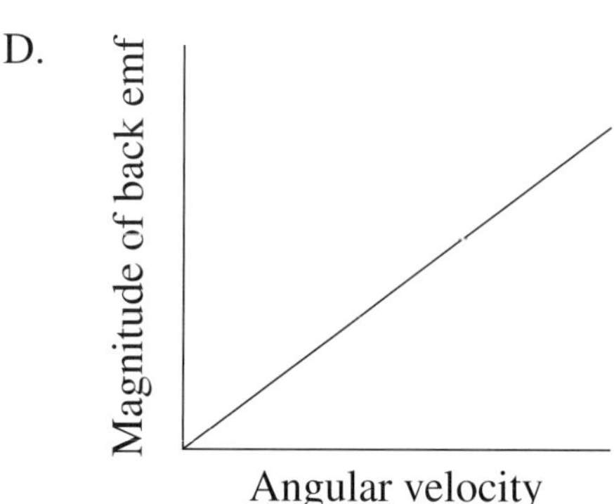

13 The diagram shows electron transitions in a Bohr-model hydrogen atom.

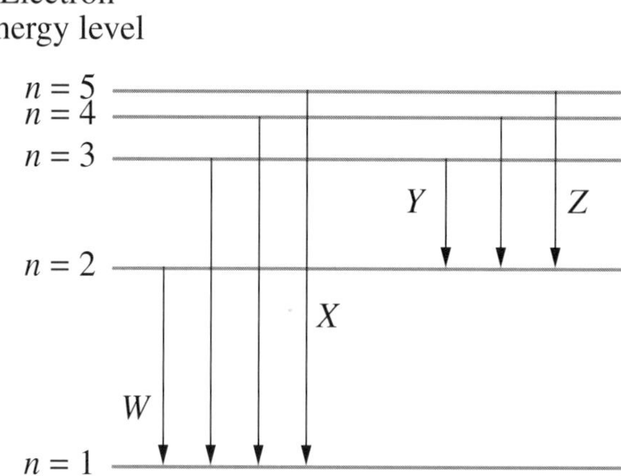

Which transition would produce the shortest wavelength of light?

A. W

B. X

C. Y

D. Z

14 Which of the following statements correctly describes the gravitational interaction between the Earth and the Moon?

A. The Earth accelerates towards the Moon.

B. The net force acting on the Earth is zero.

C. The Moon and Earth experience equal and opposite accelerations.

D. The force acting on the Moon is smaller than the force acting on the Earth.

15 Unpolarised light is incident upon two consecutive polarisers as shown. The second polariser has a fixed transmission axis which cannot be rotated. I_1 is the intensity of light after the first polariser, and I_2 is the intensity of light after the second polariser.

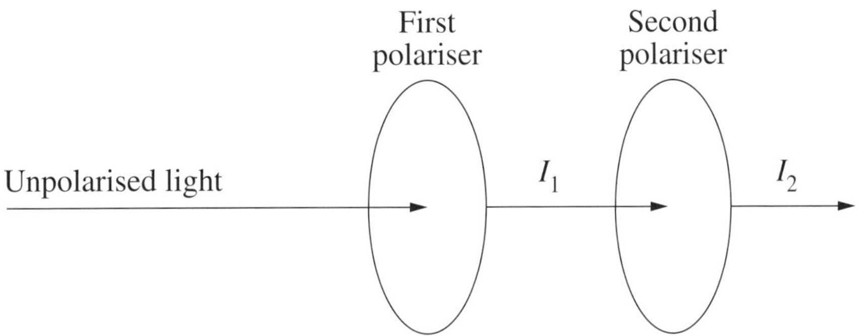

How would I_1 and I_2 be affected if the transmission axis of the first polariser was rotated?

A. Both would change.

B. Only I_1 would change.

C. Only I_2 would change.

D. Neither would change.

16 The Sun has an energy output of 3.85×10^{28} W.

By how much does the Sun's mass decrease each minute?

A. 4.28×10^{11} kg

B. 2.57×10^{13} kg

C. 1.28×10^{20} kg

D. 7.70×10^{21} kg

17 Two long, parallel conductors *X* and *Y* are connected to a light bulb and an AC power supply. The conductors are suspended horizontally from fixed points using sensitive spring balances. *X* is positioned directly below *Y*.

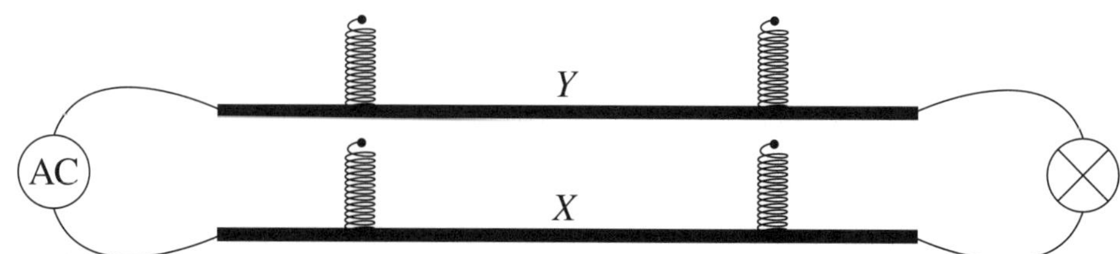

Which statement correctly compares the forces measured by the spring balances?

A. The forces measured on *X* and *Y* will always be equal.

B. The force measured on *Y* will be greater than or equal to that on *X*.

C. The force measured on *X* will be greater than or equal to that on *Y*.

D. There will be a continuous reversal of which measured force is greater.

18 An evacuated chamber contains a pair of parallel plates connected to a power supply and a switch which is initially closed.

A positively charged mass (•) falls within the chamber, under the influence of gravity, from the position shown.

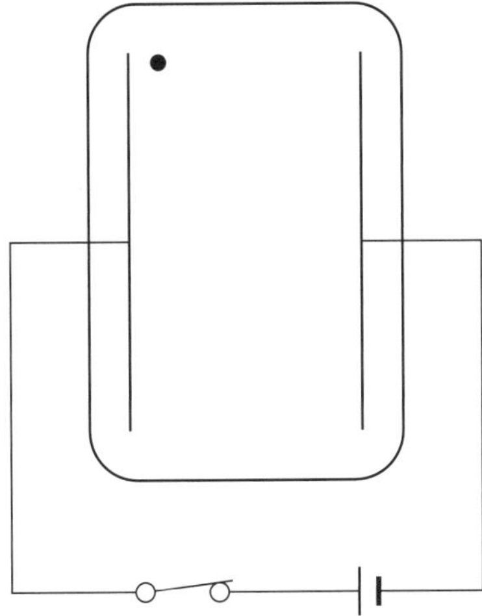

When the mass has fallen half the height of the chamber, the switch is opened.

Which of the following correctly shows the trajectory of the mass?

A.

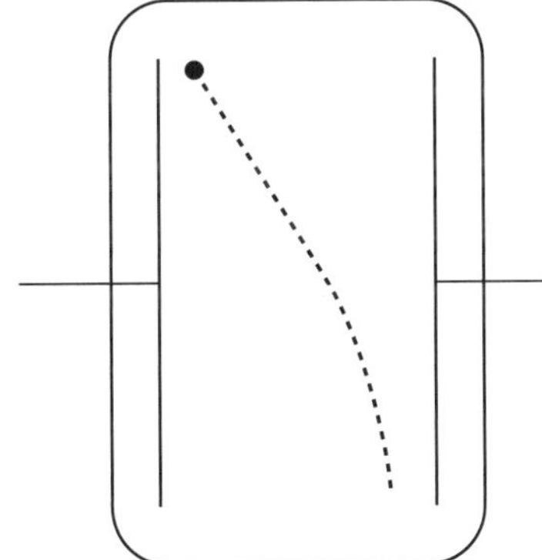

B.

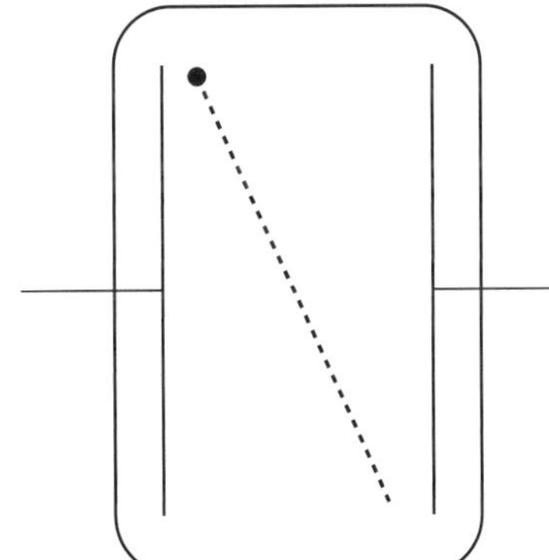

C.

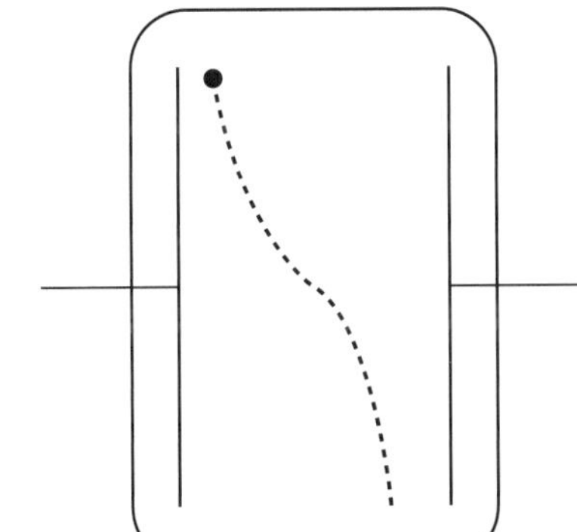

D.

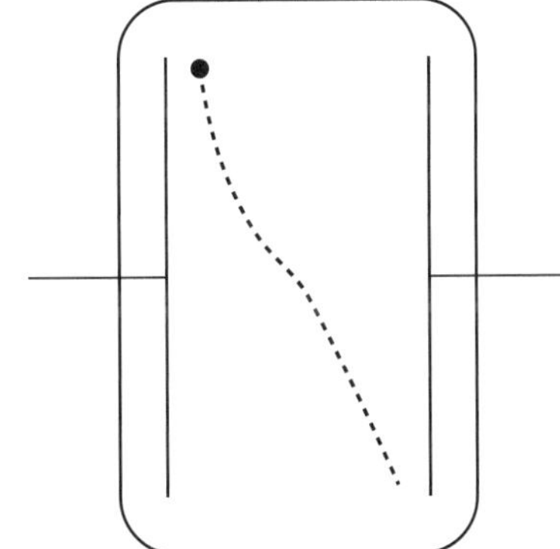

19 Rh-106 is a metallic, beta-emitting radioisotope with a half-life of 30 seconds.

A sample of Rh-106 and an electrode are placed inside an evacuated chamber. They are connected to a galvanometer and a variable DC power supply.

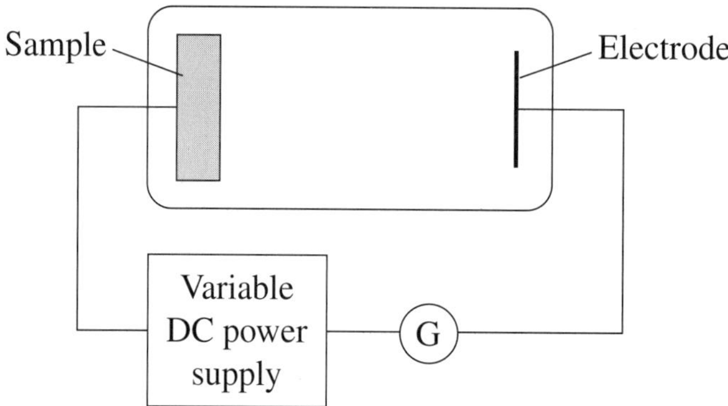

A student measures the current, I, when the power supply is set to zero. They then measure the stopping voltage, V_s. The stopping voltage is the minimum voltage needed to prevent current flowing.

A few minutes later, these measurements are repeated.

How do the TWO sets of measurements compare?

A. Only I changes.

B. Only V_s changes.

C. Both I and V_s change.

D. Neither I nor V_s changes.

20 A metal cylinder is located in a uniform magnetic field. The work function of the metal is ϕ.

Photons having an energy of 2ϕ strike the side of the cylinder, liberating photoelectrons which travel perpendicular to the magnetic field in a circular path. The maximum radius of the path is r.

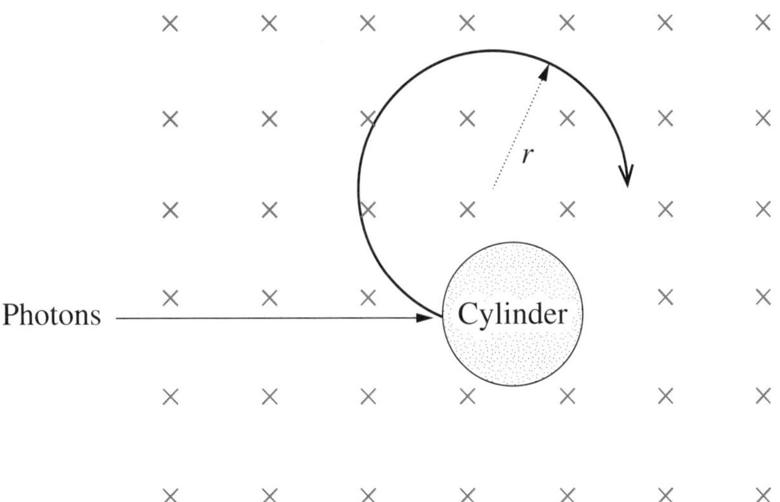

If the photon energy is doubled, what will the maximum radius of the path become?

A. $2r$

B. $3r$

C. $\sqrt{2}r$

D. $\sqrt{3}r$

2021 HIGHER SCHOOL CERTIFICATE EXAMINATION

Centre Number

Student Number

Physics

Section II Answer Booklet

80 marks
Attempt Questions 21–35
Allow about 2 hours and 25 minutes for this section

Instructions

- Write your Centre Number and Student Number at the top of this page.
- Answer the questions in the spaces provided. These spaces provide guidance for the expected length of response.
- Show all relevant working in questions involving calculations.

Please turn over

Question 21 (4 marks)

A DC motor is constructed from a single loop of wire with dimensions 0.10 m × 0.07 m. The magnetic field strength is 0.40 T and a current of 14 A flows through the loop.

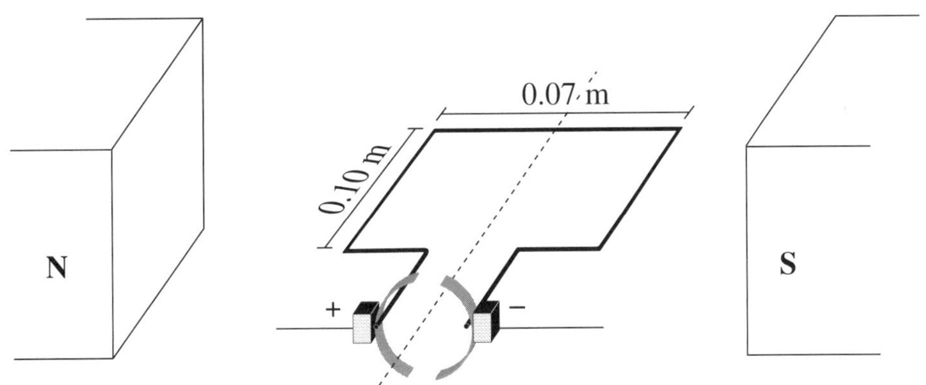

(a) Calculate the magnitude of the maximum torque produced by the motor. **2**

....

....

....

....

(b) Describe how the magnitude of the torque changes as the loop moves through half a rotation from the position shown. **2**

....

....

....

....

Question 22 (3 marks)

A horizontal disc rotates at a constant rate as shown. Two points on the disc, *X* and *Y*, are labelled. *X* is twice as far away from the centre of the disc as *Y*. **3**

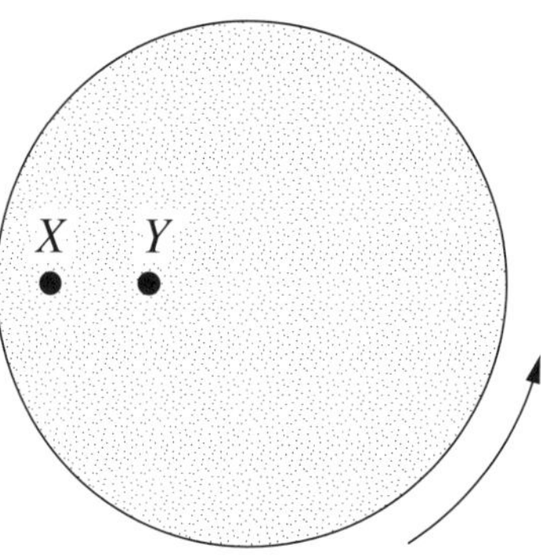

Compare the angular and instantaneous velocities of *X* with those of *Y*.

....

....

....

....

....

....

Question 23 (4 marks)

Describe how Millikan and Thomson each used fields to determine properties of the electron. **4**

....

....

....

....

....

....

....

....

Question 24 (3 marks)

A stationary coil of 35 turns and cross-sectional area of 0.02 m^2 is placed between two electromagnets, and connected to a voltmeter as shown. The electromagnets produce a uniform magnetic field of 0.15 T through the coil. **3**

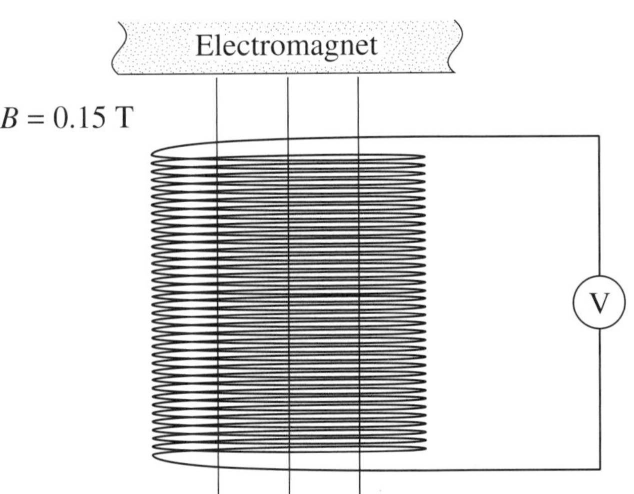

The magnitude of the magnetic field is then reduced to zero at a constant rate over a period of 0.4 s.

Calculate the magnitude of the emf induced in the coil.

....

....

....

....

....

....

Question 25 (5 marks)

A satellite is launched from the surface of Mars into an orbit that keeps it directly above a position on the surface of Mars.

Mass of Mars = 6.39×10^{23} kg

Length of Martian day = 24 hours and 40 minutes

(a) Identify TWO energy changes as the satellite moves from the surface of Mars into orbit. **2**

....

....

....

(b) Calculate the orbital radius of the satellite. **3**

....

....

....

....

....

....

Question 26 (6 marks)

A student performs an experiment to measure Planck's constant, h, using a device that emits specific frequencies of light when specific voltages are applied.

The voltage, V, needed to produce each frequency, f, is given by

$$V = \frac{hf}{q_e}$$

where q_e is the charge on an electron.

Data from the experiment is shown.

Data point	*Frequency* ($\times 10^{14}$ Hz)	*Voltage* (V)
1	3.5	1.3
2	4.8	1.7
3	5.3	1.9
4	7.0	2.6

(a) Graph this data on the axes provided. Include a line of best fit. **3**

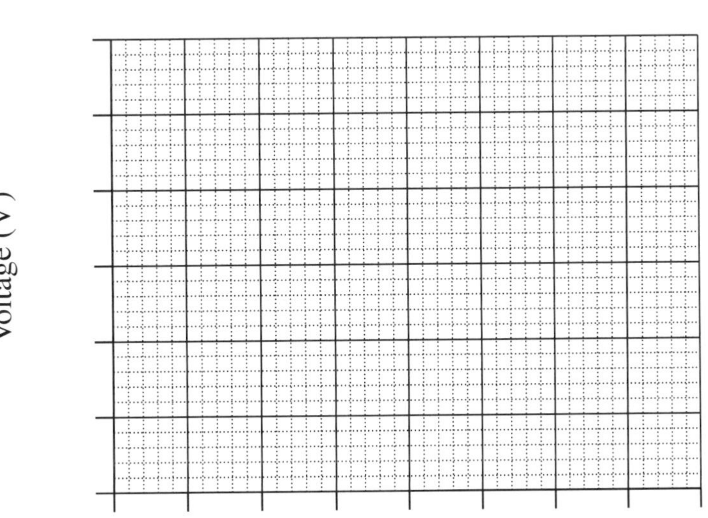

Question 26 continues on the following page

Question 26 (continued)

(b) The student proposes using data point 1 to calculate a value for Planck's constant. **3**

Justify a better method to calculate Planck's constant from the experimental data provided.

....

....

....

....

....

....

End of Question 26

Please turn over

Question 27 (6 marks)

A student is considering how to levitate a thin metal rod in a strong magnetic field of 1.2 T. The current flowing through the rod will be 2.3 A.

(a) Use a labelled diagram to show a suitable orientation of the current and the magnetic field to achieve this outcome. Include relevant forces in your diagram. **3**

(b) Explain why the maximum mass per unit length of the rod cannot exceed $0.282\ \text{kg m}^{-1}$. Support your answer with a calculation. **3**

....

....

....

....

....

....

Question 28 (5 marks)

A spaceship travels to a distant star at a constant speed, v. When it arrives, 15 years have passed on Earth but 9.4 years have passed for an astronaut on the spaceship.

(a) What is the distance to the star as measured by an observer on Earth? **3**

....

....

....

....

....

....

(b) Outline how special relativity imposes a limitation on the maximum velocity of the spaceship. **2**

....

....

....

....

Please turn over

Question 29 (5 marks)

Bohr, de Broglie and Schrödinger EACH proposed a model for the structure of the atom. **5**

How does the nature of the electron proposed in each of the three models differ?

....

....

....

....

....

....

....

....

....

....

Question 30 (5 marks)

In an experiment, a proton accelerates from rest between parallel charged plates. The spacing of the plates is 12 cm and the proton is initially positioned at an equal distance from both plates, as shown. Ignore the effect of gravity.

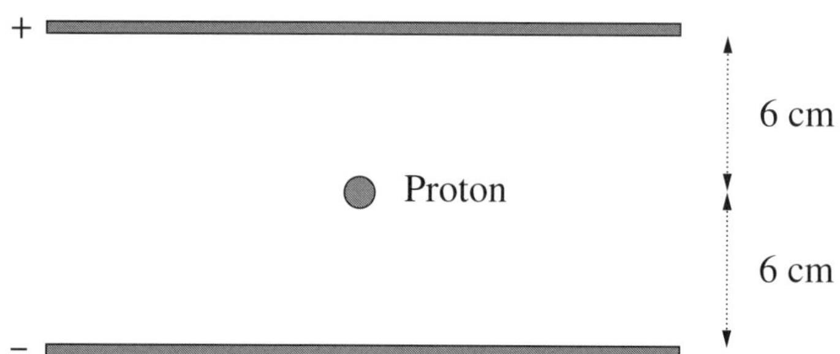

(a) The electrical potential energy of the proton is recorded in the following graph for the first 5 cm of its motion.

On the graph, sketch the corresponding kinetic energy of the proton over the same distance.

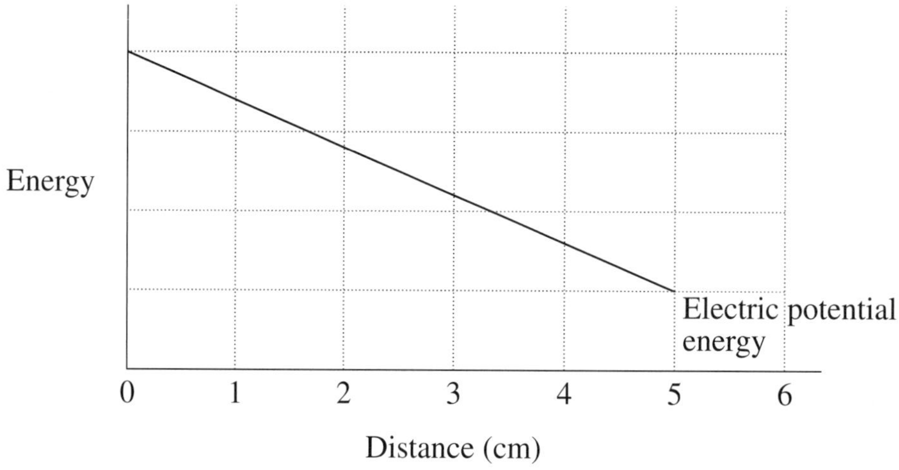

(b) The experiment is repeated using an electron in the place of the proton.

Explain how the motion of the electron would differ from that of the proton.

....

....

....

....

....

....

Question 31 (7 marks)

Two identical solenoids are mounted on carts as shown. Each solenoid is connected to a galvanometer, and the solenoid on cart 1 is also connected to an open switch and a battery. The total mass of cart 1 is twice that of cart 2. **7**

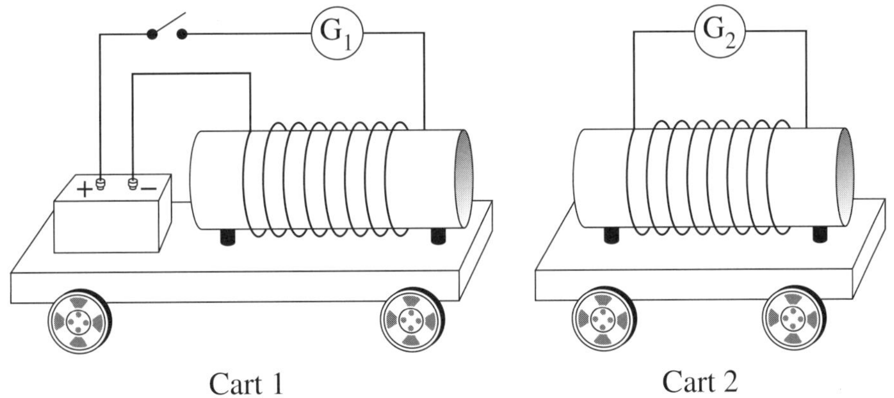

Explain what would be observed when the switch on cart 1 is closed. In your answer, refer to the current in each galvanometer and the initial movement of the carts.

....

....

....

....

....

....

....

....

....

....

....

....

....

....

....

....

Question 32 (5 marks)

Two students perform an investigation with a piece of elastic laid out straight on a table. The elastic is fixed at one end and has three markings at regular intervals. The distances from each marking to the fixed end, d_1, d_2 and d_3, are measured as shown. 5

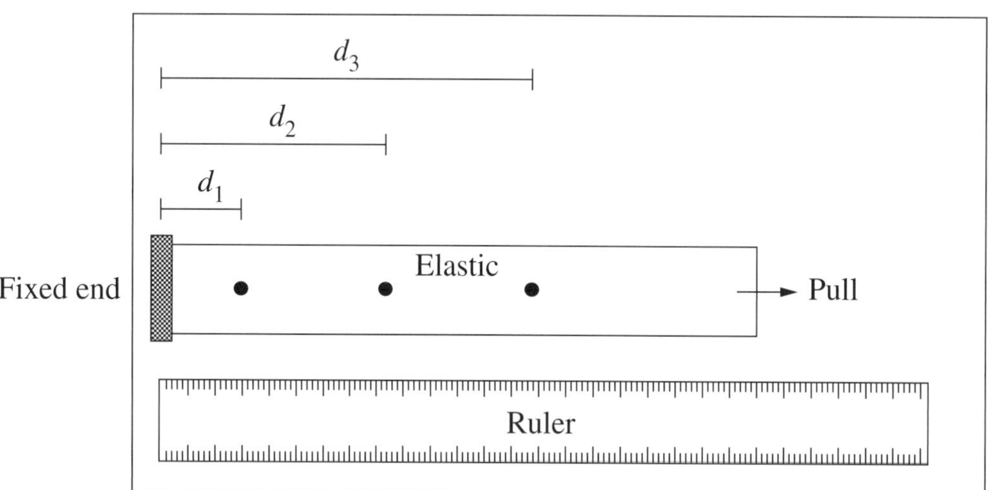

Top view of table

A student pulls the elastic to extend it, and the new values of d_1, d_2 and d_3 are measured. The student observes that each value has doubled.

How well do the observations from this investigation model the evidence that led to Hubble's discovery of the expansion of the universe? Justify your answer.

....

....

....

....

....

....

....

....

....

....

Question 33 (9 marks)

Two experiments are performed with identical light sources having a wavelength of 400 nm. **9**

In experiment *A*, the light is incident on a pair of narrow slits 5.0×10^{-5} m apart, producing a pattern on a screen located 3.0 m behind the slits.

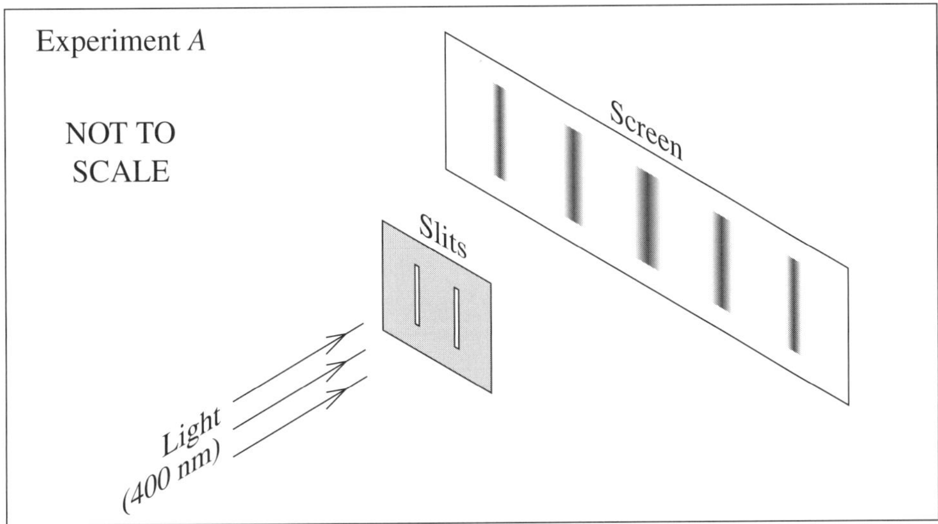

In experiment *B*, the light is incident on different metal samples inside an evacuated tube as shown. The kinetic energy of any emitted photoelectrons can be measured.

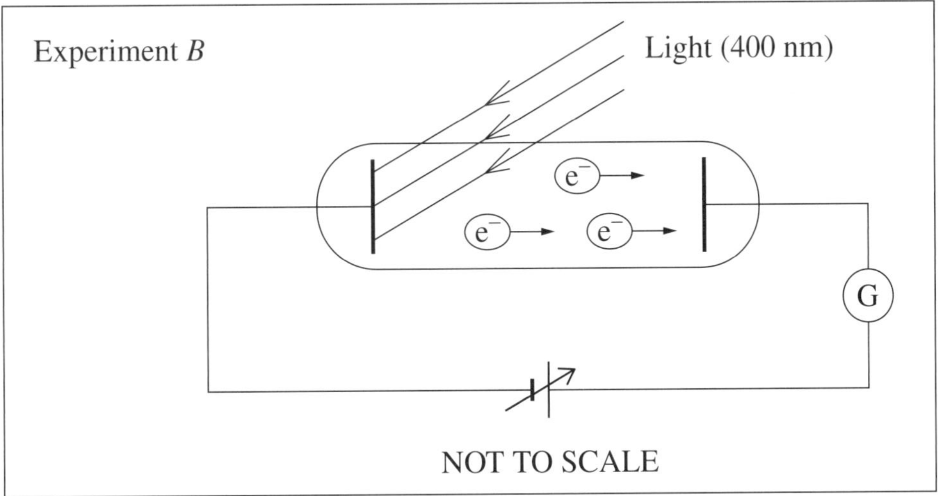

Some results from experiment *B* are shown.

Metal sample	*Work function* (J)	*Photoelectrons observed?*
Nickel	8.25×10^{-19}	No
Calcium	4.60×10^{-19}	Yes

Question 33 continues on the following page

Question 33 (continued)

How do the results from Experiment *A* and Experiment *B* support TWO different models of light? In your answer, include a quantitative analysis of each experiment.

End of Question 33

Question 34 (7 marks)

A 3.0 kg mass is launched from the edge of a cliff.

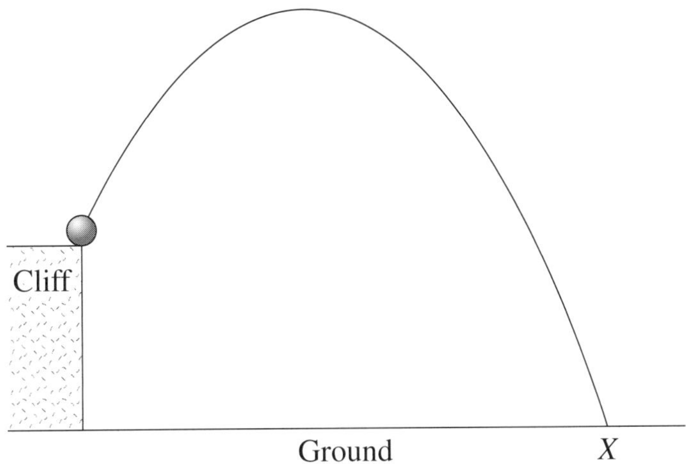

The kinetic energy of the mass is graphed from the moment it is launched until it hits the ground at *X*. The kinetic energy of the mass is provided for times t_0, t_1 and t_2.

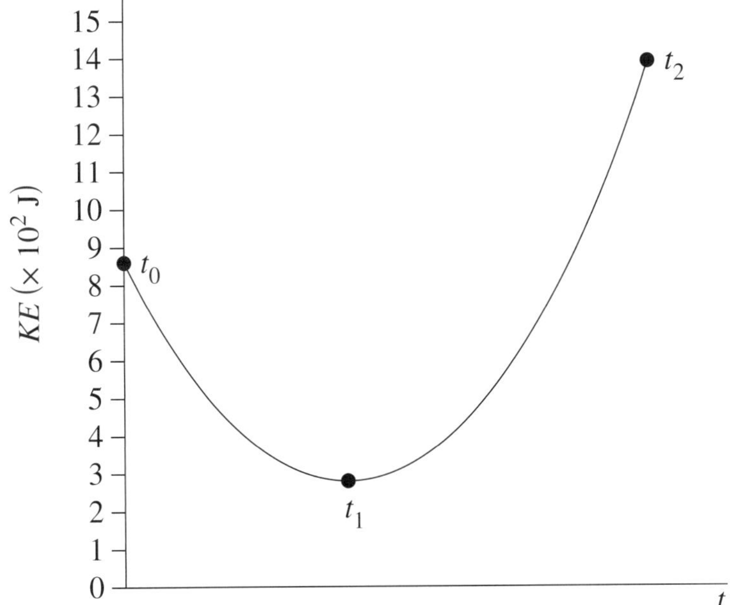

Time (t)	*KE* (J)
t_0	864
t_1	284
t_2	1393

Question 34 continues on the following page

Question 34 (continued)

(a) Account for the relative values of kinetic energy at t_0, t_1 and t_2.

....

....

....

....

....

....

....

....

....

....

(b) The horizontal component of the velocity of the mass during its flight is $13.76\ \text{m s}^{-1}$.

Calculate the time of flight of the mass.

....

....

....

....

....

....

....

....

....

....

End of Question 34

Please turn over

Question 35 (6 marks)

A spacecraft is powered by a radioisotope generator. Pu-238 in the generator undergoes alpha decay, releasing energy. The decay is shown with the mass of each species in atomic mass units, u.

$$^{238}\text{Pu} \quad \rightarrow \quad ^{234}\text{U} \quad + \quad \alpha$$

$$238.0495u \qquad 234.0409u \qquad 4.0026u$$

(a) Show that the energy released by one decay is 9.0×10^{-13} J. **3**

....

....

....

....

....

....

(b) At launch, the generator contains 9.0×10^{24} atoms of Pu-238. The half-life of Pu-238 is 87.7 years. **3**

Calculate the total energy produced by the generator during the first ten years after launch.

....

....

....

....

....

....

End of paper

2021 HSC Examination paper

Sample answers

Section I (Total 20 marks)

1 B The time of flight for a projectile is independent of the initial horizontal velocity of the projectile. It depends only on the acceleration due to gravity, the vertical displacement and the initial vertical velocity, all of which are unchanged in this example. The range is determined by the time of flight and the initial horizontal velocity ($s_x = u_x t$). Because the time of flight remains constant, reducing the initial horizontal velocity will therefore decrease the range of the projectile.

2 C Note that the charge is moving in an electric field (not a magnetic field). The electric field lines show the direction of the force that would be experienced by a positive charge if it was placed in the field. Thus a positive charge will experience a force in the same direction as the electric field.

3 D A proton is a composite particle made up of two up quarks and one down quark.

4 B The speed of light is an absolute constant that is independent of the velocity of the source or the observer. As radio waves are electromagnetic waves that travel at the speed of light, both observers will measure the waves to be travelling at the same speed.

5 A The spectrum *Y* shows a decrease in the intensity of specific wavelengths and is therefore an absorption spectrum. Discharge tubes produce emission spectra rather than absorption spectra and the overall shape of the spectrum is that of a black body. Therefore the spectrum shows the absorption lines from a star.

6 D The isotope *R* has two protons and one neutron and hence it has an atomic number of 2 and a mass number of 3. An atomic number of 2 makes it a helium nucleus and as it has three nucleons it is a He-3 isotope.

7 C In an ideal transformer the input power is equal to the output power and hence $I_P V_P = I_s V_s$ or $\frac{I_P}{I_S} = \frac{V_S}{V_P}$. Therefore, if the current is 4 times higher in the secondary than the primary, the voltage will be 4 times greater on the primary than the secondary. As the transformer reduces the voltage, it is a step-down transformer. Now, as $\frac{V_P}{V_S} = \frac{N_P}{N_S}$, we see the ratio of turns is the same as the ratio of the voltages on the primary and secondary coils. Hence, as the ratio of the voltage on the primary and secondary coils is 4 : 1, the ratio of turns must also be 4 : 1.

8 D The light diffracts around the disc and when the diffracted light from each part of the disc reaches the screen it produces an interference pattern. Refraction involves moving from one medium to another, which clearly is not occurring here. Polarisation involves light passing through a transparent polarising filter, which is also not occurring in this experiment.

9 C Gravitational potential energy at a point is a negative value, equal to the work that would be required to move the mass to an infinite distance from that point. The work required is greatest (and the potential energy most negative) when the mass is closest to the planets and decreases as the mass is moved further from the planets.

10 B The moving magnet produces a changing magnetic field in the conductor that, by Faraday's law of induction, will induce eddy currents in the copper conductor. Lenz's Law tells us that the direction of the induced currents will create magnetic fields that oppose the change that caused them. Thus the eddy currents will produce magnetic fields that will exert a force on the magnet opposing its motion and by Newton's 3rd law this will result in an equal and opposite force on the conductor in the direction of the magnet's movement.

11 A Applying Wien's Displacement Law, $\lambda_{max} = \frac{b}{T} = \frac{2.898 \times 10^{-3}}{310} = 9.3 \times 10^{-6}$ m.

12 D The back emf of a DC motor is proportional to the angular velocity and hence a graph of back emf against angular velocity will be a straight line. This is because, from Faraday's law of induction, the back emf is proportional to the rate of change of magnetic flux in the motor. As the rate of change of magnetic flux is proportional to the angular velocity, the back emf is also proportional to the angular velocity of the motor.

13 B The energy of a photon is inversely proportional to the wavelength of the photon ($E = hf = \frac{hc}{\lambda}$) and therefore the photon with the shortest wavelength will be the most energetic photon. The transition with the greatest energy difference will emit the photon with the greatest energy. Hence the shortest wavelength photon will be emitted by the transition marked X.

14 A The Moon exerts a gravitational force on the Earth which will result in the Earth accelerating towards the Moon. This centripetal acceleration enables the Earth to rotate around the centre of mass of the Earth/Moon system. We can rule out the other answers because there will be an equal force of gravitational attraction on the Earth and on the Moon but because the Moon is less massive than the Earth it will experience a greater acceleration than the Earth.

15 C When unpolarised light is incident on an ideal polariser, the emerging polarised light has half the intensity of the incident polarised light. The position of the transmission axis has no effect on the intensity of the emerging polarised light. Therefore rotating the transmission axis of the first polariser will not change the intensity of the polarised light marked I_1. The light intensity emitted from the second polariser, however, is related to the angle between the transmission axes of the two polarisers by Malus Law ($I = I_0\cos^2\theta$). Hence rotating the first polariser will only change the intensity of the light marked I_2.

16 B The energy emitted per minute is given by $E = Pt = 3.85 \times 10^{28} \times 60$ Joules min^{-1}.

Now, applying Einstein's mass/energy equation we can calculate the mass equivalent for this amount of energy by

$E = \Delta mc^2$ and hence $\Delta m = \frac{E}{c^2} = \frac{(3.5 \times 10^{28}) \times 60}{(3 \times 10^8)^2} = 2.57 \times 10^{13}$ kg.

17 C Because the current is AC, the current in the circuit will alternate and momentarily be zero twice each cycle but the current in the parallel wires will always be flowing in opposite directions. As the current in the parallel wires is flowing in opposite directions, the magnetic fields around the wires will produce an opposing force between the wires. The force on wire X will be directed downwards and the force on wire Y will be directed upwards. Therefore the force measured by the spring balance on wire X will be greater than the force measured by the spring balance on wire Y.

18 A For the first half of the flight, two forces act on the particle—the force of gravity downwards and an electrostatic force towards the right. The net force on the particle when released will therefore be diagonally downwards and the particle will accelerate in a straight line in the direction of the net force. When the electric field is switched off, only the downwards force of gravity acts on the particle and it will accelerate downwards from this point. But as the particle already has a horizontal component of velocity at the midway point, the downwards acceleration will result in a curved projectile path.

19 A The current is proportional to the number of beta particles emitted from the sample per second. Because several half-lives had elapsed before the second measurement was taken, the number of beta particles emitted per second would have decreased significantly. This will be reflected in a lower current reading.

The stopping voltage is proportional to the maximum energy of the emitted beta particles. The energy of the emitted beta particles remains constant because it is related to the mass defect in the nuclear decay that produced the particles. The stopping voltage will therefore remain constant because the energy of the emitted particles is not related to the rate at which they are emitted. (Note that an evacuated chamber is used to prevent electrons and ions being produced in the air by the emitted beta particles.)

20 D If the photon energy is doubled to 4ϕ, the emitted electron will have energy 3ϕ (as an energy ϕ is required to remove an electron from the surface of the cylinder). Thus the emitted electrons will have 3 times more energy than the electrons emitted before the photon energy was increased. As the velocity of the electrons is related to the kinetic energy of the electrons by $K = \frac{1}{2}mv^2$, the velocity of the emitted electrons will change from v to $(\sqrt{3})v$ when the incident photon energy is increased. Now, the radius of curvature is related to the velocity by $qvB = \frac{mv^2}{r}$ or $r = \frac{mv}{qB}$ and hence the radius is proportional to the velocity. If the velocity is increased to $(\sqrt{3})v$ then the radius of curvature will also increase by $(\sqrt{3})r$.

Section II

Question 21 (Total 4 marks)

(a) The maximum torque can be calculated using $\tau = nIABsin\theta$ with $\theta = 90°$.

Thus the maximum torque is
$\tau = nIABsin90° = nIAB = 14 \times (0.1 \times 0.07) \times 0.4 = 0.0392$ Nm. *(2 marks)*

(b) The torque produced is related to the angle of rotation *(θ)* by $\tau = nIABsin\theta$. Therefore the torque is proportional to $sin\theta$. In the starting position shown, $\theta = 90°$ and the torque is maximum anticlockwise. As the rotor coil rotates, the torque decreases continually until after a quarter of a rotation the plane of the rotor coil is perpendicular to the applied magnetic field, $\theta = 0°$, and the torque becomes zero. (At this point the commutator reverses the direction of current flow in the coil ensuring that when the rotor coil rotates further, the torque remains in an anticlockwise direction.) As the rotor coil moves beyond a quarter of a cycle, the torque increases continually until it again reaches a maximum after it has turned half a cycle (180°) from the initial position (i.e. when $\theta = 90°$ again). *(2 marks)*

Question 22 (Total 3 marks)

Points X and Y will have the same angular velocity ($\omega = \frac{\Delta\theta}{t}$) as both are rotating at the same number of radians per second but point X will have twice the instantaneous velocity ($v = \omega r$) of Y because it is twice as far from the centre of the circle as Y but is moving with the same angular velocity as Y. *(3 marks)*

Question 23 (Total 4 marks)

Millikan used the electric field between two horizontal, charged parallel plates to levitate microscopic, charged oil droplets. By adjusting the voltage (V) on the plates until a charged oil droplet remained stationary, he was able to equate the gravitation (mg) and electrostatic forces ($qE = \frac{qV}{d}$) on the oil droplet to determine the charge on the droplet ($\frac{qV}{d} = mg$ and hence $q = \frac{mgd}{V}$). By repeating the experiment many times Millikan was able to find the charge on an electron.

Thomson used an electric field to produce a high-speed beam of electrons. He then passed the beam through perpendicular magnetic and electric fields which exerted forces in opposite directions perpendicular to the beam's velocity. Only those electrons which had the same electric and magnetic force applied to them would not be deflected (i.e. $v = \frac{E}{B}$). He then passed these electrons with known velocity into a magnetic field directed perpendicularly to the electrons' velocity. By measuring the radius of curvature of the electrons' path in the beam and equating the magnetic force on the electron to the centripetal force, Thomson was able to find the charge to mass ratio of the electron (i.e. $qvB = \frac{mv^2}{r}$ or $\frac{q}{m} = \frac{v}{Br}$).

(4 marks)

Question 24 (Total 3 marks)

We can calculate the induced emf by applying Faraday's Law of induction.

The change in emf is $\Delta\Phi = 0 - 0.15 \times 0.02 = 0.003$ Wb

$$\varepsilon = N\frac{\Delta\Phi}{\Delta t} = 35 \times \left(\frac{0.003}{0.4}\right) = 0.26V$$

(3 marks)

Question 25 (Total 5 marks)

(a) The gravitational potential energy will remain negative but have a smaller absolute value when the satellite is in orbit ($U = -\frac{GMm}{r}$) than it would have on the surface (i.e. it increases). This is because it would require less work to remove a satellite from the gravitational influence of the planet if the satellite was in orbit than if the satellite was on the surface. The kinetic energy ($K = \frac{1}{2}mv^2$) of the satellite would also increase when it was moved into orbit. This is because it would have the same angular velocity on the surface as it would have in a geostationary orbit, but because it has a much greater orbital radius in orbit it would have a much greater linear velocity in orbit than on the surface. Thus it will have much greater kinetic energy in orbit than it has on the surface. *(2 marks)*

(b) Applying the satellite equation, $\frac{r^3}{r^2} = \frac{GM}{4\pi^2}$

and hence

$$r = \sqrt[3]{\frac{GMT^2}{4\pi^2}} = \sqrt[3]{\frac{6.67 \times 10^{-11} \times 6.39 \times 10^{23} \times (24 \times 60 \times 60 + 40 \times 60)^2}{4\pi^2}} = 2.04 \times 10^7 \text{ m}$$

(3 marks)

Question 26 (Total 6 marks)

(a)

Voltage (V)

Frequency ($\times 10^{14}$ Hz)

(3 marks)

(b) Calculating Plank's constant from a single data point will be less accurate than using all the data gathered by the student, because the single point used may be an outlier. This is indeed the case here because the first point, which the student intended to use, does not fall on the line of best fit. Using all the data also minimises the impact of random measurement errors.

A better method which uses all the data would be to find the gradient of the line of best fit. As the graph represents the relationship $v = \frac{hf}{q_e}$, the gradient (m) of the graph will be $m = \frac{h}{q_e}$ and hence Plank's constant $h = mq_e$. Thus we can obtain a more accurate value of Plank's constant by multiplying the gradient of the line of best fit by the charge on an electron. *(3 marks)*

Question 27 (Total 6 marks)

(a)

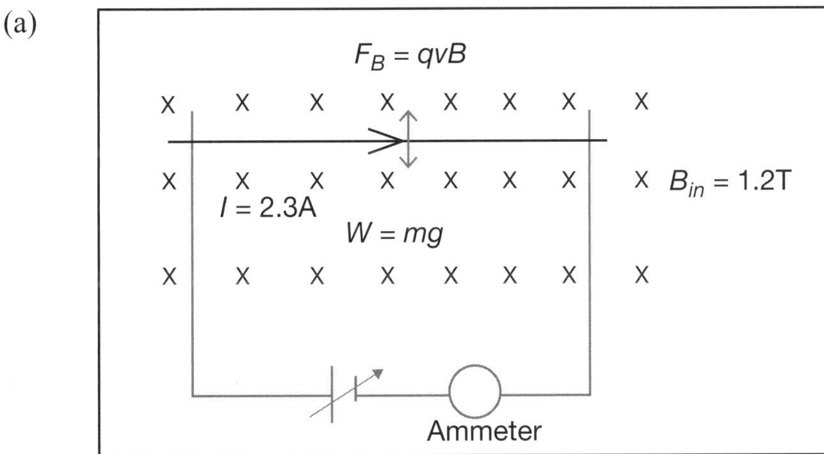

Alternatively, students may choose to have the magnetic field out of the page and the current in the rod moving towards the left. *(3 marks)*

(b) To levitate the rod, the magnetic force ($F_B = ILBsin\theta$) must be equal to the weight ($W = mg$) of the rod. Now, as the magnetic field is perpendicular to the current, we can write $ILB = mg$ and hence mass per unit length ($\frac{m}{L}$) that could be levitated is given by

$$\frac{m}{L} = \frac{BI}{g} = \frac{1.2 \times 2.3}{9.8} = 0.282 \text{ kgm}^{-1}.$$

If the mass per unit volume is greater than this value, the upwards magnetic force on the rod will be less than the weight of the rod and hence the rod will accelerate downwards rather than levitate. *(3 marks)*

Question 28 (Total 5 marks)

(a) The distance to the star can be determined once the velocity of the spaceship is determined.

We can calculate the velocity of the spacecraft by applying Einstein's time dilation with the proper time set at $t_0 = 9.4$ yrs:

$$t = \frac{t_0}{\sqrt{1 - \frac{v^2}{c^2}}}$$

Rearranging this equation to make the velocity the subject of the equation yields

$$\sqrt{1 - \frac{v^2}{c^2}} = \frac{t_0}{t} \text{ and hence } \frac{v^2}{c^2} = 1 - \left(\frac{t_0}{t}\right)^2 \text{ and finally}$$

$$v = c\sqrt{1 - \left(\frac{t_0}{t}\right)^2} = 3 \times 10^8 \sqrt{1 - \left(\frac{9.4}{15}\right)^2} = 2.34 \times 10^8 \text{ ms}^{-1}$$

The distance to the star in light years is therefore

$$s = vt = \left(\frac{2.34 \times 10^8}{3 \times 10^8}\right) \times 15 = 11.7 \text{ light years.}$$

Students may also answer using SI units, 11.7 light years = 1.11×10^{17} m. *(3 marks)*

(b) The speed of light according to the special theory of relativity is the fastest any object can travel with respect to any other observer in the universe. The theory prohibits faster-than-light travel because it predicts that the momentum and relativistic mass of a body will increase as it moves faster. Special relativity predicts the momentum of a body is related to its velocity relative to an observer by

$$p_v = \frac{m_0 v}{\sqrt{1 - \frac{v^2}{c^2}}}$$

From this equation we see that as an object is accelerated towards the speed of light, v approaches c and the momentum and hence the relativistic mass approach infinity. As an infinite force would be required to accelerate an infinite mass ($a = \frac{F}{m}$), it is impossible according to the special theory of relativity for any object to be accelerated to the speed of light and hence it is impossible for an object to travel faster than the speed of light.

Thus special relativity limits the speed of this or any other spaceship to velocities that are lower than the speed of light. *(2 marks)*

Question 29 (Total 5 marks)

All three physicists accepted that electrons had a specific charge and mass that could be measured but they had different views about the nature of the electron.

In Bohr's atomic model, electrons orbited a positive nucleus in specific orbitals. For the model to reproduce the hydrogen emission spectrum, Bohr had to assume that only orbitals with angular momentum $mvr = \frac{nh}{2\pi}$ could exist. His model treated the electron as a simple negatively charged particle that occupied a specific point in space as it orbited.

In contrast, de Broglie proposed that electrons had both wave and particle characteristics with an associated wavelength ($\lambda = \frac{h}{p}$). De Broglie's atomic model explained Bohr's postulate ($mvr = \frac{nh}{2\pi}$) by assuming that the wave nature of the electron forced the electrons to only occupy specific orbitals that corresponded to electron standing waves. In other words, the circumference of the orbital was equal to a whole number of electron wavelengths. His atomic model envisioned the electron as a particle guided by pilot waves.

Schrödinger's model of the atom treats the electron as a quantum identity. He combined de Broglie's matter/wave equation with classical wave theory to create an equation that explained the propagation of electron waves in the electric field of the nucleus in terms of a complex wave function. The wave function can then be used to generate a three-dimensional probability distribution that describes the probability of finding an electron at any point in the atom. His model shows that the electrons do not move in orbitals and have no specific location in the atom but can only be imagined as existing as a cloud in the atom. *(5 marks)*

Question 30 (Total 5 marks)

(a) The charge is accelerating freely in an electric field and hence the sum of its kinetic and potential energy will remain constant, as shown in the graph below.

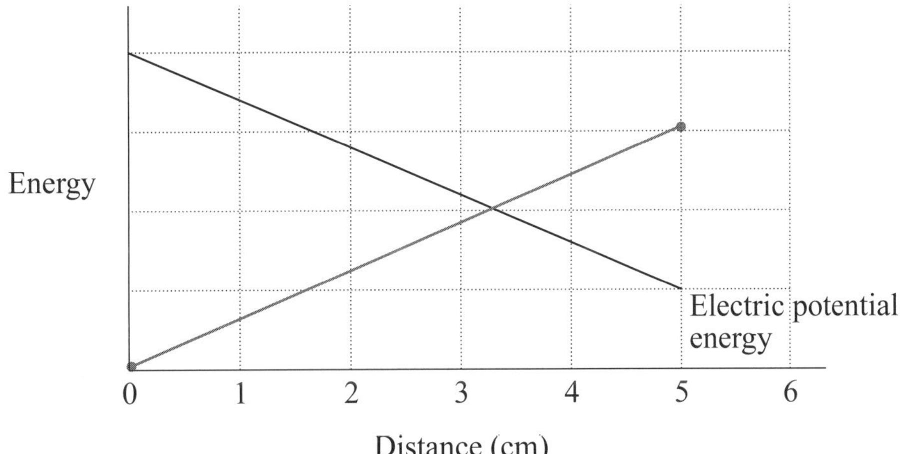

(2 marks)

(b) Because the magnitude of the charge on a proton and on an electron are equal, the magnitude of the force ($F = qE$) on each will be the same. But because the proton has a positive charge and the electron has a negative charge, the force on the proton will be downwards, while the force on an electron would be upwards. The electron has a much lower mass than the proton and this will cause it to accelerate much more rapidly (as $F = ma$ and $F = qE$ the acceleration is given by $a = \frac{qE}{m}$) than the proton. As the mass of the proton is 1837 times greater than the mass of the electron, the electron will accelerate with a magnitude 1837 times greater than the proton. Hence the electron would accelerate upwards at a rate 1837 times greater than the proton accelerates downwards. Because the electric field is uniform between the plates, both charges accelerate at a constant rate until they reach the plates. The electron will, however, reach the top plate much sooner than the proton will reach the bottom plate. *(3 marks)*

Question 31 (7 marks)

When the switch is closed, the current passing from left to right in G_1 will increase rapidly until it reaches a specific *dc* value and will then remain constant. The change in the current will not be instantaneous because the changing magnetic field produced by the current being switched on in the solenoid will induce an emf that will oppose the applied voltage (self-inductance). Once the current stabilises, the first solenoid will contain a constant magnetic field directed towards the left inside the solenoid (by the right-hand rule). That is, the solenoid on cart 1 will have a south pole on the right-hand side.

During the time the magnetic field of the first solenoid is changing from zero to its final value, the second solenoid will experience a changing magnetic flux that will induce an emf and produce a momentary current flow in the solenoid on cart 2 (Faraday's law of electromagnetic induction). The direction of the emf will be such that the current it produces will oppose the change in magnetic field that caused the emf to be induced (Lenz's law). This momentary current flow from right to left through G_2 will produce a magnetic field directed to the right in the second solenoid producing a momentary south pole on the left-hand side of the solenoid. G_2 will only register a current while the magnetic field is changing in the first solenoid. When the current in the first solenoid reaches its equilibrium *dc* value, the induced current will return to zero in G_2 and the magnetic field in the solenoid will rapidly return to zero.

During the time the magnetic field is changing in the first solenoid, the induced magnetic fields in the solenoids will produce two opposing south poles and hence an opposing force between the carts. This force will be the same magnitude on each cart and cause the carts to momentarily accelerate away from one another. As cart 1 has twice the mass of cart 2, cart 1 will accelerate at half the rate of cart 2 (as $a = \frac{F}{m}$). Both carts will have accelerated for the same time but because they are accelerating at different rates, cart 1 will not move as far or have a final velocity as great as cart 2. *(7 marks)*

Question 32 (Total 5 marks)

In the 1920s Hubble used the intensity of cepheid variable stars and several other methods to estimate the distance to 24 other galaxies. He then examined the light from the galaxies and found that all but the closest galaxies exhibited a redshift indicating that these galaxies were moving away from our galaxy. Measuring the size of the redshift showed that the recession velocity of all but the closest galaxies increased with distance. Hubble concluded that the universe was expanding and used his data to propose a law (Hubble's law) that related the recession velocity (v) of the galaxies to the distance (d) to the galaxy:

$$v = H_0 d$$

The constant of proportionality (H_0), called the Hubble constant, gives a measure of the rate at which the universe is expanding. Hubble's discovery confirmed an earlier theoretical prediction from general relativity that the universe was expanding because the space between the galaxies was expanding.

The evidence Hubble collected was that distant galaxies are moving away from us with a velocity that is proportional to their distance.

If we imagine that each dot in the student's model is representing a galaxy, time being represented by stretching the elastic and the observer being at the left-hand side of the elastic, the model shows that expanding space would result in the more distant galaxies moving away faster than the closer galaxies (as Hubble observed). The model also shows that the distance to the closer galaxies will increase less than the distance to further galaxies as time passes, which also reflects Hubble's observations.

The student's model is also consistent with Hubble's law. We can see this by examining the distance moved by each dot (galaxy) when the elastic is stretched. Assume stretching the elastic represents one interval of time (t). To account for the distances moved by each point in the time t, the galaxies at each point must be moving with velocities $v_1 = \left(\frac{d_1}{t}\right)$, $v_2 = 3\left(\frac{d_1}{t}\right)$ and $v_3 = 5\left(\frac{d_1}{t}\right)$. As the original distance (by measurement) to d_2 is three times the distance d_1 and the original distance d_3 is five times the distance to d_1, the velocities are proportional to the distance to the galaxies, as Hubble's law predicts.

The student's model is limited to one dimension, while the actual universe has three spatial dimensions (and one time dimension) but it does model Hubble's law well. Though it was an inference rather than one of Hubble's observations, the model also shows that the expansion of space (in this case the elastic), rather than the motion of galaxies through space, would account for Hubble's observations. *(5 marks)*

Question 33 (Total 9 marks)

Experiment A is a modern version of Young's famous double slit experiment that was used to support the wave model of light. When Young first demonstrated the experiment in the early 1800s, it could not be explained by Newton's corpuscular theory of light and was one of a series of experiments that led scientists at the time to abandon Newton's particle model of light and adopt a wave model of light.

Young used wave theory to provide a qualitative and quantitative explanation of the experiment. He explained that if light was a wave it would diffract from each of the slits and light from both slits would superimpose at each point on the screen producing the observed interference pattern. As the light travels a different distance to the screen from each slit, the light intensity at each point will be determined by the path difference travelled by the light from each slit (as this produces a phase difference between the waves). Young showed that the light waves from each slit would produce an interference pattern on the screen with alternate light and dark bands. He showed the bright bands (interference maxima) would occur at an angle (θ) from the central bright maxima and would be related to the wavelength of the light and slit separation by $sin\theta = m\lambda$, where m is the number of the bright band from the central bright band (the order of interference).

We can calculate the angle between the central and first (m = 1) maxima of experiment A as follows:

$$dsin\theta = m\lambda \text{ and hence } \theta = sin^{-1}\left(\frac{\lambda}{d}\right) = sin^{-1}\left(\frac{400 \times 10^{-9}}{5.0 \times 10^{-5}}\right) = 0.458^0$$

We can use this angle and the distance to the screen (L) to calculate the separation between the central and first maxima on the screen (x):

$$x = L\ tan\theta = 3\ tan\theta\ (0.458) = 0.024 \text{ m} = 2.4 \text{ cm}$$

Experiment B demonstrates the photoelectric effect, which was an experiment used by Einstein to support his photon model of light. In the photon model, electromagnetic waves (light) are quantised into small packets of energy called photons. Light in this model consists of a stream of photons rather than a continuous wave. The energy of the photons is related to the frequency of the light by $E = hf$. Einstein showed that the photoelectric effect could not be explained by classical electromagnetic wave theory but could be explained simply by the photon model of light.

The photoelectric effect occurs when light falls on a metal surface and electrons (photoelectrons) are ejected. Only light with a frequency above a specific threshold frequency for each metal eject electrons and the ejected electrons have a maximum kinetic energy that is proportional to the frequency of the incident light. By assuming one photon ejects one electron and that it requires a specific amount of energy (called the work function (ϕ)) to release an electron from the surface, Einstein applied conservation of energy to write:

Kinetic energy of photoelectron = Initial photon energy – Work function

$$K_{max} = hf - \phi$$

If the incident photon energy is less than the work function of the metal, no photoelectrons will be ejected from the surface and if the photon energy is greater than the work function, the maximum kinetic energy of the photoelectrons will be proportional to the incident light frequency. Neither of these observations can be explained by electromagnetic wave theory.

Experiment B demonstrates the requirement for the photon energy to be greater than the work function before electrons are ejected from the surface. We can calculate the incident photon energy using:

$$E = hf = \frac{hc}{\lambda} = \frac{(6.626 \times 10^{-34})(3 \times 10^{8})}{400 \times 10^{-9}} = 4.97 \times 10^{-19}\,\text{J}$$

This photon energy is less than the energy required to eject an electron from Nickel (8.25×10^{-19} J) and hence no photoelectrons are observed. But this photon energy is greater than the energy required to remove electrons from the surface of calcium (4.60×10^{-19} J), which explains why photoelectrons are ejected from calcium in the experiment. The maximum energy of the photoelectrons emitted from calcium in this experiment will be:

$$K_{max} = hf - \phi = 4.97 \times 10^{-19} - 4.60 \times 10^{-19} = 3.7 \times 10^{-20}\,\text{J}$$

(9 marks)

Question 34 (Total 7 marks)

(a) Students could answer this question using kinematics or energy.

The kinetic energy of the particle is related to the mass and velocity of the particle by $K = \frac{1}{2}mv^2$. As t_0 is the time of launch we see the kinetic energy must be due to the initial velocity of the projectile. As the projectile has a mass of 3.0 kg, we can use the kinetic energy equation to find the initial launch velocity, $v = \sqrt{\frac{2K}{m}} = \sqrt{\frac{2 \times 864}{3.0}} = 24\ \text{ms}^{-1}$.

This velocity will have a vertical and horizontal component.

The horizontal component of the initial velocity will remain constant throughout the flight but the vertical component will decrease because of downwards acceleration due to gravity until the vertical component of the velocity is zero at the top of the flight. This is also the point where the kinetic energy will be a minimum because the velocity of the particle is a minimum at the top of the flight. This point corresponds to time t_1 and we can use the energy at this point to calculate the horizontal component of the velocity,

$$v = \sqrt{\frac{2K}{m}} = \sqrt{\frac{2 \times 284}{3.0}} = 13.76\ \text{ms}^{-1}.$$

As the projectile falls from its maximum height, the vertical component of the velocity increases downwards until it reaches a maximum when the projectile reaches the ground. As the projectile's velocity is the vector sum of the horizontal and vertical components of the velocity, the velocity will increase until it reaches a maximum at time t_2 when the particle reaches the ground. The final velocity will be greater than the initial velocity because the ground in this case is significantly below the launch point. The final velocity can be calculated from the final kinetic energy,

$$v_{Final} = \sqrt{\frac{2K}{m}} = \sqrt{\frac{2 \times 1393}{3.0}} = 30.47\ \text{ms}^{-1}.$$

OR

Because the projectile is in free flight in a gravitational field, the total energy of the projectile remains constant throughout its flight. That is, the sum of the kinetic and potential energy remains constant throughout the flight.

The first time t_0 refers to the launch and initial kinetic energy of the particle when it was launched. This energy is related to the initial velocity of the projectile (u) by $K = \frac{1}{2}mu^2$. This was the energy transferred to the particle when it was launched.

As the projectile moves higher it does work on the gravitational field, its potential energy ($\Delta u = mg\Delta h$) increases and its kinetic energy decreases until at the top of the flight (time t_1), the potential energy is a maximum and the kinetic energy is a minimum. This explains why the kinetic energy at this point is least at t_1.

When the particle falls downwards from the top of its flight, gravity does work on the projectile and it loses potential energy and gains kinetic energy until it reaches the ground at time t_2. That is why the kinetic energy is a maximum at time t_2. The kinetic energy is greater at t_2 than at t_1 because the particle reaches the ground at a point that is below the starting point, where its potential energy is lower and its kinetic energy higher. *(4 marks)*

(b) The time of flight can be determined in several ways.

For example, the energy at time t_1 is due entirely to the horizontal velocity of the particle. As the horizontal component of the velocity remains constant throughout the flight, the contribution that the horizontal velocity makes to the particle's kinetic energy also remains constant. We can use this fact to determine the initial velocity and final velocity of the projectile as follows:

$K_{\text{Due to vertical component}} = K_{\text{Total}} - K_{\text{Due to horizontal component}}$

Apply this to time t_0, $K_{\text{Due to vertical component}} = 864 - 284 = 580$ J.

The initial vertical velocity is therefore $u_y = \sqrt{\frac{2K}{m}} = \sqrt{\frac{2 \times 580}{3.0}} = 19.66 \text{ ms}^{-1}$ upwards.

The final vertical component of the velocity at t_2 can be calculated in the same way:

$K_{\text{Due to vertical component}} = 1393 - 284 = 1109$ J.

And the final vertical component of the velocity will be:

$v_y = \sqrt{\frac{2K}{m}} = \sqrt{\frac{2 \times 1109}{3.0}} = 27.19 \text{ ms}^{-1}$ downwards.

Calling upwards positive and downwards negative, we can now find the time of using

$v = u + at$ or $t = \frac{v - u}{a} = \frac{-27.19 - 19.66}{-9.8} = 4.78$ seconds

OR

the initial velocity from the initial kinetic energy is $v = \sqrt{\frac{2K}{m}} = \sqrt{\frac{2 \times 864}{3.0}} = 24 \text{ ms}^{-1}$.

Knowing the horizontal component of the velocity is 13.67 ms^{-1}, the initial vertical component will be given by $u_y = \sqrt{24^2 - 13.67^2} = 19.62$ ms^{-1} upwards.

Similarly, the final velocity will be given by: $v_{Final} = \sqrt{\frac{2K}{m}} = \sqrt{\frac{2 \times 1393}{3.0}} = 30.47$ ms^{-1}.

And the vertical component of the final velocity will be given by:

$u_y = \sqrt{30.47^2 - 13.67^2} = 27.23$ ms^{-1} downwards.

Calling upwards positive and downwards negative, we can now find the time of using:

$v = u + at$ or $t = \frac{v - u}{a} = \frac{-27.23 - 19.62}{-9.8} = 4.78$ seconds

(3 marks)

Question 35 (Total 6 marks)

(a) We first find the mass defect in the nuclear decay reaction:

$$\Delta m = (234.0409 + 4.0026) - 438.0495 = -0.006\ u$$

The negative sign tells us that mass is converted into energy in this reaction.

Hence the energy released is $E = (0.006) \times 931.5 = 5.589$ MeV

or $E = (5.589 \times 1.602 \times 10^{-19}) \times 10^6 = 8.954 \times 10^{-13}$ J

or to two significant figures $E = 9.0 \times 10^{-13}$ Joules, as expected.

Alternatively, students may convert the mass defect in atomic mass units to kilograms and then apply Einstein's mass/energy equation ($E = mc^2$) to find the energy released in the reaction. *(3 marks)*

(b) We must find how many atoms have undergone radioactive decay.

The decay constant for this reaction will be given by $\lambda = \frac{ln2}{t_{\frac{1}{2}}} = \frac{ln2}{87.7} = 7.9 \times 10^{-3}\ yr^{-1}$

The number of isotopes that have not undergone radioactive decay after 10 years will be then given by $N_t = N_0e^{-\lambda t} = (9 \times 10^{24})e^{-(0.079 \times 10)} = 8.316 \times 10^{24}$ isotopes.

Now, the number of uranium isotopes that have undergone radioactive decay from the original sample will be given by $N = 9 \times 10^{24} - 8.316 \times 10^{24} = 6.84 \times 10^{23}$ isotopes.

The energy released in 10 years will be $E = (6.84 \times 10^{23})(9.0 \times 10^{-13}) = 6.16 \times 10^{11}$ Joules.

(3 marks)

CHAPTER 8

NSW Education Standards Authority

2022 HIGHER SCHOOL CERTIFICATE EXAMINATION

Physics

General Instructions

- Reading time – 5 minutes
- Working time – 3 hours
- Write using black pen
- Draw diagrams using pencil
- Calculators approved by NESA may be used
- A data sheet, formulae sheet and Periodic Table are provid the back of this paper

Total marks: 100

Section I – 20 marks

- Attempt Questions 1–20
- Allow about 35 minutes for this section

Section II – 80 marks

- Attempt Questions 21–35
- Allow about 2 hours and 25 minutes for this section

Section I

20 marks
Attempt Questions 1–20
Allow about 35 minutes for this section

Use the multiple-choice answer sheet for Questions 1–20.

---

1 An ideal transformer has 20 turns on the primary coil and an input voltage of 100 V.

How many turns are there on the secondary coil if the output voltage is 400 V?

A. 4

B. 5

C. 80

D. 400

2 The absorption lines in a star's spectrum are shown.

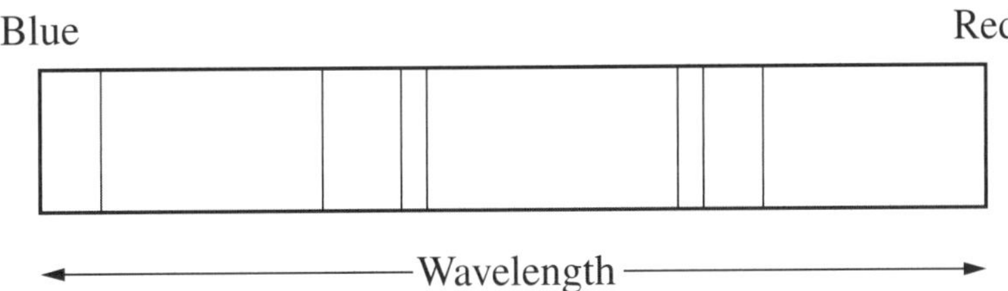

What feature of the star is directly responsible for these absorption lines?

A. Size

B. Colour

C. Distance from Earth

D. Chemical composition

3 A radioisotope emits radiation which is deflected by an electric field, as shown.

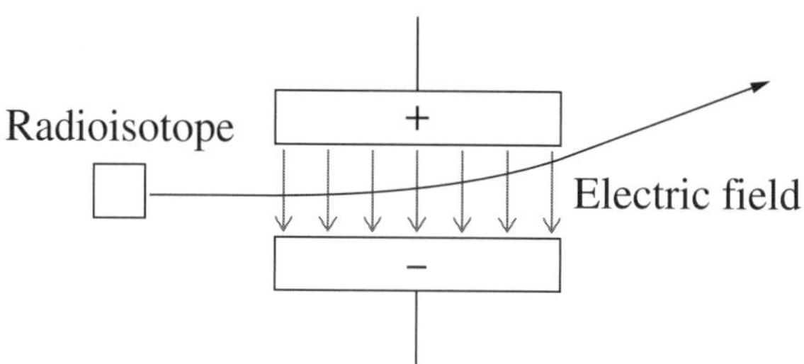

What type of radiation is this?

A. Alpha

B. Gamma

C. Beta positive (positron)

D. Beta negative (electron)

4 A current-carrying wire is in a magnetic field, as shown.

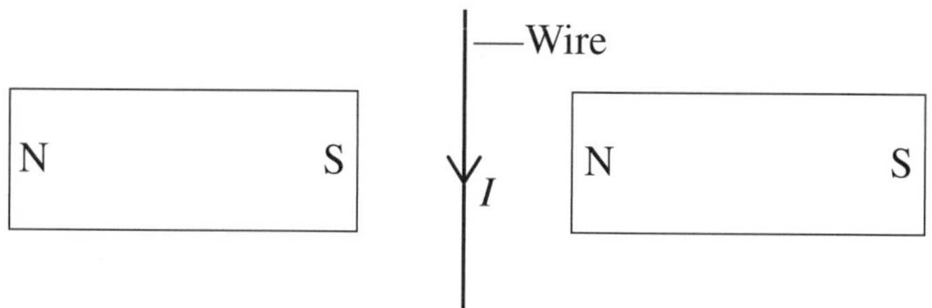

What is the direction of the force on the wire?

A. Left

B. Right

C. Into the page

D. Out of the page

5 Protons and neutrons are made up of quarks. The table shows the charges of these quarks.

Quark	*Charge*
Up	$+\frac{2}{3}$
Down	$-\frac{1}{3}$

What combination of quarks forms a neutron?

A. 1 up, 1 down

B. 1 up, 2 down

C. 2 up, 1 down

D. 2 up, 2 down

6 The elliptical orbit of a planet around a star is shown.

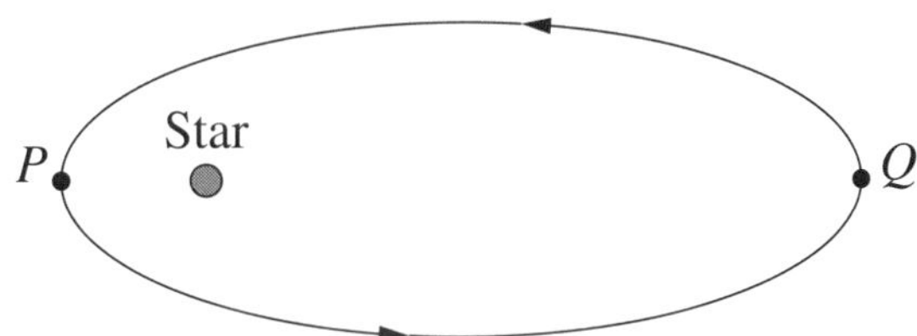

Which type of energy is greater at position P than at Q?

A. Kinetic

B. Nuclear

C. Potential

D. Total

7 A photon has an energy of 9.0×10^{-24} J.

What is the frequency of this radiation?

A. 1.00×10^{-40} Hz

B. 7.36×10^{-11} Hz

C. 1.36×10^{10} Hz

D. 5.97×10^{11} Hz

8 An object is launched with an initial velocity, u, and hits a wall with a final velocity, v.

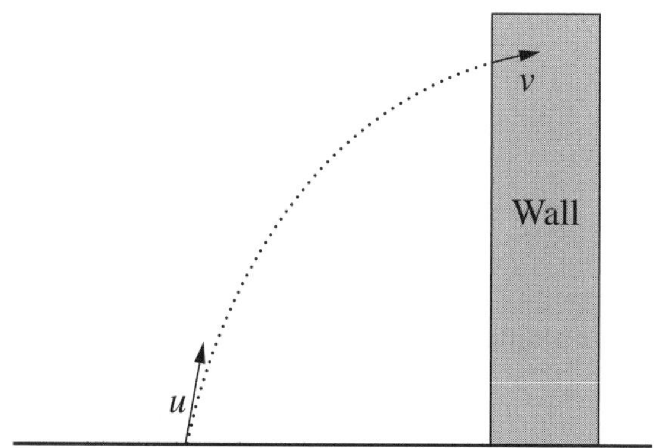

Which statement correctly compares components of u and v?

A. The vertical component of v is less than the vertical component of u.

B. The vertical component of v is greater than the vertical component of u.

C. The horizontal component of v is less than the horizontal component of u.

D. The horizontal component of v is greater than the horizontal component of u.

9 The radiation emitted by a black body has a peak wavelength of 5.8×10^{-7} m.

What is its temperature?

A. 3000 K

B. 4500 K

C. 5000 K

D. 5500 K

10 The orbital velocity, v, of a satellite around a planet is given by $v = \sqrt{\dfrac{GM}{r}}$.

Which graph is consistent with this relationship?

A.

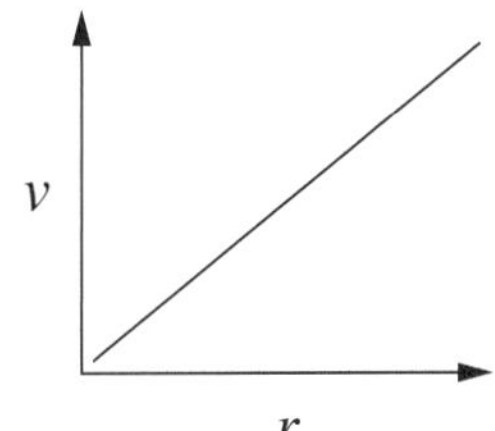

B.

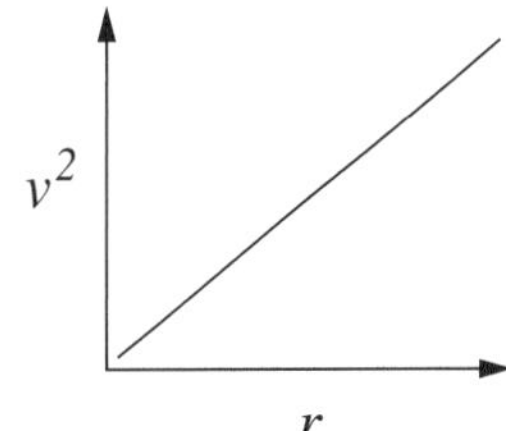

C.

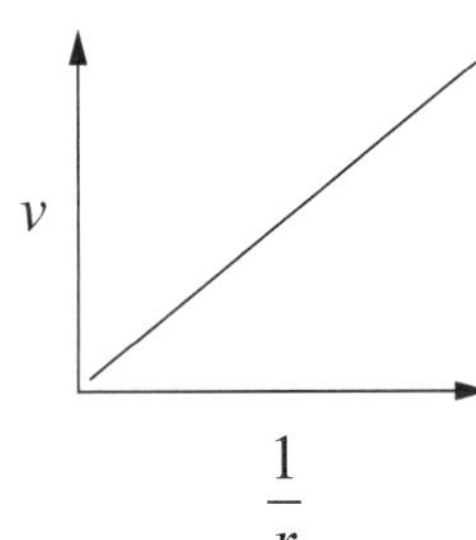

D.

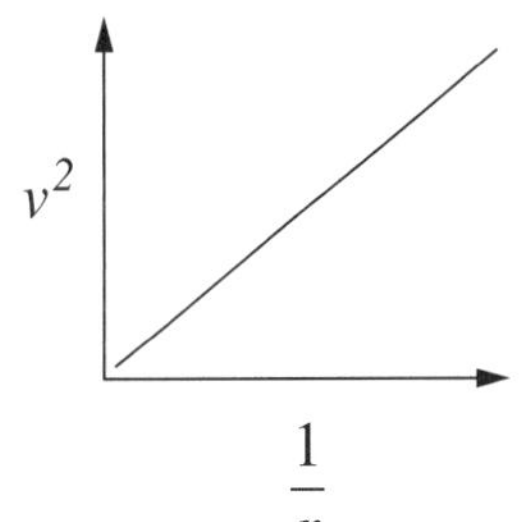

11 A projectile is launched vertically upwards. The displacement of the projectile as a function of time is shown.

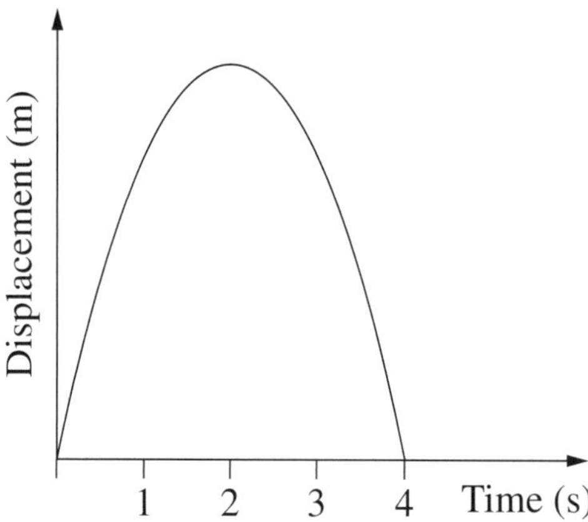

Which velocity–time graph corresponds to this motion?

A.

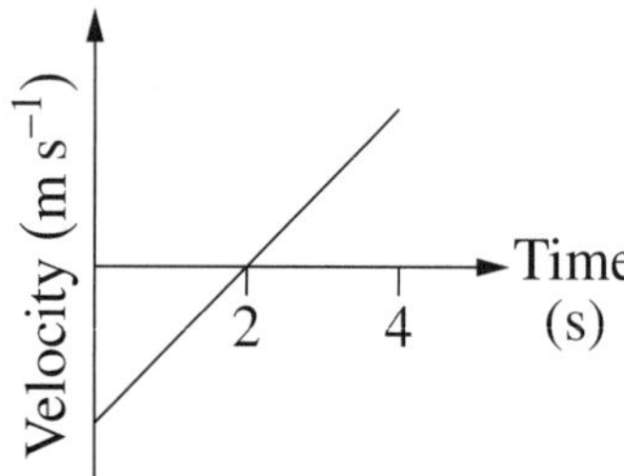

B.

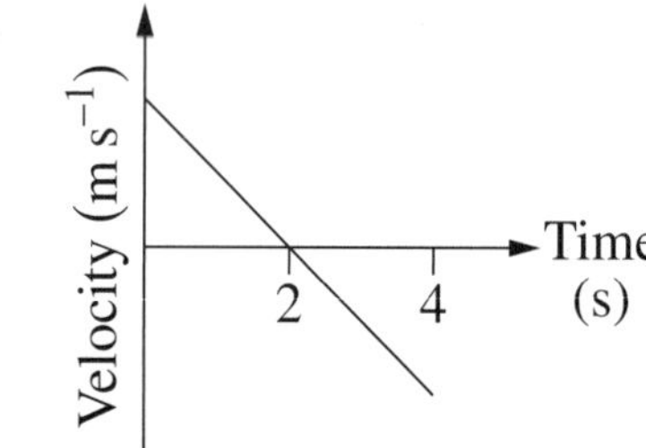

C.

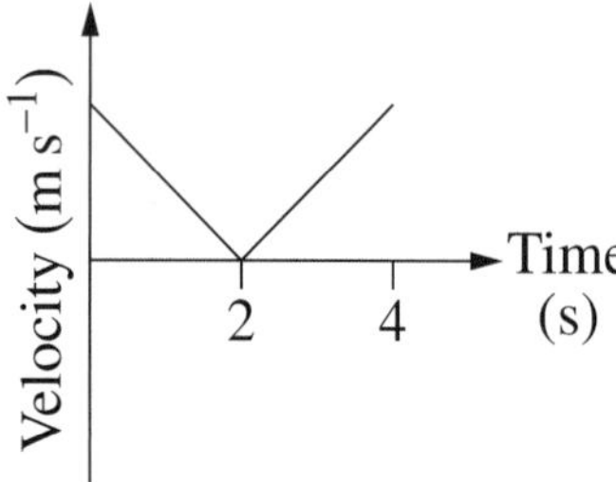

D.

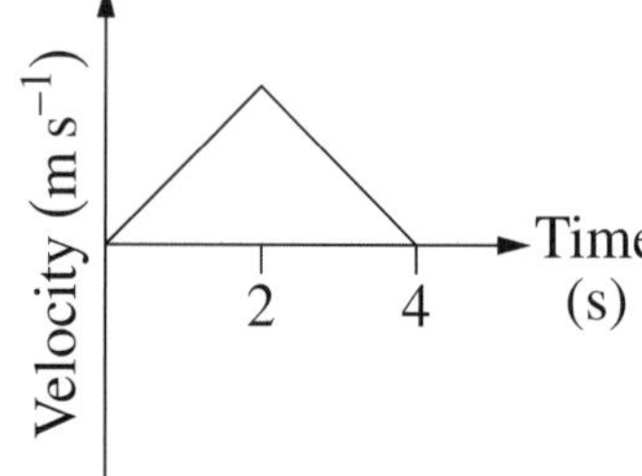

12 The diagram shows a region in which there are uniform electric and magnetic fields. A positively charged particle moves in the region at constant velocity.

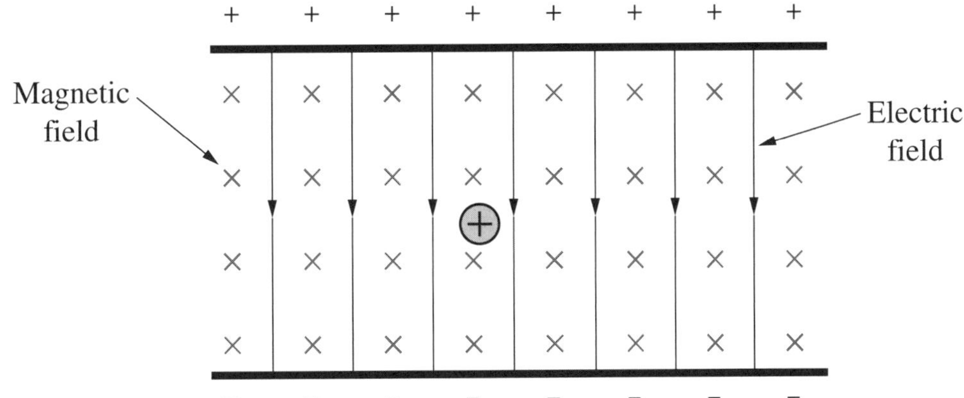

What is the direction of the particle's velocity?

A. Up the page

B. Down the page

C. To the left

D. To the right

13 Two satellites share an orbit around a planet. One satellite has twice the mass of the other.

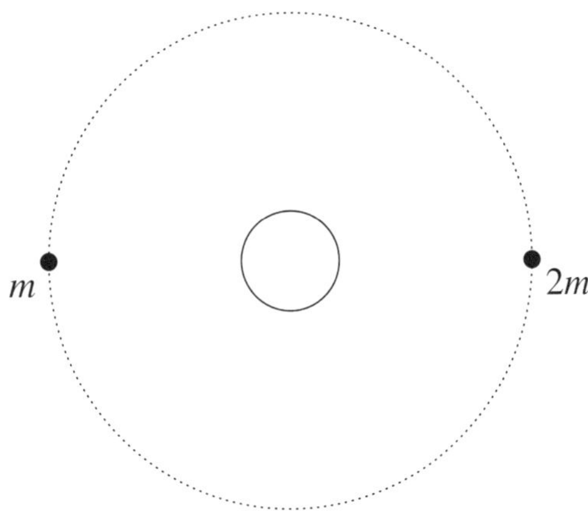

Which quantity would be different for the two satellites?

A. Speed

B. Momentum

C. Orbital period

D. Centripetal acceleration

14 Line X shows the results of an experiment carried out to investigate the photoelectric effect.

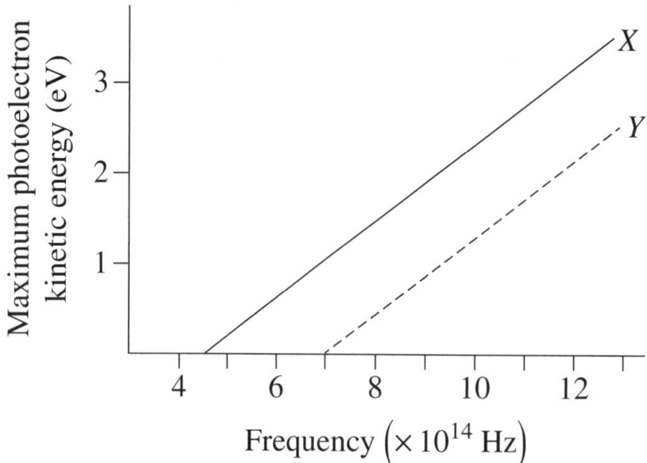

What change to this experiment would produce the results shown by line Y?

A. Increasing the frequency of the radiation

B. Using a metal that has a greater work function

C. Decreasing the intensity of the incident radiation

D. Decreasing the maximum energy of photoelectrons

15 Two wires separated by a distance, d, carry equal electric currents producing a magnetic force between them.

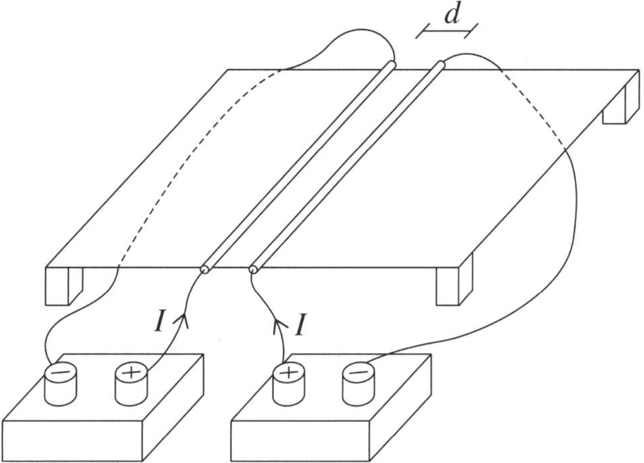

The separation between the wires is increased to $4d$ and the current in each wire is doubled.

What happens to the magnetic force between the wires, compared to the original force?

A. It does not change.

B. It increases by a factor of 4.

C. It decreases by a factor of 4.

D. It decreases by a factor of 8.

16 The binding energy of helium-4 (He-4) is 28.3 MeV and the binding energy of beryllium-6 (Be-6) is 26.9 MeV.

Which of the following rows in the table is correct?

A.	He-4 requires more energy to separate into individual protons and neutrons	He-4 is less massive than Be-6
B.	He-4 requires less energy to separate into individual protons and neutrons	He-4 is less massive than Be-6
C.	He-4 requires more energy to separate into individual protons and neutrons	He-4 is more massive than Be-6
D.	He-4 requires less energy to separate into individual protons and neutrons	He-4 is more massive than Be-6

17 Unpolarised light of intensity I_0 is incident upon a vertically polarised filter. The filtered light then passes through a pair of glasses. The glass have polarising filters, with one side polarised vertically and the other horizontally.

The filter undergoes one complete 360° rotation around point P, as shown.

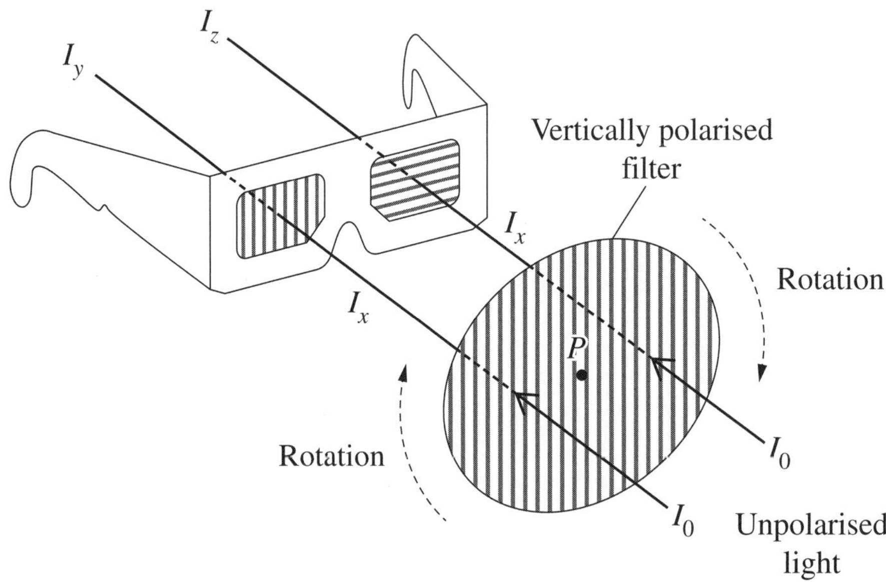

Which of the following correctly compares I_y to the intensity at other positions?

A. I_y never equals I_x

B. I_y never equals I_z

C. I_y sometimes equals I_z

D. I_y sometimes equals I_0

18 A charged oil droplet was observed between metal plates, as shown.

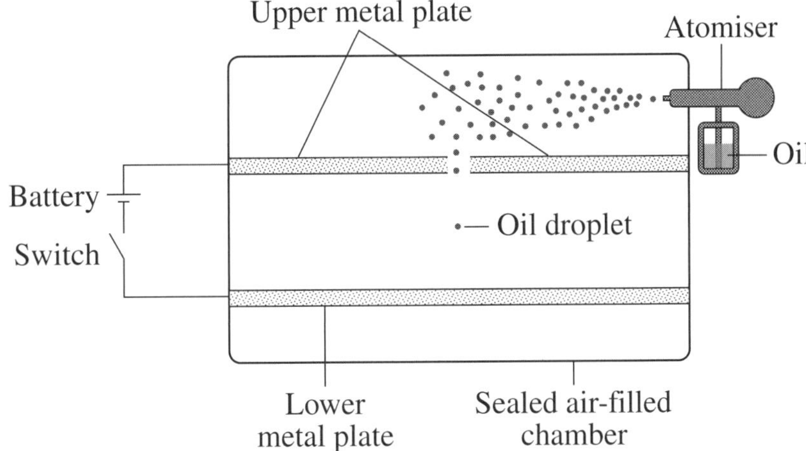

While the switch was open, the oil droplet moved downwards at a constant speed. After the switch was closed, the oil droplet moved upwards at the same constant speed.

Assume that the only three forces that may act on the oil droplet are the force of gravity, the force due to the electric field and the frictional force between the air and the oil droplet. The magnitudes of these forces are F_G (due to gravity), F_E (due to the electric field) and F_F (due to the frictional force).

Which row of the table shows all the forces affecting the motion of the oil droplet in the direction indicated, and the relationship between these forces?

	Downwards motion	*Upwards motion*
A.	$F_G > F_F$	$F_E > F_F$
B.	$F_G > F_F$	$F_E > F_G + F_F$
C.	$F_G = F_F$	$F_G = F_E$
D.	$F_G = F_F$	$F_E = F_G + F_F$

19 An AC generator is operated by turning a handle, which rotates a coil in a magnetic field.

The handle is turned at a constant speed and the AC voltage output of the generator causes a light globe connected to it to light up, as shown in Circuit 1.

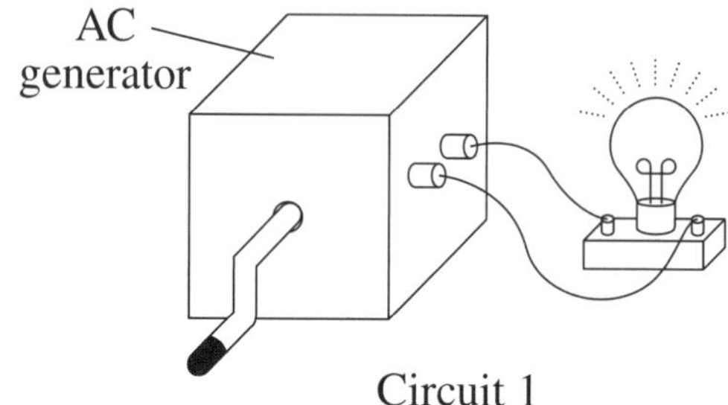

Circuit 1

A second identical light globe is then connected in series to the generator output, as shown in Circuit 2. The handle is turned at the same constant speed.

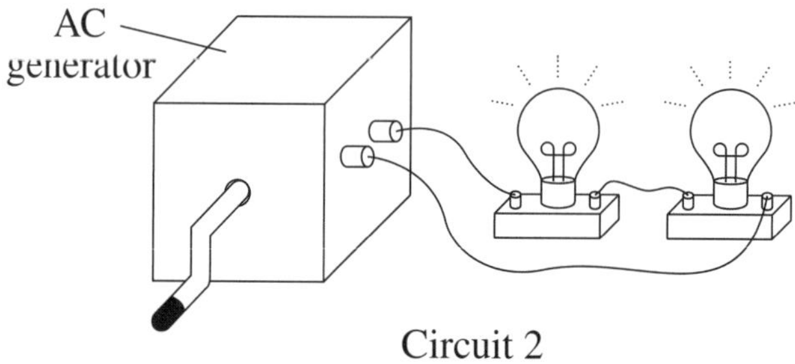

Circuit 2

Which statement describes and explains the effort required to turn the handle in Circuit 2, compared to Circuit 1?

A. The handle in Circuit 2 is easier to turn because the smaller current in Circuit 2 produces less opposing torque.

B. The handle in Circuit 2 is easier to turn because the voltage output is shared equally across the two identical light globes.

C. The handle in Circuit 2 is more difficult to turn because the larger current in Circuit 2 produces more opposing torque.

D. The handle in Circuit 2 is more difficult to turn because it takes more power to operate the two identical globes than it does to operate the single globe.

20 In a thought experiment, a car is travelling at a uniform velocity of $0.4c$. The diagram shows one of the car's wheels as it rolls past a stationary observer at X.

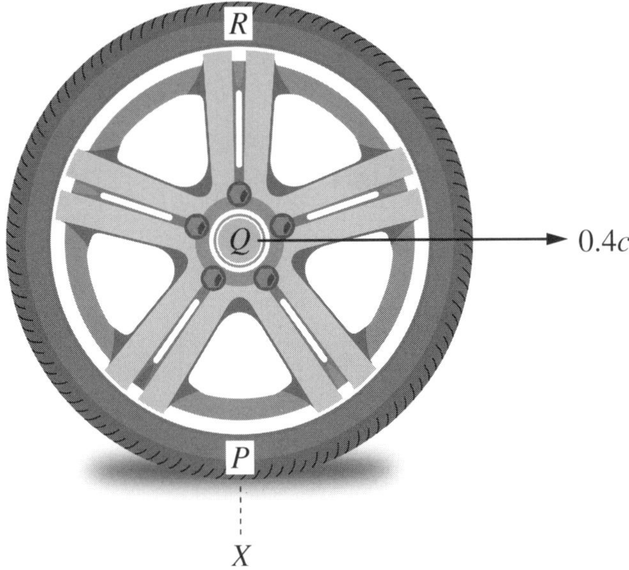

Consider the instantaneous velocity of different points on the car's wheel relative to the ground. Assume that there is no slippage of the tyre on the road.

At the instant the centre of the wheel, Q, passes X, how would the observer describe the relativistic length contraction at points P, Q and R?

A. It is the same at P, Q and R.

B. It is zero at P and greatest at R.

C. It is equal at P and R, and least at Q.

D. It is zero at P and the same value at Q and R.

2022 HIGHER SCHOOL CERTIFICATE EXAMINATION

Centre Number

Student Number

Physics

Section II Answer Booklet

80 marks
Attempt Questions 21–35
Allow about 2 hours and 25 minutes for this section

Instructions

- Write your Centre Number and Student Number at the top of this page.
- Answer the questions in the spaces provided. These spaces provide guidance for the expected length of response.
- Show all relevant working in questions involving calculations.

Please turn over

Question 21 (4 marks)

The positions of two stars, *X* and *Y*, are shown in the Hertzsprung–Russell diagram.

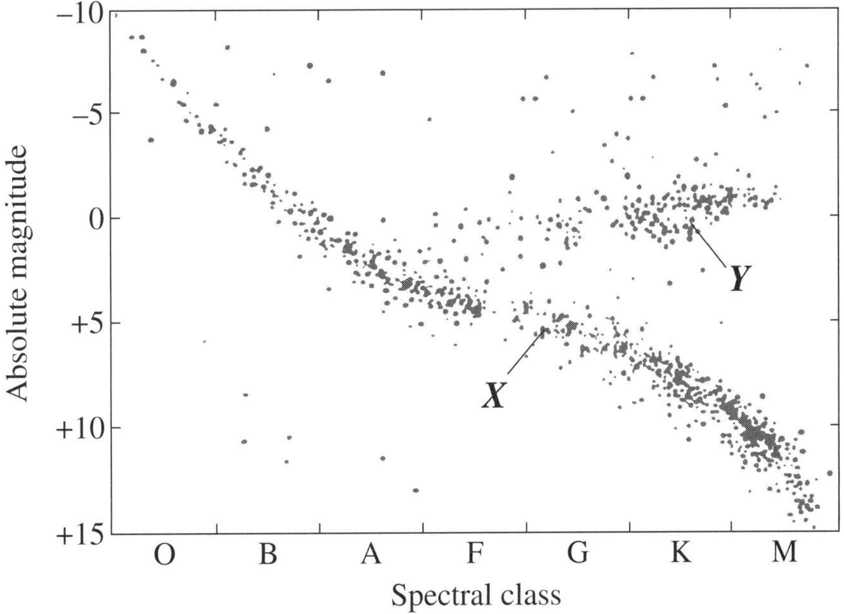

(a) Compare qualitatively the surface temperature and luminosity of *X* and *Y*. **2**

Surface temperature: ....

....

Luminosity: ....

....

(b) Identify the elements undergoing fusion in the core of each star, *X* and *Y*. **2**

....

....

....

....

Question 22 (4 marks)

The diagram shows features of a transformer. **4**

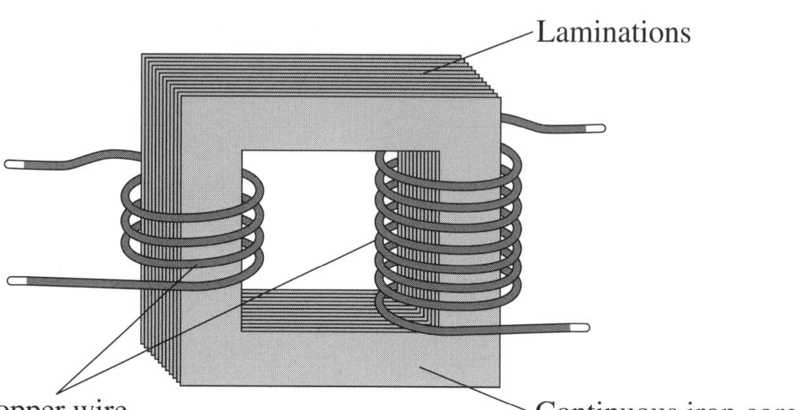

For TWO features of the transformer, describe how each contributes to the transformer's efficiency.

....

....

....

....

....

....

....

....

Question 23 (4 marks)

Outline a method that could be used to determine a value for the speed of light. In your answer, identify ONE factor that would limit the accuracy of the experimental data. **4**

Question 24 (4 marks)

The radioactive decay curve for americium-242 is shown.

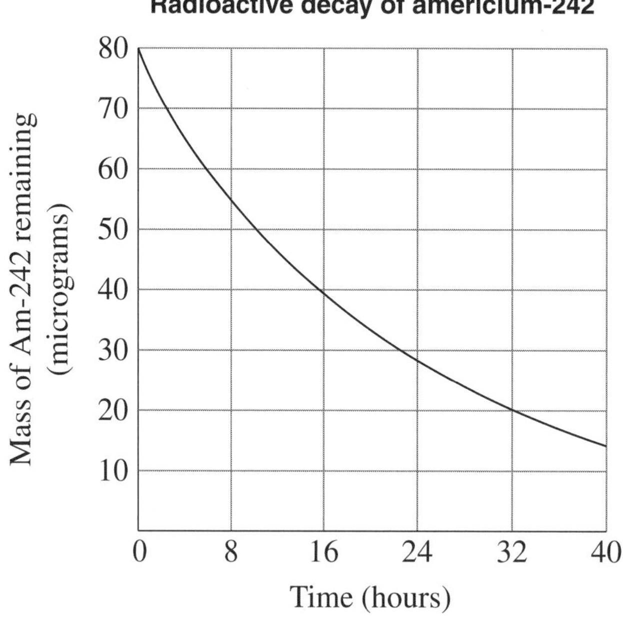

(a) Use the graph to find the half-life of Am-242 and hence show that the decay constant, λ, is 0.043 h^{-1}. **2**

....

....

....

....

(b) Calculate how long it takes until the mass of Am-242 is 8 micrograms. **2**

....

....

....

....

Question 25 (5 marks)

A rocket is launched vertically from a planet of mass M. After it leaves the atmosphere, the rocket's engine is turned off and it continues to move away from the planet. From this time the rocket's mass is 200 kg. The rocket's speed, v, at two different distances from the planet's centre, R, is shown.

Point	R (m)	v (m s^{-1})
1	4.3×10^6	5500
2	2.5×10^7	2900

(a) Show that the magnitude of the change in kinetic energy from point 1 to point 2 is 2.2×10^9 J. **2**

....

....

....

....

(b) Determine the mass M of the planet using the law of conservation of energy. **3**

....

....

....

....

....

....

Question 26 (6 marks)

Light of frequency 7.5×10^{14} Hz is incident on a calcium metal sheet which has a work function of 2.9 eV. Photoelectrons are emitted.

The metal is in a uniform electric field of 5.2 NC^{-1}, perpendicular to the surface of the metal, as shown.

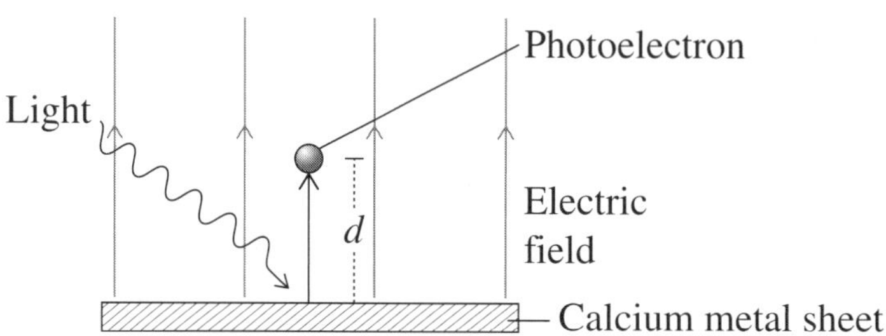

(a) Show that the maximum kinetic energy of an emitted photoelectron is 3.2×10^{-20} J. **3**

....

....

....

....

....

....

(b) Calculate the maximum distance, d, an emitted photoelectron can travel from the surface of the metal. **3**

....

....

....

....

....

....

Question 27 (7 marks)

A laser producing red light of wavelength 655 nm is directed onto double slits separated by a distance, $d = 5.0 \times 10^{-5}$ m. A screen is placed behind the double slits.

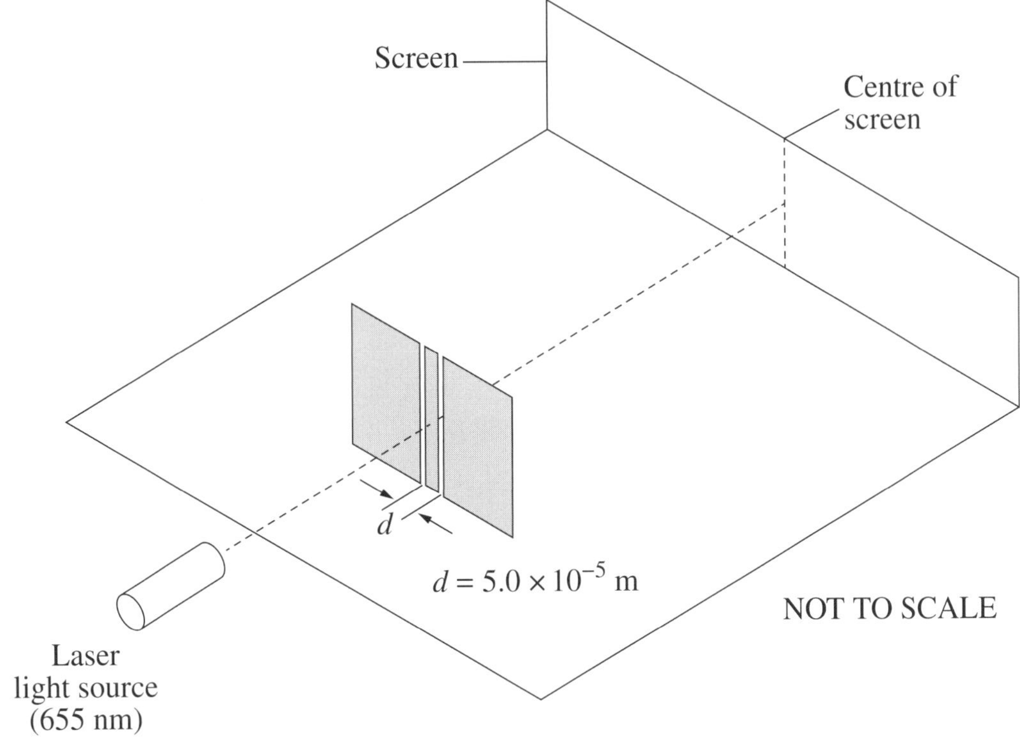

(a) Newton proposed a model of light. Use a labelled sketch to show the pattern on the screen that would be expected from Newton's proposed model. **2**

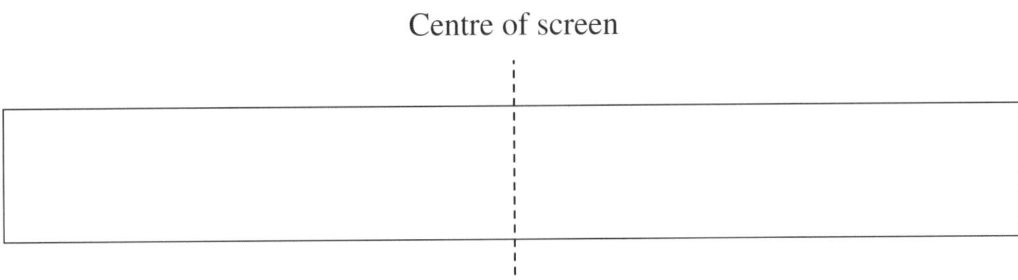

Question 27 continues on the following page

Question 27 (continued)

When the laser light is turned on, a series of vertical bright lines are seen on the screen.

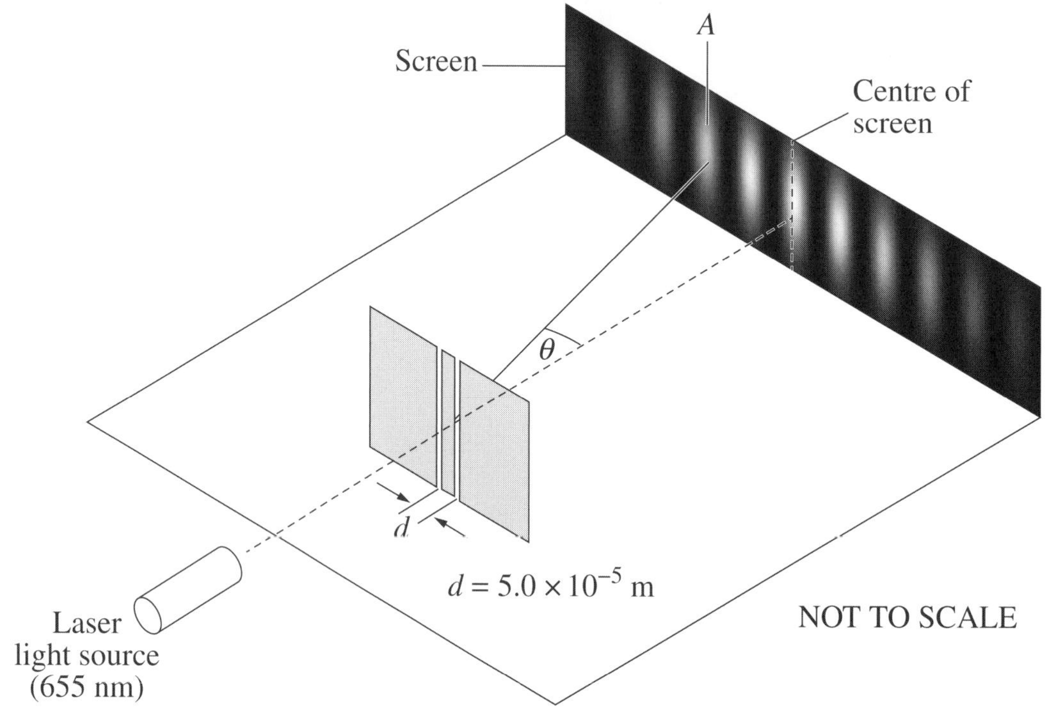

(b) Calculate the angle, θ, between the centre line and the bright line at A. **3**

....

....

....

....

....

....

(c) The laser is replaced with one producing green light of wavelength 520 nm. **2**

Explain the difference in the pattern that would be produced.

....

....

....

....

End of Question 27

Question 28 (3 marks)

Two steps in the CNO cycle of nuclear fusion are shown. 3

Step *X*	$^{13}N \rightarrow {}^{13}C + e^{+} + \nu$
Step *Y*	$^{13}C + {}^{1}H \rightarrow {}^{14}N$

Step *X* releases 1.20 MeV.

The masses in Step *Y* are shown in the table.

Isotope	*Mass* (*u*)
Carbon-13	13.003
Proton	1.007
Nitrogen-14	14.003

Propose a reason why Step *Y* releases more energy than Step *X*. Support your answer with calculations.

....

....

....

....

....

....

Question 29 (4 marks)

An apple was thrown horizontally to the east from the window of a car which was moving with a uniform velocity to the north. **4**

Explain the horizontal and vertical components of the apple's motion during its flight.

....

....

....

....

....

....

....

....

Please turn over

Question 30 (6 marks)

In a thought experiment, light travels from *X* to a mirror *Y* and back to *X* on a moving train carriage. The path of the light relative to an observer on the train is shown.

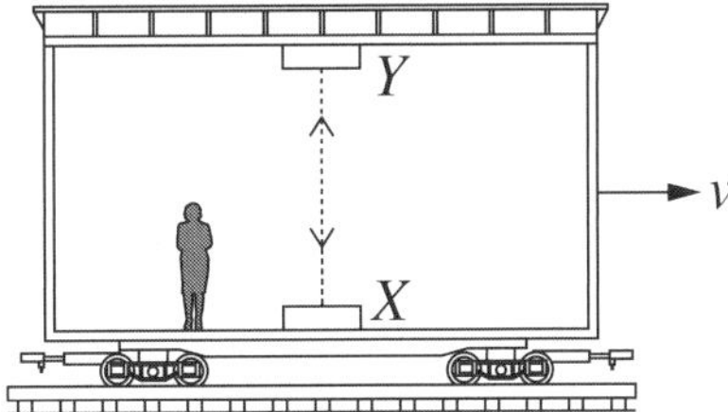

Relative to an observer outside the train, the path of the light is shown below, at three consecutive times as the train carriage moves along the track.

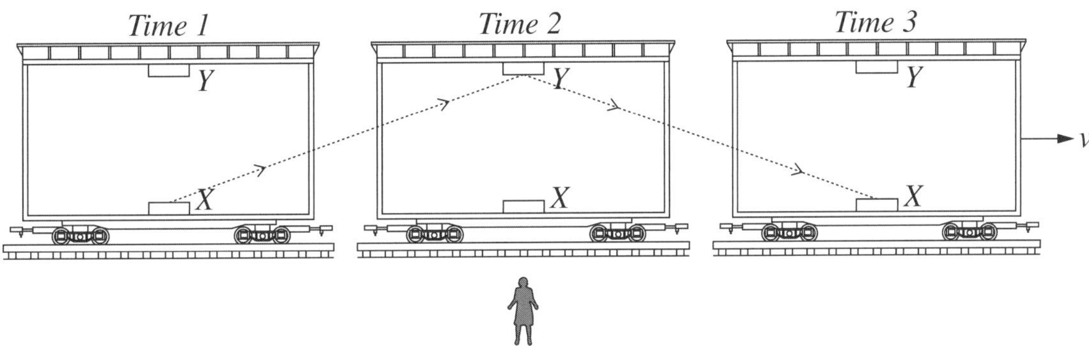

(a) Describe qualitatively how the constancy of the speed of light and the thought experiment above led Einstein to predict time dilation. **3**

....

....

....

....

....

....

Question 30 continues on the following page

Question 30 (continued)

(b) The train is travelling with a velocity $v = 0.96c$. To the observer inside the train, the return journey for the light between X and Y takes 15 nanoseconds. **3**

How long would this return journey take according to the observer outside the train?

....

....

....

....

....

....

End of Question 30

Please turn over

Question 31 (9 marks)

Following the Geiger–Marsden experiment, Rutherford proposed a model of the atom. **9**

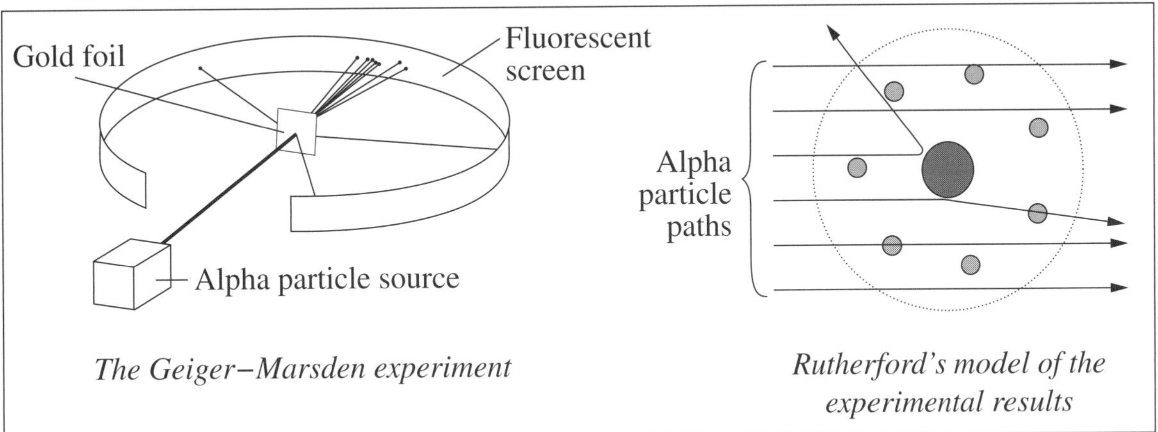

The Geiger–Marsden experiment

Rutherford's model of the experimental results

Bohr modified this model to explain the spectrum of hydrogen observed in experiments.

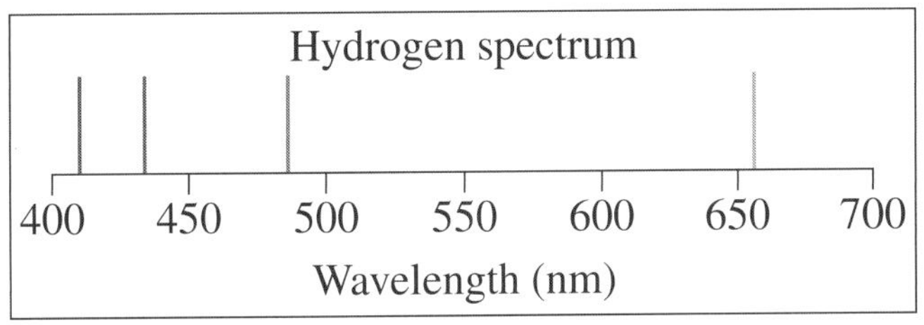

The Balmer series

The Bohr–Rutherford model of the atom consists of electrons in energy levels around a positive nucleus.

How do features of this model account for all the experimental evidence above? Support your answer with a sample calculation and a diagram, and refer to energy, forces and photons.

....

....

....

....

....

....

....

....

....

Question 31 continues on the following page

Question 31 (continued)

End of Question 31

Question 32 (6 marks)

One type of stationary exercise bike uses a pair of strong, movable magnets placed on opposite sides of a thick, aluminium flywheel to provide a torque to make it harder to pedal.

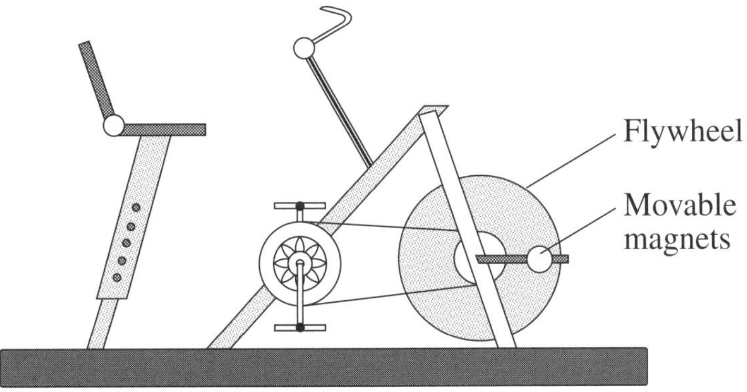

(a) Explain the principle by which these magnets make it harder to pedal. **3**

....

....

....

....

....

....

(b) The bike rider wants to increase the opposing torque on the flywheel. Justify an adjustment that could be made to the magnets to achieve this. **3**

....

....

....

....

....

....

Question 33 (6 marks)

In a hammer throw sport event, a 7.0 kg projectile rotates in a circle of radius 1.6 m, with a period of 0.50 s. It is released at point *P*, which is 1.2 m above the ground, where its velocity is at 45° to the horizontal.

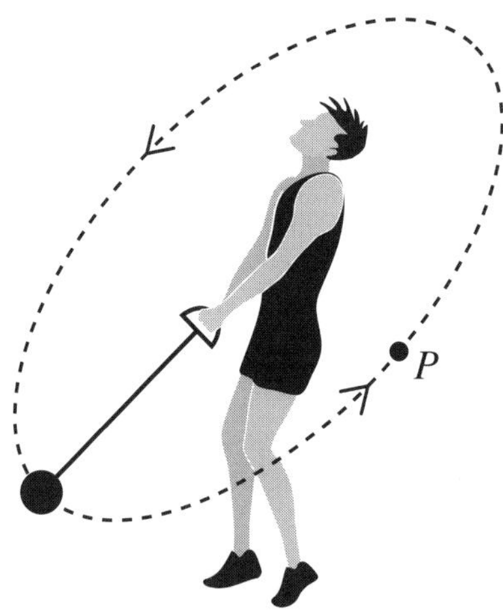

(a) Show that the vertical component of the projectile's velocity at *P* is 14.2 m s^{-1}. **2**

....

....

....

....

(b) Calculate the horizontal range of the projectile from point *P*. **4**

....

....

....

....

....

....

....

....

....

....

Question 34 (7 marks)

Three charged particles, *X*, *Y* and *Z*, travelling along straight, parallel trajectories at the same speed, enter a region in which there is a uniform magnetic field which causes them to follow the paths shown. Assume that the particles do not exert any significant force on each other. **7**

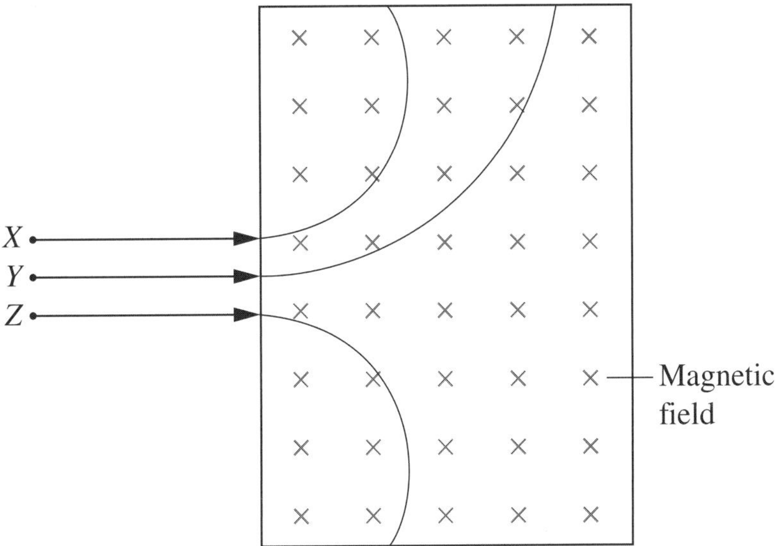

Explain the different paths that the particles follow through the magnetic field.

....

....

....

....

....

....

....

....

....

....

....

....

....

....

....

....

Question 35 (5 marks)

A capsule travels around the International Space Station (ISS) in a circular path of radius 200 m as shown. **5**

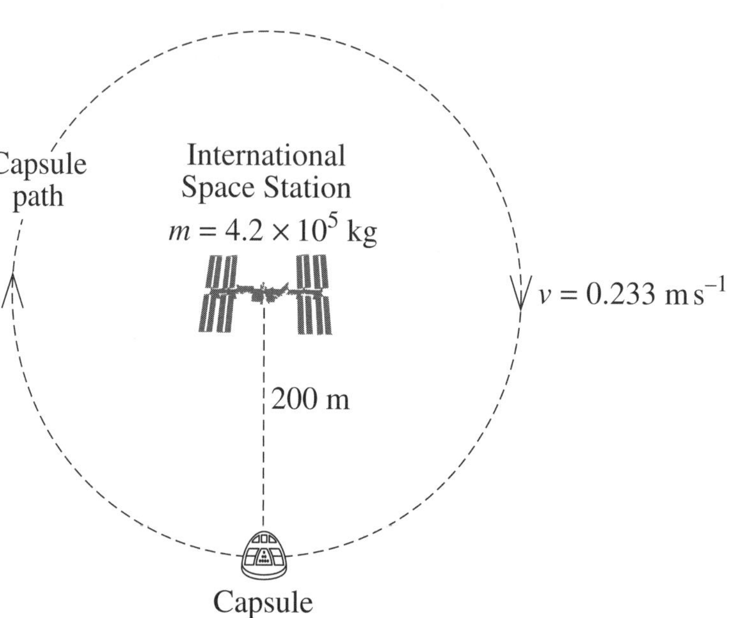

Analyse this system to test the hypothesis below.

The uniform circular motion of the capsule around the ISS can be accounted for in terms of the gravitational force between the capsule and the ISS.

....

....

....

....

....

....

....

....

....

....

End of paper

2022 HSC Examination paper

Sample answers

Section I (Total 20 marks)

1 C The ratio of turns on the primary and secondary coils of a transformer is equal to the ratio of the voltages on the coils.

Now, as $\frac{V_P}{V_S} = \frac{N_P}{N_S}$ we can write $N_S = \frac{N_P V_S}{V_P} = \frac{20 \times 400}{100} = 80V$.

2 D The continuous black body radiation emitted from a star passes through the outer layers of the star where specific wavelengths are absorbed by the atoms and molecules in the outer layers of the star. The absorption spectra produced is, therefore, related to the chemical composition of the star.

3 D The electric field lines indicate the direction of the force that a positively charged particle would experience in the field. Because the radioisotope experiences a force in the opposite direction to the electric field lines, it must be a negatively charged particle. Alpha particles and positrons are positively charged and gamma radiation has no charge; hence the radiation must be beta negative (i.e. electrons).

4 C By the right-hand palm rule, the force on the wire must be into the page.

5 B A neutron has no net charge. The only combination of quarks from the choices given that would result in a zero net charge is 1 up and 2 down.

That is $+\frac{2}{3} + 2(-\frac{1}{3}) = 0$.

6 A The sum of the potential (U) and kinetic (K) energy of a planet in orbit is a constant. At the position P the planet is closest to the star and hence the potential energy is a minimum (maximum negative value, as $U = -\frac{GMm}{r}$) and therefore the planet must have its maximum kinetic energy at this point. At point Q the planet is furthest from the star, the potential energy is maximum and hence the kinetic energy of the planet must be least at this point.

7 C The energy of a photon is related to its frequency by $E = hf$ and, hence,

$f = \frac{E}{h} = \frac{9.0 \times 10^{-24}}{6.626 \times 10^{-34}} = 1.36 \times 10^{10}\,Hz$.

8 A Because the acceleration due to gravity is directed downwards, the vertical component of the velocity decreases as the vertical displacement increases. The vertical velocity is given by $v^2 = u^2 + 2as$ and as the acceleration is negative the quantity $2as$ is also negative. As the initial velocity (u) is constant, the vertical velocity (v) will decrease as the vertical displacement (s) increases.

9 C The relationship between the peak emission wavelength and the temperature of a black body is given by Wien's displacement law, $\lambda_{max} = \frac{b}{T}$.

Hence $T = \frac{b}{\lambda_{max}} = \frac{2.898 \times 10^{-3}}{5.8 \times 10^{-7}} = 4997$ K.

10 D By squaring both sides of the equation we obtain $v^2 = (GM)(\frac{1}{r})$. As the quantity (GM) is a constant, a plot of v^2 against $(\frac{1}{r})$ will be a straight line that passes through the origin with gradient equal to *GM*.

11 B At $t = 0$ s, the launch velocity must be positive; at $t = 4$ s, the velocity just before landing must be negative; and at $t = 2$ s, at the maximum height the velocity must be zero. Only graph B meets these requirements.

12 D As the particle moves with a constant velocity, the net force on the particle must be zero. For this to occur the downwards force produced by the electric field on the positive particle must be balanced by an upwards magnetic force of the same magnitude. Using the right-hand palm rule we see that the magnetic field would produce an upwards force on the particle if the particle was moving towards the right.

13 B As the satellites have the same orbital radius, they must be travelling with the same velocity $\left(v = \sqrt{\frac{GM}{r}}\right)$. Now, as momentum is given by $p = mv$ and the planets are travelling at the same velocity, the satellite with mass $2m$ will have twice the momentum of the satellite of mass m. The direction of the momentum will also be different for each satellite.

14 B The work function of a metal is the minimum energy required by an incident photon to remove a photoelectron from the metal's surface. As the energy of a photon is proportional to the frequency of the photon ($E = hf$), the work function is also proportional to the minimum frequency of the incident photon that will eject a photoelectron from the surface.

Because a higher frequency is required to release photoelectrons in experiment *Y*, the metal used for experiment *Y* must have had a greater work function than the metal used for experiment *X*.

15 A The force between two parallel current-carrying wires is given by $F = \frac{\mu_0}{2\pi}\frac{I_1 I_2 l}{r}$.

Hence increasing the separation (r) between the wires by a factor of 4 would decrease the force by the same factor (i.e. $\frac{F}{4}$). As the force is proportional to the product of the currents, doubling the current in each wire would increase the force between the wires by a factor of 4. Thus the net effect of the combined changes would be that the force between the parallel wires would not change.

16 A The binding energy of a nucleus is the energy required to separate the nucleus into its constituent nucleons (i.e. into individual protons and neutrons). Helium-4 has the greater binding energy and hence requires the greatest energy to separate it into its constituent nucleons.

The mass of a proton is only slightly greater than the mass of a neutron. Now, as He-4 has 4 nucleons (2 protons and 2 neutrons) and Be-6 has 6 nucleons (4 protons and 2 neutrons), Be-6 will be the heavier nucleus.

17 C In the initial position, the polarisation axis of both the initial polarising filter and the *Y* filter of the glasses are in the same direction and hence $I_y = I_x$. Now, I_y can never equal I_0 because polarising unpolarised light decreases the intensity by half. At several times during the rotation (45°, 135°, 225° and 315°) of the polarising filter, the angle between the filter and the polarising axis of both lenses will be the same (45°) and at these times the intensity of $I_y = I_x$. None of the other options are possible.

18 D Because the droplets move down with a constant velocity, there is no net force on the droplets and hence the forces acting on the droplets must sum to zero. Therefore when the droplets are moving downwards $F_G = F_F$.

Again, because the droplets moved upwards at a constant velocity the net force on the droplets must be zero. As the velocity is the same in the upwards direction as the downwards direction, the force of friction on the droplets must be the same as when the droplet was travelling downwards. The electric force acts upwards while gravity and friction act downwards (when the droplet is moving upwards). To ensure the net force on the drop is zero, the upwards force must balance the downwards forces, which we can write as $F_E = F_G + F_F$.

19 A The voltage produced by an electric generator is determined by its rate of rotation (from Faraday's law of induction the rate of change of flux is proportional to the EMF induced). Therefore the generator voltage produced will be equal in both experiments because the handle is turned at a constant rate.

Adding an identical light globe in series will increase the total resistance (it would double it if the globes operated at the same temperature in both experiments but, as the voltage is shared, they will operate at a slightly lower temperature so it may not increase the resistance this much) and hence decrease the current drawn from the generator. The opposing torque is proportional to the current drawn and therefore the handle will be easier to turn when the second light globe is added.

20 B With respect to the centre of the wheel, point *R* is moving with velocity 0.4*c* to the right and point *P* is moving 0.4*c* to the left. Because the centre of the wheel is moving at 0.4*c* to the right with respect to a stationary observer, the observer will see point *R* moving at 0.8*c* to the right, while point *P* will be at rest. As length contraction is related to the relative velocity of the points to a stationary observer, the observer will see no length contraction for *P* and a maximum length contraction for the point *R*.

Section II

Question 21 (Total 4 marks)

(a) Surface temperature: The spectral class of a star is related to the surface temperature of a star. As star *Y* has a K spectral class and star *X* has a G spectral class, star *X* will have a higher surface temperature than star *Y*.

Luminosity: Absolute magnitude is a measure of the luminosity of a star. The smaller the absolute magnitude, the more luminous the star. As star *Y* has a smaller absolute magnitude than star *X*, star *Y* must have a greater luminosity than star *X*. (*2 marks*)

(b) Star *X* is a main sequence star and hence will be fusing hydrogen to helium in its core. Star *Y* is a red giant which has finished fusing hydrogen to helium in its core. Star *Y* would be fusing hydrogen to helium in a shell around the core and fusing helium to carbon in the core. (*2 marks*)

Question 22 (Total 4 marks)

The efficiency of the transformer is improved by using a laminated iron core. The AC current in the input coil produces a changing magnetic flux in the iron core, which induces eddy currents in the iron core. These eddy currents heat the core via resistance heating and reduce the efficiency of the transformer. By splitting the iron core into thin sheets (laminates) and insulating the sheets from one another, eddy currents cannot flow between the sheets. This reduces the eddy currents in the core and significantly increases the efficiency of the transformer.

Using a continuous iron core also improves the efficiency of a transformer. A continuous iron core minimises magnetic flux leakage and hence improves the flux linkage between the primary and secondary coils.

OR

Students could explain how the use of a soft iron in the core minimises losses caused by magnetising and demagnetising the core each half cycle (hysteresis losses) or how using thick copper wires minimises resistance heating losses in the coils. (*4 marks*)

Question 23 (Total 4 marks)

We could use an experiment like the one performed by Fizeau in 1851 and illustrated below. Light is passed from an intense light source through a gap in a toothed wheel that is spun at high velocity. The light passes through one of the gaps in the wheel to a mirror several kilometres away and is reflected on the same path to the toothed wheel. The speed of the wheel is increased until the returning light passes through the next gap in the wheel when it has rotated to the initial position of the first gap. By knowing the distance to the mirror and the rate of rotation of the toothed wheel, the speed of light can be calculated. One of the problems that limits the accuracy of this experiment is measuring the exact distance from the toothed wheel to the mirror.

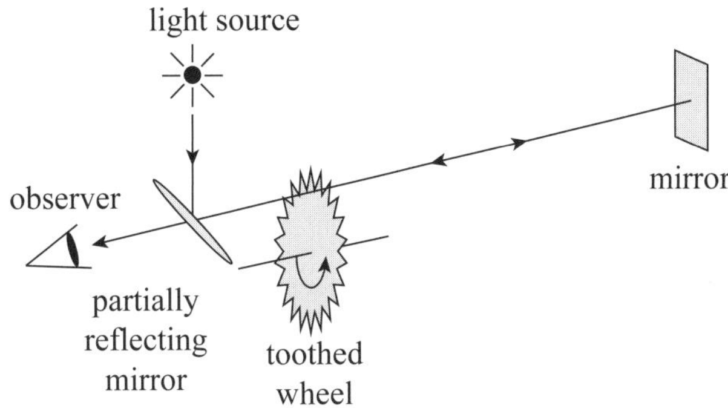

Students could outline any of several other experiments, such as the astronomical method used by Romer or the experiment conducted by Michelson. *(4 marks)*

Question 24 (Total 4 marks)

(a) Reading from the graph the time for half of the sample to undergo radioactive decay is $t_{\frac{1}{2}} = 16$ h.

The decay constant is given by $\lambda = \frac{ln2}{t_{\frac{1}{2}}} = \frac{ln2}{16} = 0.043\ \text{h}^{-1}$. *(2 marks)*

(b) Taking the natural log of $N_t = N_0 e^{-\lambda t}$ gives $ln\left(\frac{N_1}{N_0}\right) = -\lambda t$ and, hence,

$$t = \left(\frac{1}{\lambda}\right)\left(ln\frac{N_0}{N_t}\right) = \left(\frac{1}{0.043}\right)ln\left(\frac{80}{8}\right) = 53.55\ \text{h}.$$ *(2 marks)*

Question 25 (Total 5 marks)

(a) The change in kinetic energy will be given by

$$\Delta K = \frac{1}{2}mv_f^2 - \frac{1}{2}mv_i^2 = \frac{1}{2}(200)(2900)^2 - \frac{1}{2}(200)(5500)^2 = 2.18 \times 10^9\ \text{J}.$$

The negative sign tells us the kinetic energy of the rocket (to two significant figures) has decreased by 2.2×10^9 J as expected. *(2 marks)*

(b) When the rocket is in free flight, the total energy of the rocket remains constant. Therefore if the kinetic energy has decreased by 2.2×10^9 J, the gravitational potential energy of the rocket must have increased by the same amount.

The change in potential energy can be calculated using

$$\Delta U = -GMm\left(\frac{1}{r_f} - \frac{1}{r_i}\right) = GMm\left(\frac{1}{r_i} - \frac{1}{r_f}\right).$$

Hence $$M = \frac{\Delta U}{Gm\left(\frac{1}{r_i} - \frac{1}{r_f}\right)} = \frac{(2.18 \times 10^9)}{(6.67 \times 10^{-11})(200)\left(\frac{1}{4.3 \times 10^6} - \frac{1}{2.5 \times 10^7}\right)}$$

$$= 8.50 \times 10^{23}\ \text{kg}.$$ *(3 marks)*

Question 26 (Total 6 marks)

(a) The maximum kinetic energy will be given by

$K_{max} = hf - \phi = (6.626 \times 10^{-34})(7.5 \times 10^{14}) - (2.9 \times 1.602 \times 10^{-19}) = 3.24 \times 10^{-20}$ J

(3 marks)

(b) The work done by the field against the motion of the photoelectron must be equal to the initial kinetic energy of the photoelectron.

$K_{max} = W = qEd$, and hence, $d = \dfrac{K_{max}}{qE} = \dfrac{(3.24 \times 10^{-20})}{(1.602 \times 10^{-19})(5.2)} = 0.039$ m.

(3 marks)

Question 27 (Total 7 marks)

(a) This diagram shows the intensity of light that would reach the screen according to Newton's corpuscular model of light.

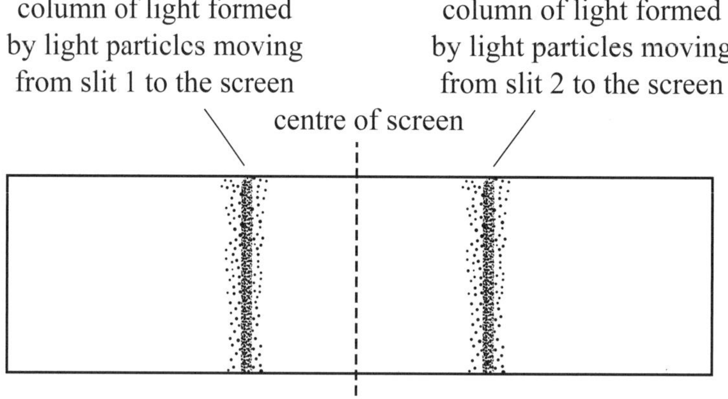

(2 marks)

(b) The angle between interference maxima and the central axis is given by

$$d\sin\theta = m\lambda.$$

Now, as A is the second interference maximum from the centre of the screen, $m = 2$.

Hence $= \theta = \sin^{-1}\left(\dfrac{m\lambda}{d}\right) = \sin^{-1}\left(\dfrac{2(655 \times 10^{-9})}{5.0 \times 10^{-5}}\right) = 1.5^0$. *(3 marks)*

(c) The central maxima would remain in the same position but the distance between the maxima on the screen would be reduced, compressing the pattern inwards. Using light with a shorter wavelength will also change the colour of the pattern from red to green.

Students may also support their answer quantitatively by using the equation in part (b) to show the new angle to the second order of interference will be smaller:

$\theta = \sin^{-1}\left(\dfrac{m\lambda}{d}\right) = \sin^{-1}\left(\dfrac{2(520 \times 10^{-9})}{5.0 \times 10^{-5}}\right) = 1.2^0$.

(2 marks)

Question 28 (Total 3 marks)

The beta plus decay (reaction *X*) will release less energy than the fusion reaction (reaction *Y*) because more mass is converted into energy in reaction *Y* than in reaction *X*.

The mass converted to energy in reaction *Y* is given by

$$\Delta m = \text{mass of products} - \text{mass of reactants}$$
$$= 14.003 - (13.003 + 1.007) = -0.007\text{u}$$

The negative sign indicates that mass is lost (i.e. converted into energy) in this reaction. This mass defect corresponds to an energy of $E = 0.007 \times 931.5 = 6.52$ MeV.

The mass defect for reaction *X* is given by $\Delta m = \frac{1.2}{931.5} = 0.0013\text{u}$.

As expected, less mass is converted into energy in reaction *X* (0.0013u) than in reaction *Y* (0.007u) and hence more energy will be released in reaction *Y* (6.52 MeV) than in reaction *X* (1.2 MeV). *(3 marks)*

Question 29 (Total 4 marks)

The initial vertical velocity is zero. The apple will fall with the acceleration due to gravity (g) and hence the vertical velocity (v_y) will increase in the downwards direction as time (t) passes according to $v_y = gt$ (from $v = u + at$). This will result in the vertical component of the velocity increasing in a downwards direction until the apple reaches the ground. If we let the height at which the apple was released be $s_y = 0$, the vertical distance fallen (s_y) from this point as time (t) passes will be given by

$s_y = \frac{gt^2}{2}$ (from $s = ut + \frac{1}{2}at^2$).

The direction of the horizontal component of the velocity (v_x) of the apple could be determined by vectorially adding the northern velocity of the car to the eastern horizontal launch velocity of the apple. The vector sum of these two velocities will give the direction and magnitude of the initial horizontal velocity (u_x). The direction and magnitude of the initial horizontal component of the velocity of the apple will remain constant throughout the flight (i.e. $v_x = u_x$). The horizontal displacement (s_x) from the launch position during the flight will be given by ($s_x = u_x t$). *(4 marks)*

Question 30 (Total 6 marks)

(a) Because the speed of light is a constant for both observers, and the distance between *X* and *Y* is shorter for the observer in the train, the observer in the train will measure a shorter time for a pulse of light to travel to *Y* and to return to *X*. Therefore if the observers synchronised their clocks at time t_1 (when the pulse left *X*) and then compared their clocks at time t_3 (when the pulse returns to *X*), more time would have passed for the stationary observer than for the observer on the train. Einstein resolved this logical inconsistency by suggesting that two observers could only measure different times for the same event if time passed at a different rate for each observer. The outside observer would conclude that time for the observer on the moving train was passing more slowly than time passed for the stationary observer outside the train. Einstein concluded that time was not a universal constant but a relative quantity and that an observer would always see time running more slowly in an inertial reference frame moving with respect to the observer. He called this effect 'time dilation'. *(3 marks)*

(b) Applying Einstein's time dilation equation with the proper time (moving frame's time interval) set at $t_0 = 15$ ns

$$t = \frac{t_0}{\sqrt{\left(1 - \frac{v^2}{c^2}\right)}} = \frac{15 \times 10^{-9}}{\sqrt{1 - 0.96^2}} = 5.36 \times 10^{-8} \text{ s.}$$ *(3 marks)*

Question 31 (Total 9 marks)

In the Geiger–Marsden experiment shown, thin gold foil was bombarded by alpha particles. The vast majority of the alpha particles were observed to pass through the foil with little or zero deflection but a small fraction of the alpha particles (about 1 in 10 000) was deflected by large angles. The Bohr–Rutherford model explains this result by assuming that most of the atom was empty space, with all the positive charge and almost all the atom's mass concentrated in a tiny, central, positively charged nucleus. The model had minute negative electrons orbiting the nucleus, held in orbit by the electrostatic force of attraction between the negative electrons and positive nucleus. Almost all the positively charged alpha particles would pass straight through because the atoms were almost entirely empty space. A tiny fraction of the alpha particles, however, would come close enough to the intense electric field near the central positive nucleus to be scattered by a large angle by the electrostatic force of repulsion between positive alpha particle and the positive nucleus. This is illustrated in Rutherford's model of the experimental results shown in the question.

When hydrogen gas is excited in an electric discharge, it emits specific wavelengths of light over a broad region of the electromagnetic spectrum. The series of specific wavelengths emitted in the visible region of the electromagnetic spectrum from excited hydrogen atoms shown is called the Balmer series. The Bohr–Rutherford model explained the spectrum of hydrogen by postulating that the negatively charged electrons orbiting the nucleus in Rutherford's initial model could only occupy orbits with specific energies (i.e. the energy of the orbiting electron was quantised). Bohr proposed that electrons could move between these allowed energy levels, provided energy was conserved. In an unexcited (ground state) hydrogen atom the electron would occupy the lowest energy orbital. But when the hydrogen atom was in a gas discharge it could absorb energy by exciting the electron to a higher energy orbital. These energised electrons rapidly move (relax) to a less energetic orbital by emitting a photon of electromagnetic radiation with an energy equal to the energy difference between the orbitals. Using Einstein's relationship $E = hf = \frac{hc}{\lambda}$, Bohr showed that when an electron moved to a lower energy orbit it emitted a photon with a specific energy and wavelength (see Figure 1a). Bohr also showed that hydrogen atoms could absorb a photon if the photon energy was equal to the energy required to excite an electron from its current orbital to a higher energy orbital (see Figure 1b). Thus, the Bohr–Rutherford model could explain how the emission and absorption spectra of hydrogen was produced.

An energy level diagram for the hydrogen orbitals is shown schematically in Figure 2 below. Note that each orbital is denoted by the principal quantum number (n), where $n = 1$ is the lowest energy (ground state orbital), $n = 2$ is the first excited orbital, $n = 3$ is the next more energetic orbital, etc. When electrons are excited to different higher energy orbitals and then relax to a specific lower energy orbital, they produce a series of spectral lines. Electrons that move to the $n = 2$ orbital from higher energy levels produce the Balmer series of lines shown in the question. Bohr was able to use his model to explain and derive the Rydberg equation which had previously been shown to predict the position of the wavelengths in a hydrogen spectrum.

Bohr explained the Rydberg equation $\frac{1}{\lambda} = \left(R\frac{1}{n_f^2} - \frac{1}{n_i^2}\right)$

by saying that n_f corresponded to the principal quantum number of the final electron energy level and n_i corresponded to the initial energy level. The Balmer series of wavelengths can be calculated by setting $n_f = 2$.

For example, to calculate longest wavelength (least energetic) photon that would be emitted in the Balmer series, we set $n_i = 3$ and $n_f = 2$, and see that

$\frac{1}{\lambda} = \left(R\frac{1}{n_f^2} - \frac{1}{n_i^2}\right) = 1.097 \times 10^7\left(\frac{1}{2^2} - \frac{1}{3^2}\right) = 1\,523\,611\ \text{m}^{-1}$ and, hence, $\lambda = 656$ nm.

Note that this corresponds to the longest wavelength in the Balmer series shown in the emission spectrum diagram in the question. This photon would have an energy given by

$$E = \frac{hc}{\lambda} = \frac{(6.626 \times 10^{-34}) \times (3 \times 10^8)}{656 \times 10^{-9}} = 3.03 \times 10^{-19}\ \text{J}.$$

Note that this is the energy difference between the $n = 2$ and $n = 3$ energy levels.

Figure 1 Emission and absorption of a photon

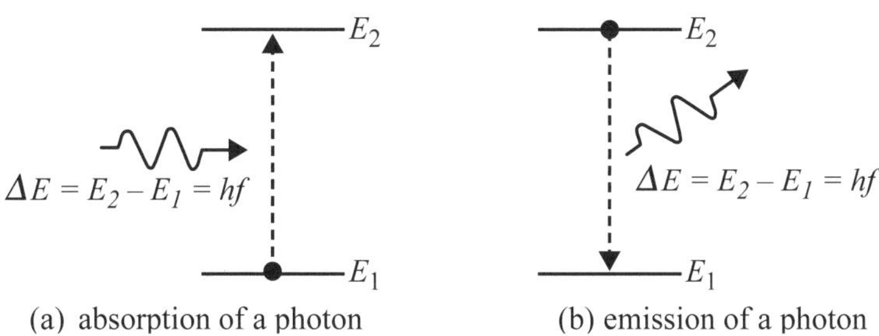

(a) absorption of a photon (b) emission of a photon

Figure 2 Bohr orbitals

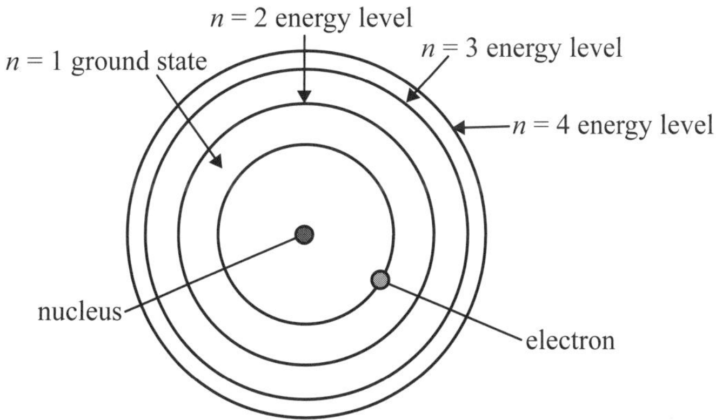

Four lowest energy levels of Bohr model of the hydrogen atom

(9 marks)

Question 32 (Total 6 marks)

(a) Because the aluminium wheel is spinning in the magnetic field produced by the magnets, it will experience a changing magnetic field and, by Faraday's law of induction, eddy currents will be induced in the aluminium wheel. By Lenz's law, these currents will be in a direction which will produce magnetic fields that will oppose the change that produced the eddy currents. In this case the eddy currents will produce magnetic fields that will produce a force and hence a torque that will oppose the rotation of the aluminium wheel. *(3 marks)*

(b) For a fixed rate of rotation (i.e. fixed angular velocity) the linear velocity of the wheel is faster the further it is from the centre of the wheel ($v = \omega r$). Hence if the magnets are moved further from the centre of the wheel, the aluminium will pass more rapidly between the magnets, inducing more intense eddy currents, which would create stronger magnetic fields and a greater opposing force on the wheel. Moving the magnets further from the centre of the wheel will also increase the distance between the applied force and the centre of rotation. As torque is the product of the force (F) and the distance (r) it is applied from the centre of the wheel (i.e. $\tau = rF$), moving the magnets further from the centre of the wheel will increase both the magnitude of the opposing force and the distance it is applied from the centre of the wheel, resulting in a greater opposing torque. *(3 marks)*

Question 33 (Total 6 marks)

(a) The velocity of the hammer head will be given by $v = \frac{2\pi r}{T} = \frac{2\pi \times 1.6}{0.5} = 20.1\ \text{ms}^{-1}$.

As the hammer is thrown at 45° to the horizontal, the vertical component of the velocity will be given by $v_x = v\sin\theta = 20.1 \times \sin 45° = 14.2\ \text{ms}^{-1}$. *(2 marks)*

(b) Calling the quantities in the upwards direction positive, the downwards direction negative and the initial vertical height of launch $s_y = 0$, we first find the final vertical velocity using

$v^2 = u^2 + 2as = 14.2^2 + 2(-9.8)(-1.2) = 225\ \text{ms}^{-1}$.

Hence $v = -15\ \text{ms}^{-1}$. (We take the negative root because the final velocity is downwards.)

Then the time of flight can be determined using

$v = u + at$ and, hence, $t = \frac{v-u}{a} = \frac{-15 - 14.2}{-9.8} = 2.98$ s.

(Or the student may solve the quadratic equation $s = ut + \frac{1}{2}at^2$, to find t.)

Now the range can be determined using the horizontal component of the motion.

As the projectile was released at 45° to the horizontal, the initial horizontal and vertical components of the motion will be equal and, hence, $u_y = 14.2\ \text{ms}^{-1}$ and finally

$Range = u_y t = 14.2 \times 2.98 = 42.6$ m. *(4 marks)*

Question 34 (Total 7 marks)

The direction of the force that the magnetic field exerts on the charged particles when they enter the magnetic field is given by the right-hand palm rule. Applying this rule indicates that particles X and Y are positively charged as they move upwards when they enter the field and particle Z is negatively charged.

As the charges are moving perpendicularly to the magnetic field, they will experience a force given by $F = qvB$ when they are in the magnetic field. Because this force is perpendicular to the velocity of the charge, it is a centripetal force which will cause the particles to follow a circular trajectory in the field.

As the particles move at the same speed in the same magnetic field, the radius of the path taken by each charged particle is proportional to the mass-to-charge ratio of the particle. By equating the centripetal and magnetic force equations we see,

$qvB = \frac{mv^2}{r}$ or $r = \frac{mv}{qB}$, and as $\frac{v}{B}$ is a constant (k) for the particles, we can write, $r = k\frac{m}{q}$.

As particles X and Z move at the same velocity and have the same radius of curvature in the field, the magnitude of their mass-to-charge ratio must be equal. Hence particles X and Z have the opposite charge but the same magnitude mass-to-charge ratio.

Particle Y also moves at the same velocity as the other particles but has a larger radius of curvature and hence a greater absolute value for its mass-to-charge ratio than either particle X or Z. (*7 marks*)

Question 35 (Total 5 marks)

To test the hypothesis, we must see if the capsule could be a satellite moving in an orbit with uniform circular motion around the ISS with the specified radius and velocity. For the capsule to undergo uniform circular motion around the ISS, the gravitational attraction between the capsule and the ISS would have to provide the centripetal force required to keep the capsule moving in a circular orbit.

The gravitational force between the ISS and the capsule is given by

$$F = \frac{GMm}{r^2} = \frac{(6.67 \times 10^{-11} \times 4.2 \times 10^5 \times 1.2 \times 10^4)}{(200)^2} = 8.4 \times 10^{-6}\text{N}$$

The centripetal force required to keep the satellite moving in the circular orbit specified would be

$$F_c = \frac{mv^2}{r} = \frac{(1.2 \times 10^4) \times (0.233)^2}{200} = 3.26\text{N}$$

The centripetal force required to keep the capsule in orbit around the ISS at the specified distance and speed is millions of times greater than the gravitational attraction between the ISS and the capsule. Hence the hypothesis is incorrect and the motion of the capsule around the ISS cannot be accounted for in terms of the gravitational attraction between the capsule and the ISS.

The hypothesis is also incorrect because the capsule would not orbit in a circle around the ISS but would orbit around the common centre of mass of the capsule/ISS system. (*5 marks*)

CHAPTER 9

NSW GOVERNMENT

Centre Number

Student Number

NSW Education Standards Authority

2023 HIGHER SCHOOL CERTIFICATE EXAMINATION

Physics

General Instructions

- Reading time – 5 minutes
- Working time – 3 hours
- Write using black pen
- Draw diagrams using pencil
- Calculators approved by NESA may be used
- A data sheet, formulae sheet and Periodic Table are provided at the back of this paper
- Write your Centre Number and Student Number at the top of this page

Total marks: 100

Section I – 20 marks

- Attempt Questions 1–20
- Allow about 35 minutes for this section

Section II – 80 marks

- Attempt Questions 21–34
- Allow about 2 hours and 25 minutes for this section

Section I

20 marks
Attempt Questions 1–20
Allow about 35 minutes for this section

Use the multiple-choice answer sheet for Questions 1–20.

---

1 The gravitational field strength acting on a spacecraft decreases as its altitude increases.

This is due to a change in the

A. mass of Earth.

B. mass of the spacecraft.

C. density of the atmosphere.

D. distance of the spacecraft from Earth's centre.

2 Which diagram best represents the transmission of energy from a power station to people's houses?

A.

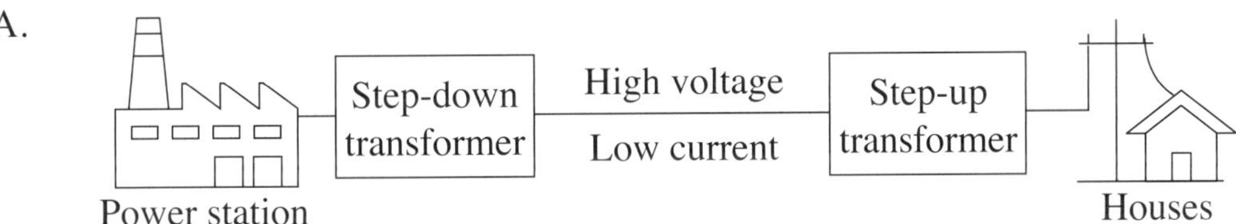

B.

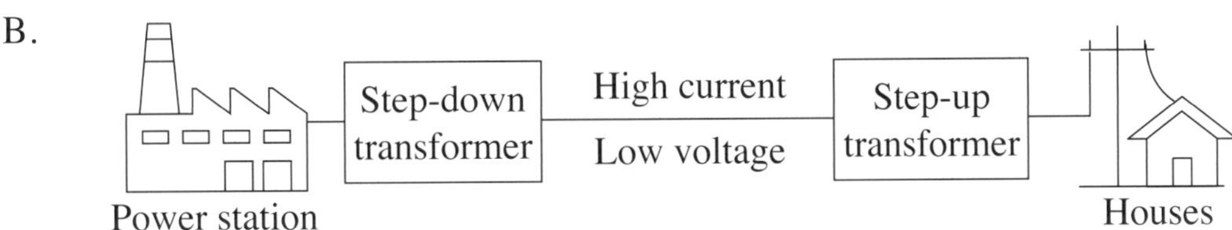

C.

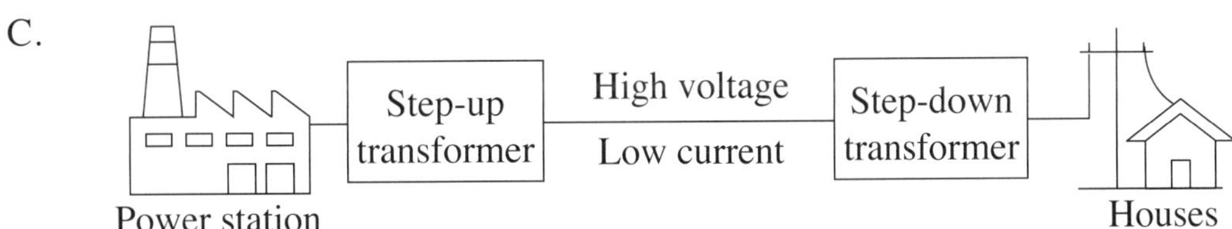

D.

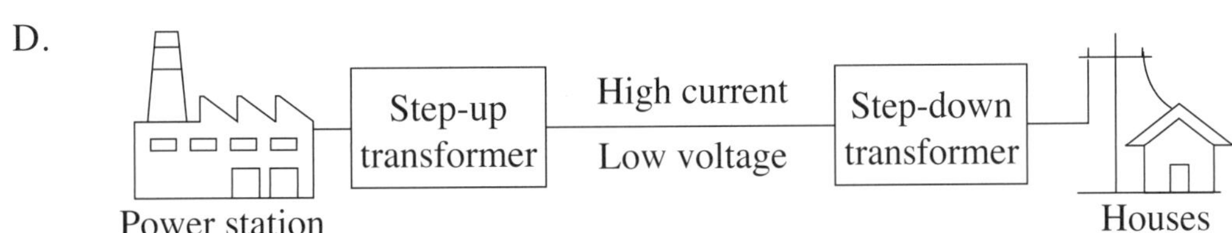

3 A diagram representing a double slit experiment using light is shown.

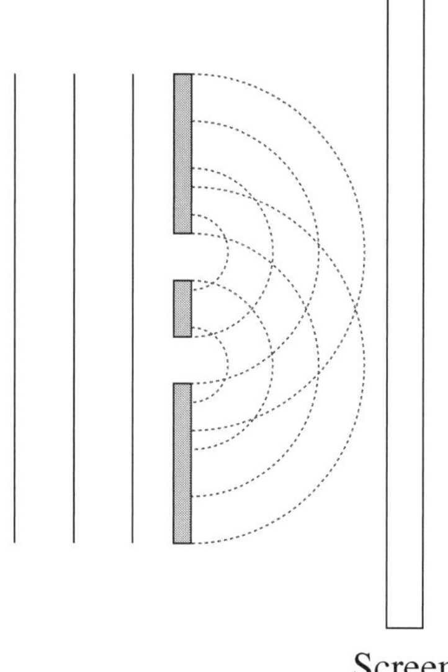

Which of the following best represents the expected pattern on the screen?

A.

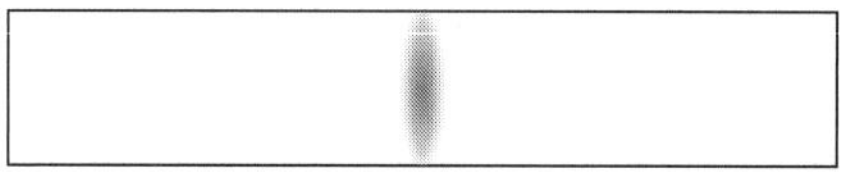

B.

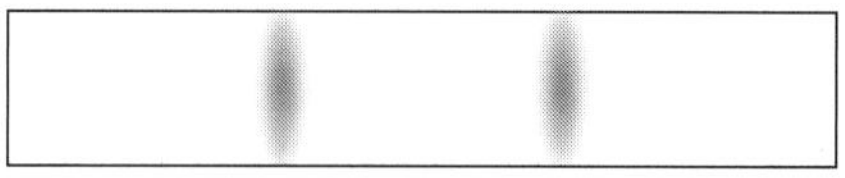

C.

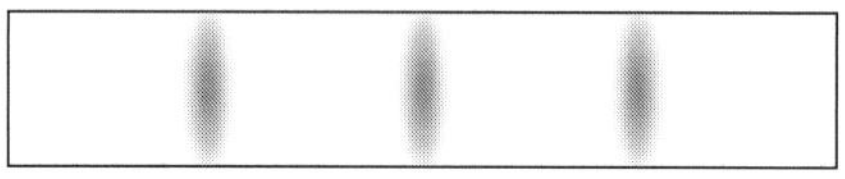

D.

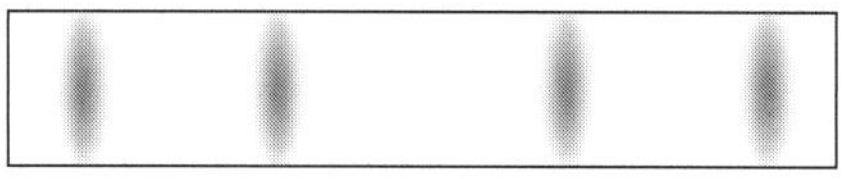

4 Caesium-137 has a half-life of 30 years.

What mass of caesium-137 will remain after 90 years, if the initial mass was 120 g?

A. 4 g

B. 15 g

C. 40 g

D. 60 g

5 An exoplanet is in an elliptical orbit, moving in the direction shown. The distances between consecutive positions P, Q, R and S are equal.

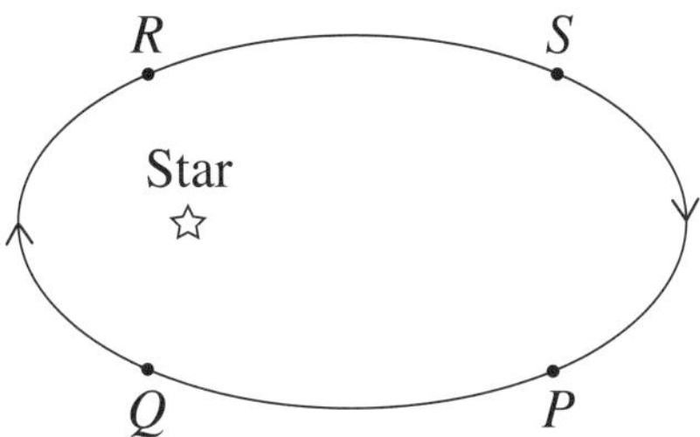

Between which two points is the exoplanet's travel time the greatest?

A. P and Q

B. Q and R

C. R and S

D. S and P

6 An electron would produce an electromagnetic wave when it is

A. stationary.

B. in a stable hydrogen atom.

C. moving at a constant velocity.

D. moving at a constant speed in a circular path.

7 A proton and a neutron travel at the same speed.

Which statement correctly explains the difference between their de Broglie wavelengths?

A. The proton has a longer wavelength because its mass is greater.

B. The proton has a shorter wavelength because its mass is smaller.

C. The neutron has a shorter wavelength because its mass is greater.

D. The neutron has a longer wavelength because its mass is smaller.

8 A ball is launched from a platform at position *A* with velocity *u*. It lands in the position shown.

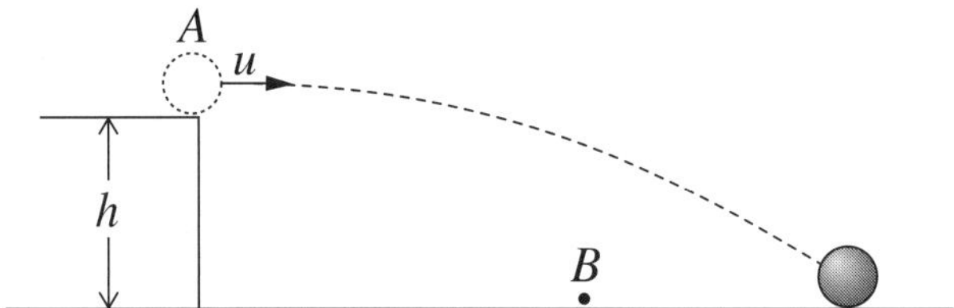

The ball could be made to land at position *B* by increasing the

A. velocity *u*.

B. launch angle.

C. mass of the ball.

D. height of the platform.

9 The graph shows the relationship between radiation intensity and wavelength for a black body at 4500 K.

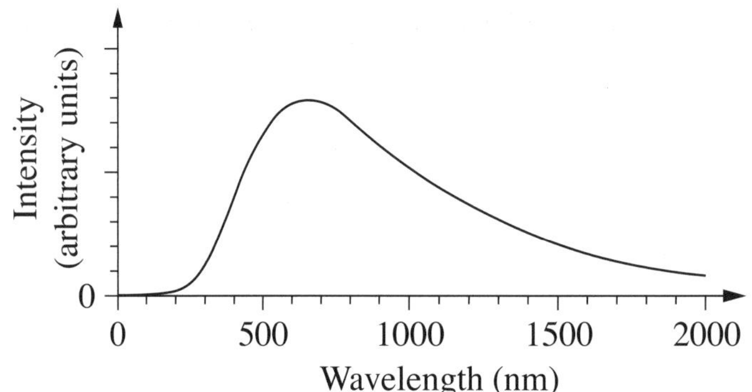

Which statement describes the expected difference in the graph for a black body at 4000 K?

A. Intensity at all wavelengths will be less.

B. Intensity at all wavelengths will be greater.

C. The peak intensity will occur at a higher frequency.

D. The peak intensity will occur at a shorter wavelength.

10 Figure I shows a current flowing through a loop of wire that is in a uniform magnetic field.

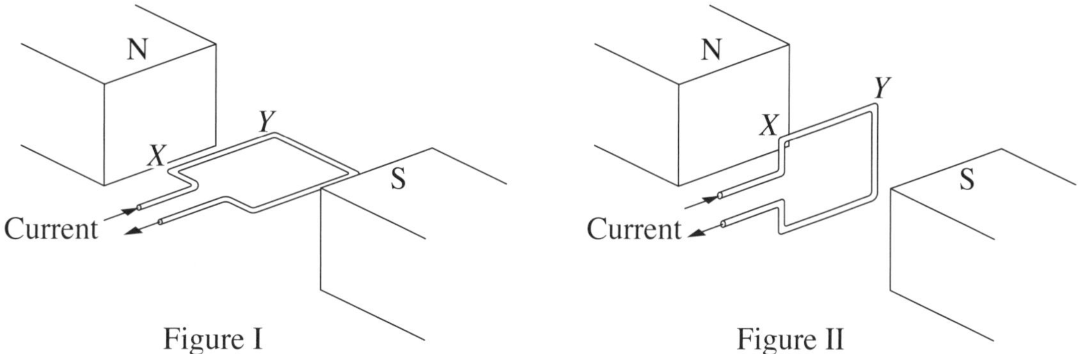

Figure I

Figure II

The loop is then rotated to the position shown in Figure II.

The magnitude of the force on the side *XY* and the magnitude of the torque on the loop in Figure II are compared to those in Figure I.

Which row of the table correctly describes the comparison?

	Force	*Torque*
A.	I > II	I = II
B.	I > II	I > II
C.	I = II	I = II
D.	I = II	I > II

11 The chart shows part of the nuclear decay series beginning with uranium.

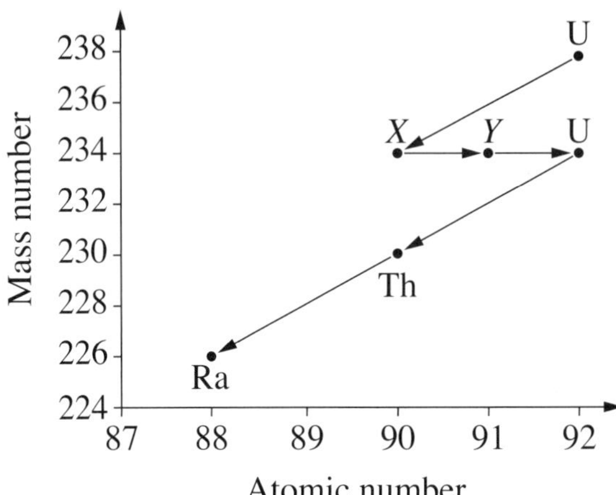

Which option correctly identifies *X* and *Y* and the process by which each was produced?

	X	*Y*
A.	$^{234}_{90}\text{Th}$ alpha decay	$^{234}_{91}\text{Pa}$ beta decay
B.	$^{234}_{90}\text{Th}$ alpha decay	$^{234}_{91}\text{Pa}$ alpha decay
C.	$^{234}_{91}\text{Pa}$ beta decay	$^{234}_{90}\text{Th}$ beta decay
D.	$^{234}_{91}\text{Pa}$ beta decay	$^{234}_{90}\text{Th}$ alpha decay

12 Figure I shows a positively charged particle accelerating freely from X to Y, between oppositely charged plates. The change in the particle's kinetic energy is W.

The distance between the plates is then doubled as shown in Figure II. The same charge accelerates from rest over the same distance from X to Y.

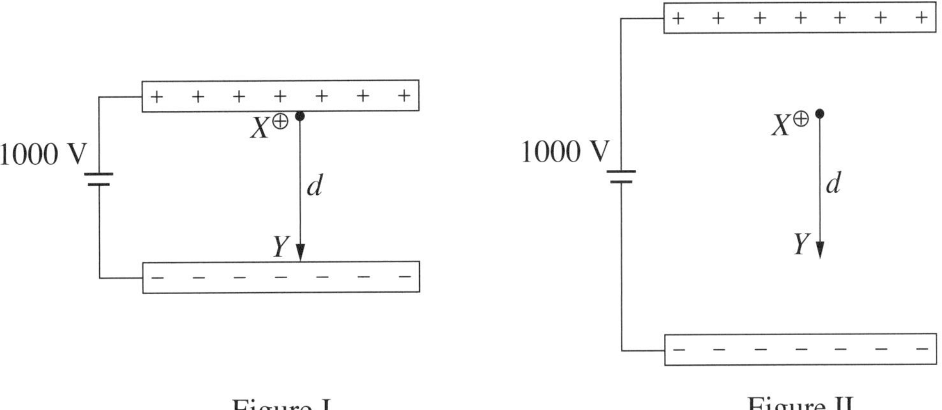

Figure I　　　　Figure II

What is the change in kinetic energy of the positively charged particle shown in Figure II?

A. W

B. $\frac{W}{2}$

C. $\sqrt{W}$

D. $2W$

13 Nucleus X has a greater binding energy than nucleus Y.

What can be deduced about X and Y?

A. X is more stable than Y.

B. Y is more stable than X.

C. X has a greater mass defect than Y.

D. Y has a greater mass defect than X.

14 Planet X has a mass 4 times that of Earth and a radius 3 times that of Earth. The escape velocity at the surface of Earth is 11.2 km s^{-1}.

What is the escape velocity at the surface of planet X?

A. 8.40 km s^{-1}

B. 9.70 km s^{-1}

C. 12.9 km s^{-1}

D. 14.9 km s^{-1}

15 What evidence resulting from investigations into the photoelectric effect is consistent with the model of light subsequently proposed by Einstein?

A. Photoelectrons were only ejected from a metal if the light was less than a specific wavelength.

B. Increasing the intensity of light on a metal increased the maximum kinetic energy of the photoelectrons.

C. If photons had sufficient energy to eject photoelectrons from a metal, the maximum kinetic energy was independent of the type of metal used.

D. The probability of photoelectrons being emitted from a metal was proportional to the duration of exposure to light for any given wavelength used.

16 In a thought experiment, two identical parallel aluminium rods, X and Y, are carrying electric currents of equal magnitude. Rod X rests on a table. Rod Y remains stationary, vertically above X, as a result of the magnetic interaction. The masses of the connecting wires are negligible.

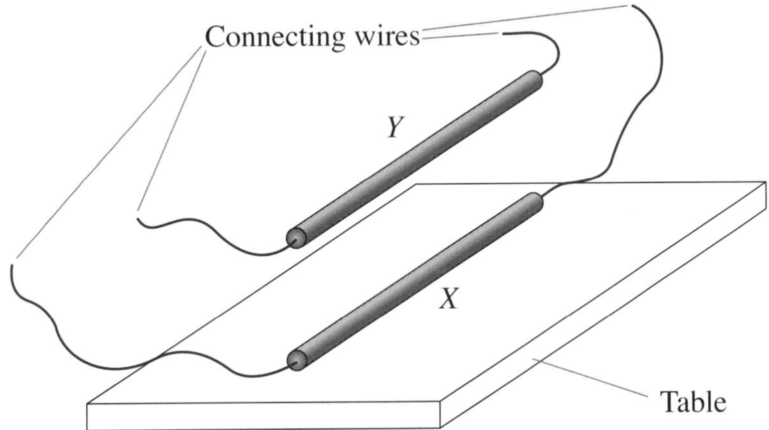

Which statement must be correct if rod Y is stationary?

A. The magnetic force acting on X is upward.

B. The currents through X and Y are in the same direction.

C. The force the table exerts on X is equal and opposite to the total weight of X and Y.

D. The force the table exerts on X is equal and opposite to the force of gravity acting on Y.

17 A mass attached to a lightweight, rigid arm hanging from point O, oscillates freely between X and Z.

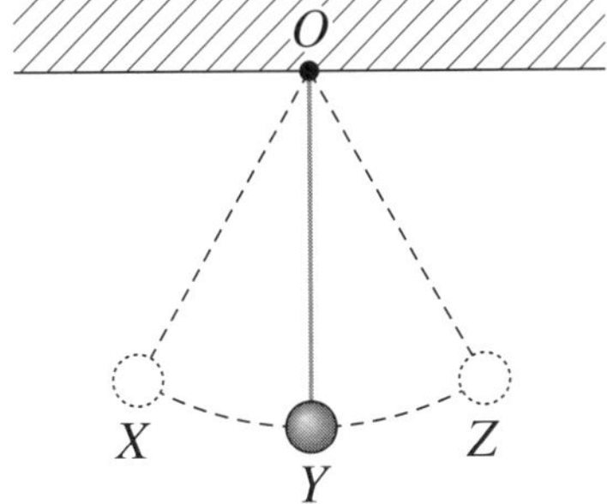

Which statement best describes the torque acting on the arm as it oscillates?

A. It is constant in magnitude and direction.

B. It is zero at Y and a maximum at X and Z.

C. It is zero at X and Z and a maximum at Y.

D. It is constant in magnitude but its direction changes.

18 The diagrams show the trajectories of two particles with the same mass and charge and which initially have the same velocity u, as shown. The subsequent motion of each particle is determined by its properties and by its interaction with the field in which it is moving.

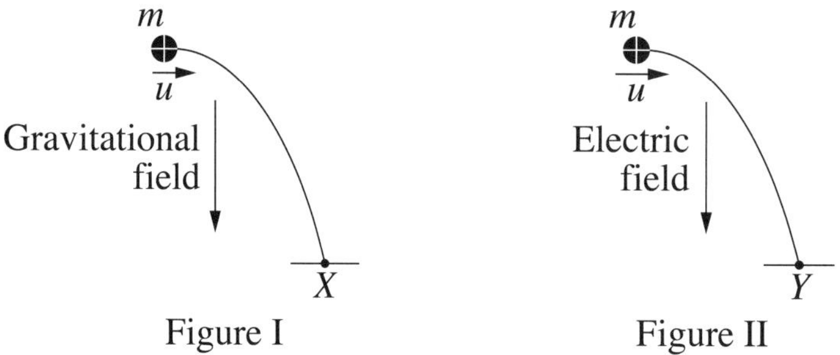

Figure I

Figure II

X and Y represent the landing points in Figures I and II.

Which row of the table shows the correct paths of the particles if the mass of each is increased by the same amount and they are given the same initial velocity u?

	Gravitational field	Electric field
A.	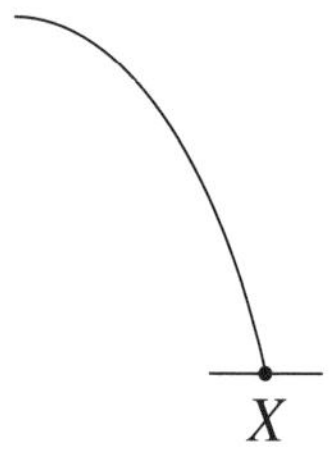 X	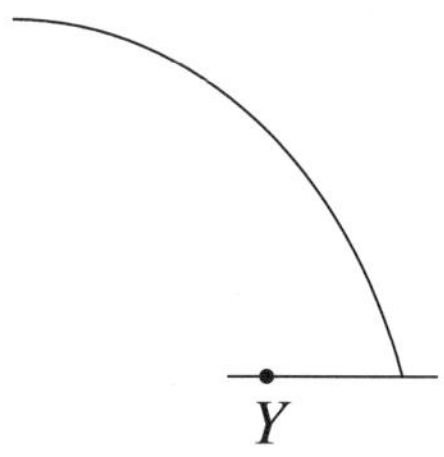 Y
B.	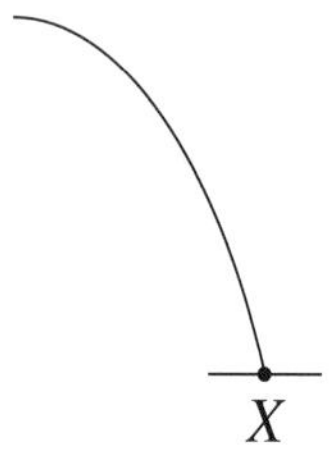 X	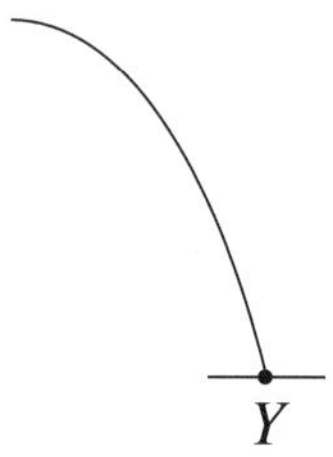 Y
C.	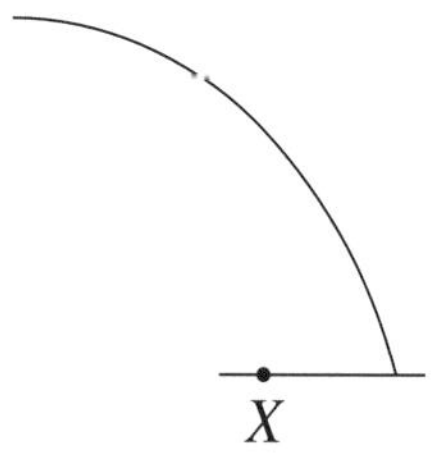 X	 Y
D.	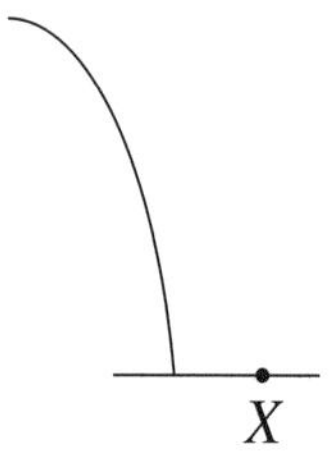 X	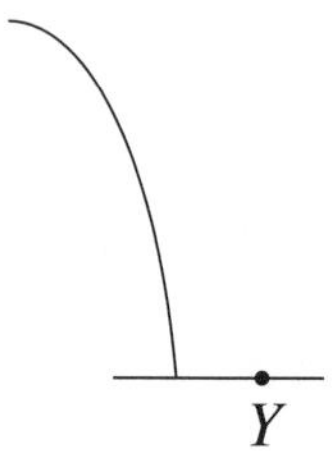 Y

19 The diagram represents the distribution of positive charges in identical wires when no current is flowing.

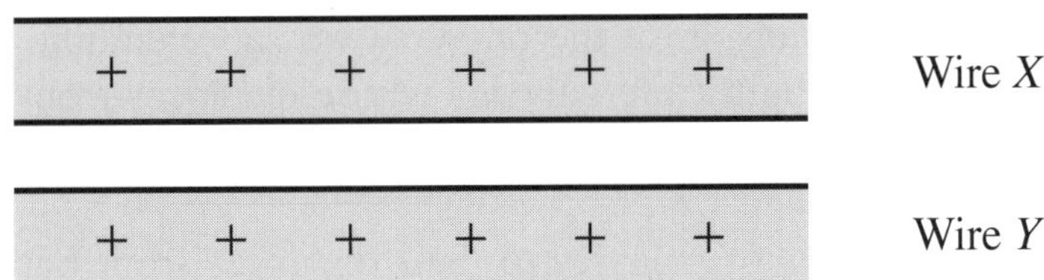

Equal currents then flow in each wire, but in opposite directions. These currents are considered conventionally as the flow of positive charge.

Which diagram represents the charge distribution in the wires, from the frame of reference of a positive charge in wire Y?

A.

+ + + Wire X

+ + + + + + Wire Y

B.

+ + + + + + + + + + + + Wire X

+ + + + + + Wire Y

C.

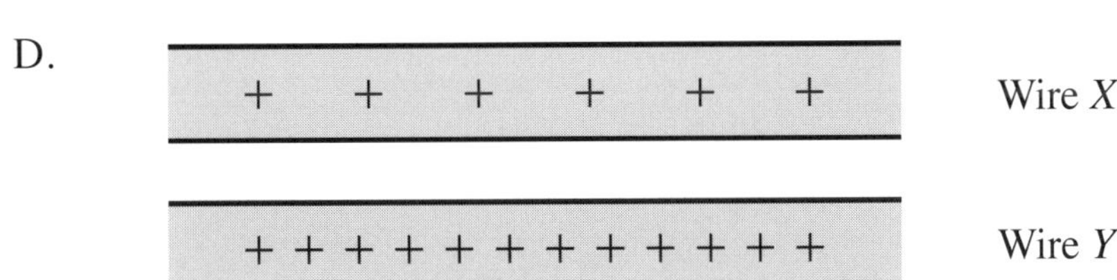

D.

+ + + + + + Wire X

+ + + + + + + + + + + + Wire Y

20 In 1995, observational evidence showed that Hubble's description of the expansion of the universe was inaccurate.

It was discovered that the expansion of the universe was accelerating. This discovery was based on observations of light from galaxies whose distances from Earth could be accurately measured, and were significantly more distant than any observed by Hubble.

Which graph relating velocities of galaxies to their distances from Earth is consistent with an accelerating rate of expansion of the universe?

A.

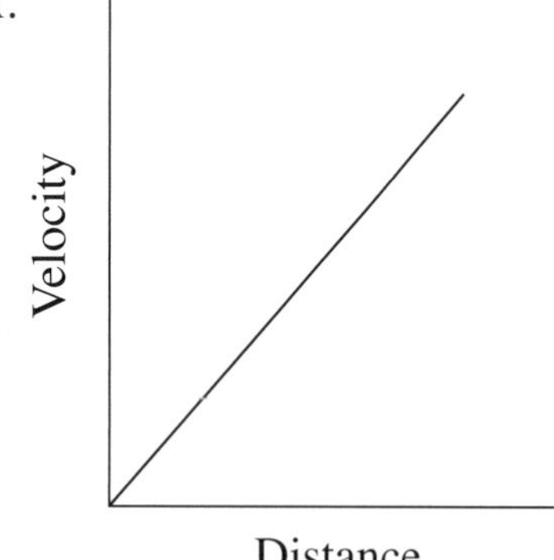

B.

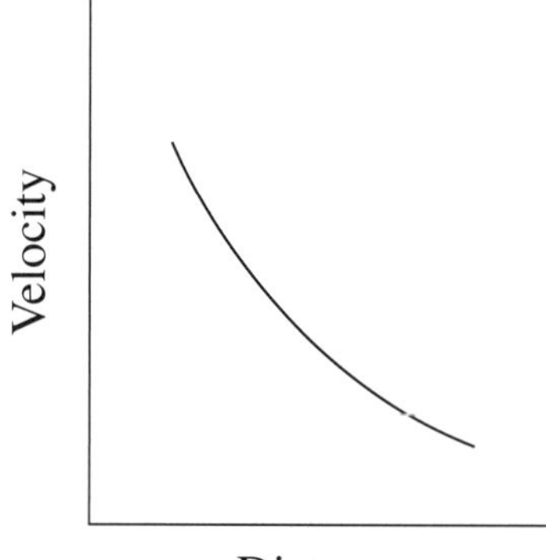

C.

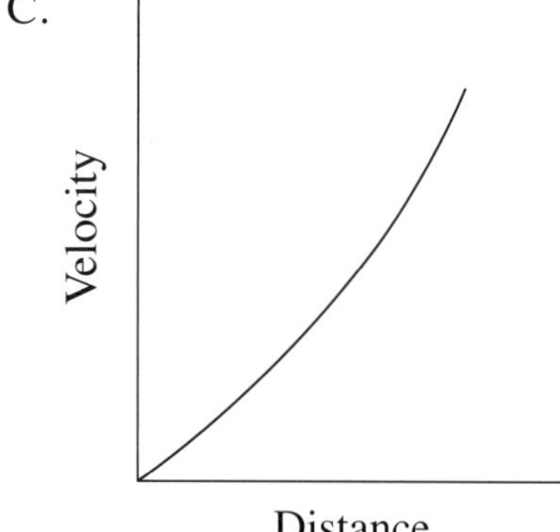

D.

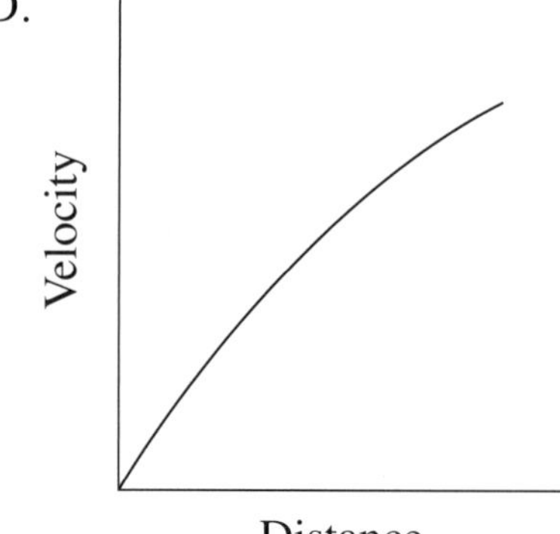

2023 HIGHER SCHOOL CERTIFICATE EXAMINATION

Centre Number

Student Number

Physics

Section II Answer Booklet

80 marks
Attempt Questions 21–34
Allow about 2 hours and 25 minutes for this section

Instructions

- Write your Centre Number and Student Number at the top of this page.
- Answer the questions in the spaces provided. These spaces provide guidance for the expected length of response.
- Show all relevant working in questions involving calculations.
- Extra writing space is provided at the back of this booklet. If you use this space, clearly indicate which question you are answering.

Please turn over

Question 21 (5 marks)

A Hertzsprung–Russell diagram is shown.

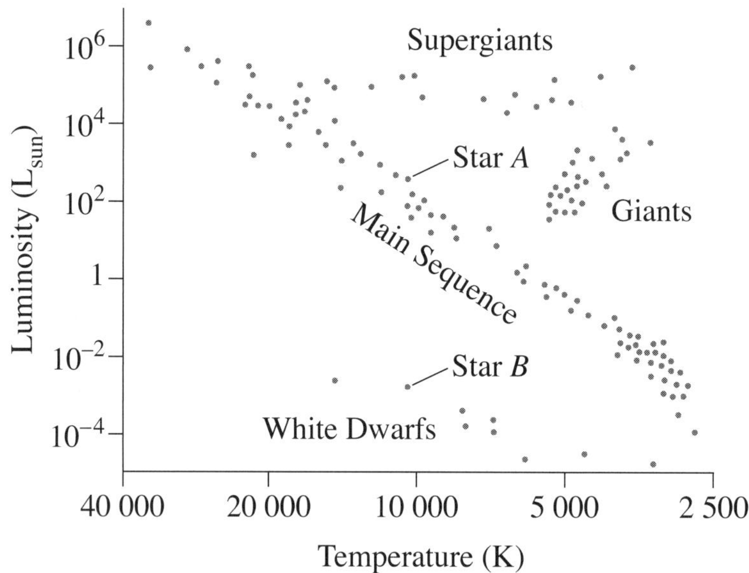

(a) Identify TWO variables that determine the luminosity of a star. **2**

....

....

....

....

(b) Describe differences between stars *A* and *B*. **3**

....

....

....

....

....

....

Question 22 (3 marks)

A spacecraft passes Earth at a speed of $0.9c$. The spacecraft emits a light pulse every 3.1×10^{-9} s, as measured by the crew on the spacecraft. **3**

What is the time between the pulses, as measured by an observer on Earth?

....

....

....

....

....

....

Please turn over

Question 23 (7 marks)

The James Webb Space Telescope (JWST) has a mass of 6.1×10^3 kg and orbits the Sun at a distance of approximately 1.52×10^{11} m.

(a) The Sun has a mass of 1.99×10^{30} kg. **2**

Calculate the magnitude of gravitational force the Sun exerts on the JWST.

....

....

....

....

(b) The telescope is sensitive to wavelengths from 6.0×10^{-7} m to 2.8×10^{-5} m. **3**

What is the minimum photon energy that it can detect?

....

....

....

....

....

....

(c) The JWST observed an exoplanet emitting a peak wavelength of 1.14×10^{-5} m. **2**

Calculate the temperature of the exoplanet.

....

....

....

....

Question 24 (3 marks)

An electron is travelling at 3.0×10^6 m s^{-1} in the path shown. 3

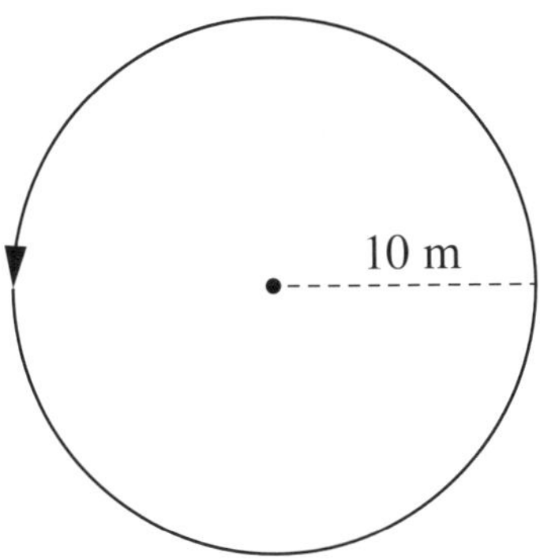

Calculate the magnetic field required to keep the electron in the path.

Please turn over

Question 25 (4 marks)

(a) The diagram represents one type of electric motor. **2**

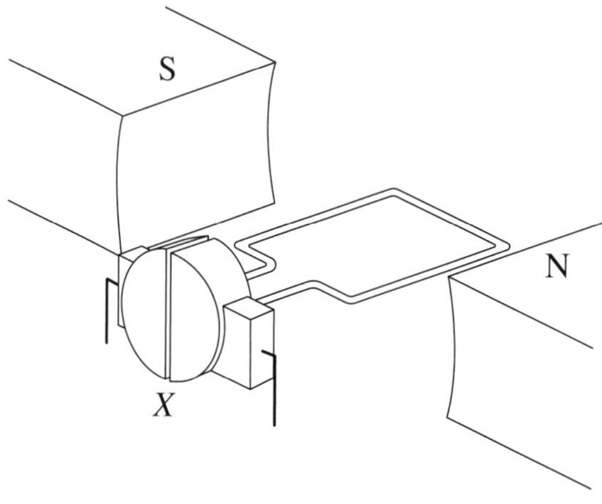

Describe the function of part *X*.

....

....

....

....

....

(b) Explain why the torque of a DC motor decreases as its rotational speed increases. **2**

....

....

....

....

....

....

Question 26 (3 marks)

Consider the following nuclear reaction. **3**

$$^{12}_{6}C + ^{1}_{1}H \rightarrow ^{9}_{5}B + ^{4}_{2}He$$

The masses of the isotopes in this process are shown in the table.

Isotope	*Mass* (*u*)
$^{12}_{6}C$	12.064
$^{9}_{5}B$	9.013
$^{4}_{2}He$	4.003
$^{1}_{1}H$	1.008

Calculate the energy released in this reaction.

....

....

....

....

....

....

Question 27 (8 marks)

(a) Explain how the composition and temperature of a star can be determined from its spectrum. **4**

....

....

....

....

....

....

....

....

....

....

Question 27 continues on the following page

Question 27 (continued)

(b) The diagram represents one hydrogen emission line from the spectrum of a star. **4**

656 nm

Explain the changes to this spectral line that would be observed as a result of the star's rotational velocity. Modify the diagram to support your answer.

....

....

....

....

....

....

....

....

End of Question 27

Question 28 (5 marks)

An ideal transformer is connected to a 240 V AC supply. It has 300 turns on the primary coil and 50 turns on the secondary coil.

It is connected in the circuit with two identical light globes, *X* and *Y*, as shown.

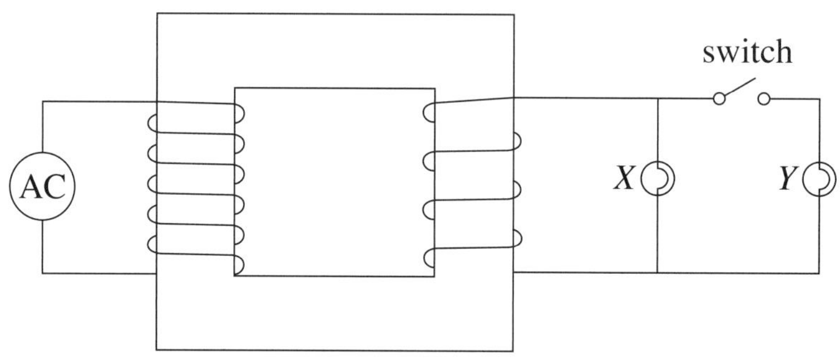

(a) Calculate the voltage across light globe *X* when the switch is open. **2**

....

....

....

....

(b) Explain why, after the switch has been closed, the current in the primary coil is different from when the switch is open. **3**

....

....

....

....

....

....

....

Question 29 (4 marks)

When light from an incandescent lamp is passed through a plane polarising filter, the intensity of the light is reduced. **4**

Explain this phenomenon.

....

....

....

....

....

....

....

....

Please turn over

Question 30 (8 marks)

The diagram shows apparatus that is used to investigate the interaction between the magnetic field produced by a coil and two copper rings *X* and *Y*, when each is placed at position *P*, as shown.

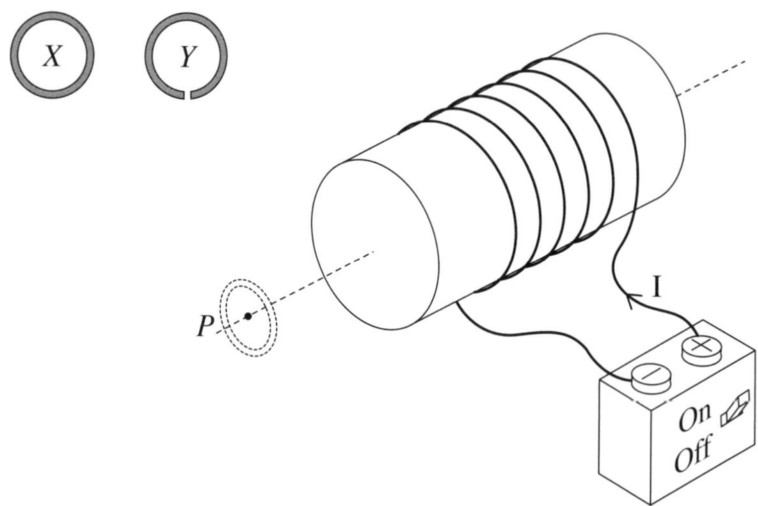

Ring *X* is a complete circular ring, and a small gap has been cut in ring *Y*. Each of the rings has a cross-sectional area of 4×10^{-4} m^2.

The power supply connected to the coil produces an increasing current through the coil in the direction shown, when the switch is turned on. This produces a magnetic field at *P* that varies as shown in the graph.

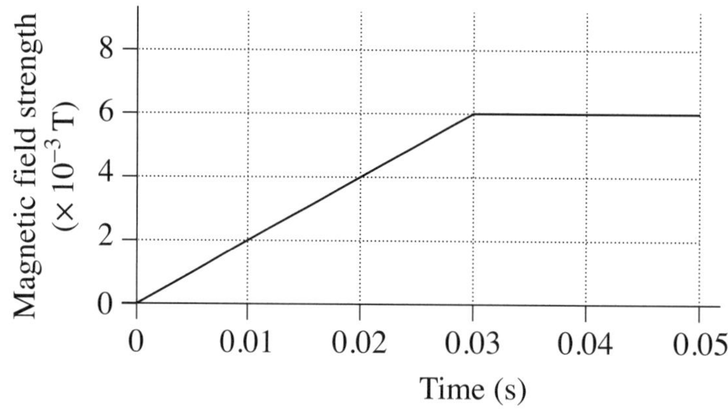

Question 30 continues on the following page

Question 30 (continued)

(a) In the first part of the investigation, ring X is held near the end of the electromagnet at position P. **4**

Account for the force acting on the ring from 0 to 0.05 seconds after the power supply is turned on.

....

....

....

....

....

....

....

(b) (i) In the second part of the investigation, ring Y is placed at P, and the power supply is turned on. **2**

Explain the behaviour of the ring.

....

....

....

....

(ii) Calculate the maximum induced emf in ring Y. **2**

....

....

....

....

End of Question 30

Question 31 (5 marks)

A roller coaster uses a braking system represented by the diagrams. **5**

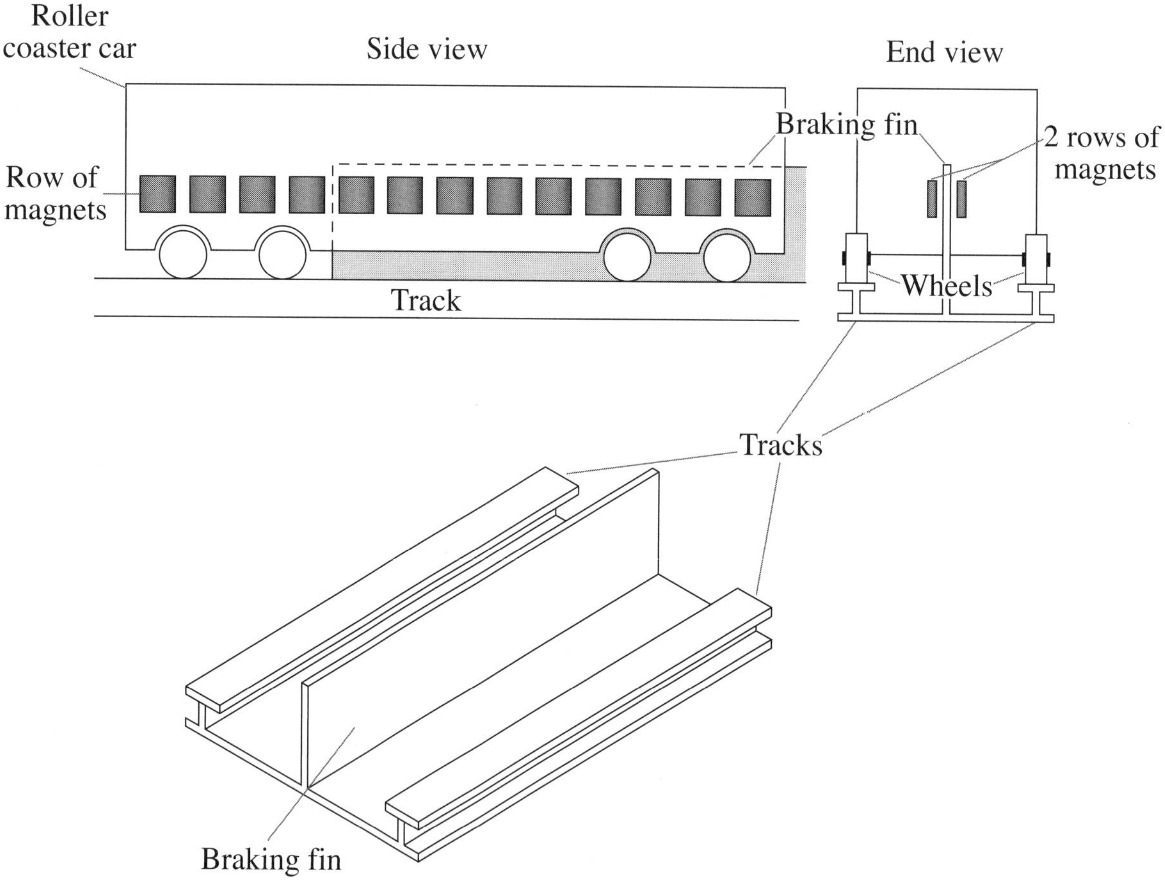

When the roller coaster car reaches the end of the ride, the two rows of permanent magnets on the car pass on either side of a thick aluminium conductor called a braking fin.

Question 31 continues on the following page

Question 31 (continued)

The graph shows the acceleration of the roller coaster reaching the braking fin at two different speeds.

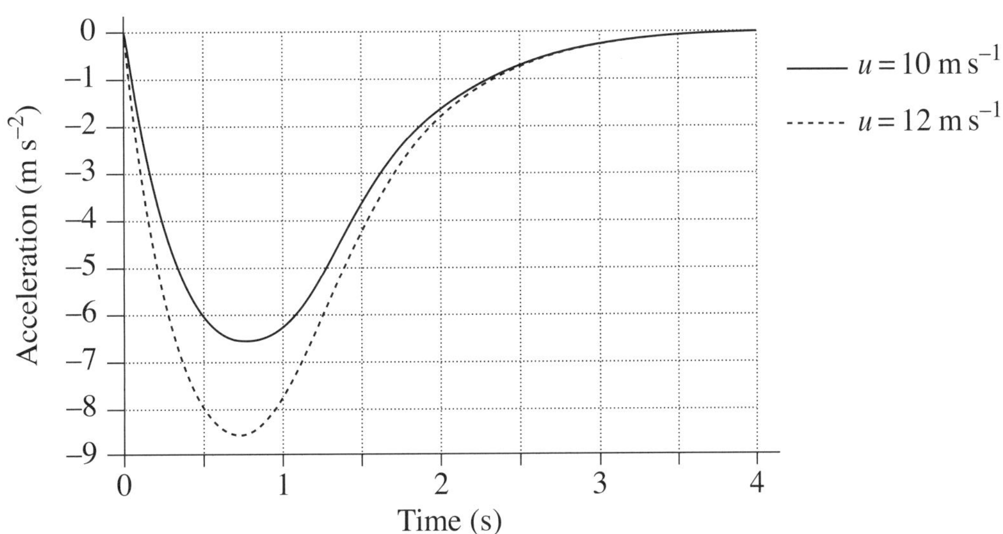

Explain the similarities and differences between these two sets of data.

....

....

....

....

....

....

....

....

....

....

....

....

End of Question 31

Question 32 (7 marks)

A horizontal disc rotates at 3 revolutions per second around its centre, with the top of the disc at ground level. **7**

At 2 m from the centre of the disc, a ball is held in place at ground level on the top of the disc by a spring-loaded projectile launcher. At position X, the launcher fires the ball vertically upward with a velocity of 5.72 m s^{-1}.

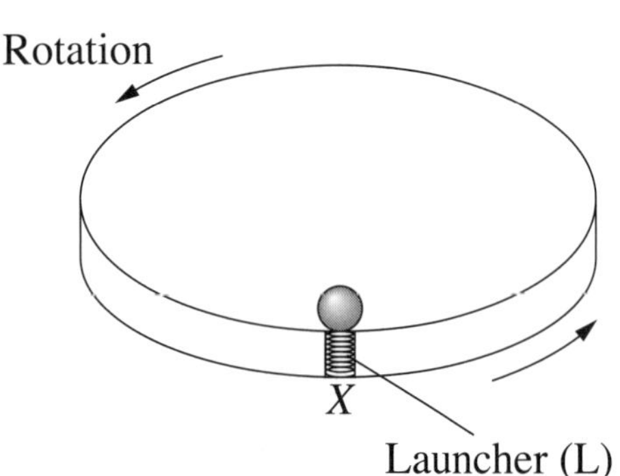

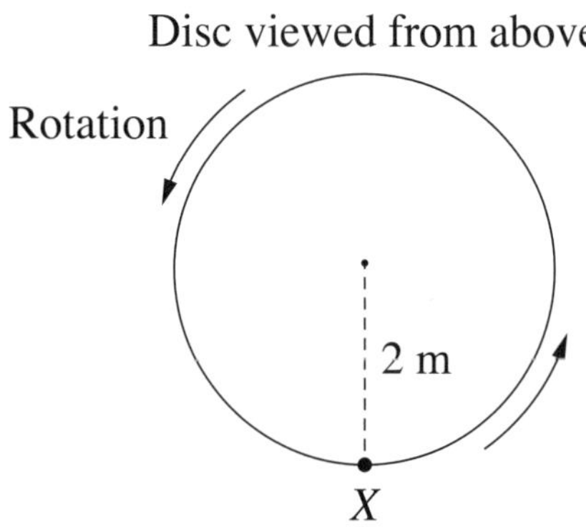

Calculate the ball's position relative to the launcher's new position, at the instant the ball hits the ground.

....

....

....

....

....

....

....

....

....

....

....

....

....

....

Question 33 (9 marks)

Consider the following statement. **9**

> The interaction of subatomic particles with fields, as well as with other types of particles and matter, has increased our understanding of processes that occur in the physical world and of the properties of the subatomic particles themselves.

Justify this statement with reference to observations that have been made and experiments that scientists have carried out.

Question 34 (9 marks)

A 400 kg satellite is travelling in a circular orbit of radius 6.700×10^6 m around Earth. Its potential energy is -2.389×10^{10} J and its total energy is -1.195×10^{10} J.

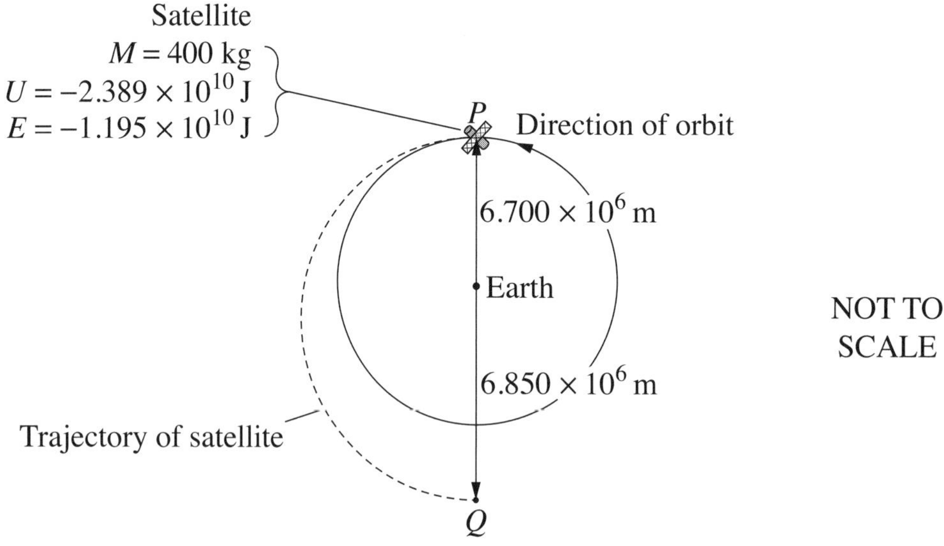

At point P, the satellite's engines are fired, increasing the satellite's velocity in the direction of travel and causing its kinetic energy to increase by 5.232×10^8 J. Assume that this happens instantaneously and that the engine is then shut down.

The satellite follows the trajectory shown, which passes through Q, 6.850×10^6 m from Earth's centre.

(a) Analyse qualitatively the energy changes as the satellite moves from P to Q. **2**

....

....

....

....

Question 34 continues on the following page

Question 34 (continued)

(b) Show that the kinetic energy of the satellite at Q is 1.194×10^{10} J. **4**

....

....

....

....

....

....

....

....

(c) Explain the motion of the satellite after it passes through Q. **3**

....

....

....

....

....

....

End of paper

2023 HSC Examination paper

Sample aswers

Section I (Total 20 marks)

1 D Gravitational field strength is given by $g = \frac{GM}{r^2}$. Because r = radius of planet + altitude, the gravitational field strength (g) will decrease as the altitude increases.

2 C Line loss is minimised by transmitting power using low current (as $P_{loss}=I^2R$) and high voltage. Answer A is incorrect as a step-up transformer is required to increase the voltage and reduce the current.

3 C The maxima will occur when $dsin\theta = m\lambda$. There must be a central maximum ($m = 0$) and so answers B and D are incorrect. The diagram in the question shows the wavefronts (crests) from each slit overlap in the middle and at each side which will produce three interference maxima on the screen as shown in answer C.

4 B As the half-life of caesium-137 is 30 years, 90 years represents three half-lives. Hence, after 30 years have elapsed, there will be 60 g remaining. After the next 30 years there will be 30 g remaining and, finally, after a further 30 years have elapsed there will be 15 g of caesium-137 remaining.

5 D The sum of the potential (U) and kinetic (K) energy of a planet in orbit is a constant. Hence, as gravitational potential energy $\left(U = \frac{-GMm}{r}\right)$ increases with the distance (r) to the central body, the kinetic energy must decrease to ensure the total energy remains constant. The path between the points S and P is furthest from the star and so the exoplanet will have a lower velocity in this section than between any of the other points shown. As the path length between consecutive points are equal, the exoplanet will take longer to travel from S to P than it will to move between any of the other consecutive points.

6 D An clcctron must accclcratc to producc an clcctromagnctic wavc. In a stablc hydrogen atom, the electron acts as a standing wave and does not emit electromagnetic radiation. An electron moving in a circle is undergoing centripetal acceleration and so will produce electromagnetic waves.

7 C The de Broglie wavelength is given by $\lambda = \frac{h}{mv}$ and a neutron is more massive than a proton. Hence a neutron will have a shorter de Broglie wavelength than a proton if both particles are travelling at the same speed.

8 B The change must result in a reduction in the range of the projectile. The range is given by Range = $u_x t$. As the project is launched horizontally, increasing the initial velocity will increase the range. Increasing the height of the platform will increase the time of flight and hence would also increase the range. Changing the mass

would have no effect as the range is independent of the mass. Changing the launch angle (which will change the time of flight and the horizontal component of the velocity) is therefore the only way the range could be reduced.

9 A The relationship between the peak emission wavelength and the temperature of a black body is given by Wien's displacement law, $\lambda_{max} = \frac{b}{T}$. Reducing the temperature of the body will increase the wavelength and decrease the frequency at which peak emission occurs. Hence answers C and D are incorrect. Reducing the temperature of a black body reduces the radiant power emitted from a black body and so the intensity at all wavelengths will be lower.

10 D As the magnetic field is perpendicular to the direction of current, the force on the side XY is given by $F = lIB$. This force on XY in position I will be equal to the force on XY in position II. The torque, however, is related to the product of the distance to the axis of rotation and the component of the applied force perpendicular to the plane of the coil. Hence the torque in position I will be at a maximum as force is perpendicular to the plane of the coil but will be zero in position II as the force is parallel to the plane of the coil.

11 A Reading the atomic and mass numbers from the graph and using the atomic number to determine the element from the periodic table, we see X is Th_{90}^{234} and Y is Pa_{91}^{234}.

As the atomic number decreased by 2 when X was formed, it must have been produced by an alpha decay. The atomic number then increases by one, indicating that Y was the product of a beta decay.

12 B The kinetic energy gained is equal to the work done, $W = qEd$. The same charge is moved across the same distance in Figure II but the electric field strength is halved as the plate separation is doubled. Hence the work done, and kinetic energy gained, is given by $\frac{qEd}{2} = \frac{W}{2}$.

13 C The binding energy is proportional to the mass defect between the nucleus and its constituent nucleons and so nucleus X has a greater mass defect than nucleus Y. Note that the stability of a nucleus is determined by the mass defect per nucleon and not by the mass defect alone.

14 C Escape velocity is given by $v = \sqrt{\frac{2GM}{r}}$ and hence the escape velocity is proportional to $\sqrt{\frac{M}{r}}$. If the mass is four times greater and the radius is three times greater than that of the Earth, the escape velocity at the surface of Planet X will be given by $v = 11.2 \times \sqrt{\frac{4}{3}} = 12.9$ km s^{-1}.

15 A Classical wave physics predicts that, provided the intensity of the incident light is great enough, photo electrons would be emitted from the surface by the light of any wavelength. In contrast, Einstein's photon model of light predicts that photoelectrons will only be ejected if the incident photon energy is sufficient to remove an electron from the surface of the metal. As photon energy is related to wavelength ($E = hf = \frac{hc}{\lambda}$), this means that photoelectrons will only be emitted when the incident light is less than a specific wavelength.

16 C As Y is stationary, the net force on Y must be zero. Hence there must be an upwards magnetic force on Y equal to the weight of Y. There will be an equal and opposite magnetic force on wire X. The total downwards force exerted on the table by X will therefore be the total weight of X and Y. As Y is stationary, the table must exert an upwards reaction force equal to the total weight of X and Y.

17 B The torque on the mass is equal to the product of the distance from the mass to the pivot (i.e. the length of the pendulum) times the component of the force perpendicular to the line joining the mass to the pivot point marked *0*. The only force on the mass is the downwards weight force. The component of the weight force perpendicular to the pivot arm is zero at Y and so the torque is zero at point Y. The perpendicular component of the weight is maximum at points X and Z and so the torque is a maximum at points X and Z.

18 A Changing the mass has no effect on the particle in Figure I, as the acceleration due to gravity is independent of the mass. The downwards acceleration of the particle in Figure II will be reduced when the mass is increased as $a = \frac{F}{m} = \frac{qE}{m}$. Therefore the particle will be in flight for a longer time. Increasing the time of flight will increase the range of the particle, as the horizontal range is $x = u_x t$.

19 B From the reference frame of a positive charge in wire Y, the other charges in the wire will appear to be stationary as they are all moving with the same velocity. The distance between the positive charges in wire Y, therefore, will not change when current flows. This is because a reference frame moving at constant velocity is the same as a reference frame at rest (i.e. all inertial frames are equivalent). An observer on a positive charge in wire Y would see the charges in wire X moving. Therefore a positive charge in wire Y would observe the distance between the positive charges in wire X to be reduced (in accordance with Einstein's length contraction equation).

20 D Hubble used the red shift of comparatively nearby galaxies to show that the velocity (red shift) of a galaxy is proportional to its distance. He assumed the universe was expanding at a constant rate and that his graph could be extrapolated to large distances. Distance, however, correlates to time in astronomy and, if the rate of expansion has been accelerating, the universe would have been expanding at a much slower rate a long time ago than it is now.

Section II

Question 21 (Total 5 marks)

(a) The absolute luminosity of a star is determined by its surface temperature and the size (radius) of the star. *(2 marks)*

(b) Because both stars have similar surface temperatures and star *A* is much more luminous than star *B*, star *A* must have a much greater radius than star *B*. Star *A* is on the main sequence, which means it produces energy by fusing hydrogen to helium in its core, while star *B* is a white dwarf star that does not produce energy and features no fusion reactions. Star *A* is at an earlier stage of its evolutionary process than star *B* because mid-range main-sequence stars like star *A* become red giants when they are no longer able to sustain H to He fusion and then become white dwarf stars when fusion reactions cease in the star. *(3 marks)*

Question 22 (Total 3 marks)

An observer on the Earth will observe time to be slowed in the moving spacecraft in accordance with Einstein's time dilation equation. Hence the time between pulses measured by the observer on Earth will be given by:

$$t = \frac{t_0}{\sqrt{\left(1 - \frac{v^2}{c^2}\right)}} = \frac{3.1 \times 10^{-9}}{\sqrt{1 - 0.9^2}} = 7.1 \times 10^{-9}\ \text{s}$$

(3 marks)

Question 23 (Total 7 marks)

(a) Applying Newton's law of gravity.

$$F = \frac{GMm}{r^2} = \frac{6.67 \times 10^{-11} \times 1.99 \times 10^{30} \times 6.1 \times 10^3}{(1.52 \times 10^{11})^2} = 35\ \text{N}$$

(2 marks)

(b) Photon energy is inversely proportional to wavelength and so the longest wavelength light will have the lowest photon energy. Applying Einstein's photon energy equation, the minimum photon energy the telescope can detect is:

$$E = hf = \frac{hc}{\lambda} = \frac{6.626 \times 10^{-34} \times 3.00 \times 10^8}{2.8 \times 10^{-5}} = 7.8 \times 10^{-21}\ \text{Joules}$$

(3 marks)

(c) Assuming the exoplanet behaves as a perfect black body, the surface temperature of the exoplanet will be given by Wein's Displacement Law:

$$\lambda_{max} = \frac{b}{T} \text{ and hence } T = \frac{b}{\lambda_{max}} = \frac{2.898 \times 10^{-3}}{1.14 \times 10^{-5}} = 254.2\ \text{K}$$

(2 marks)

Question 24 (Total 3 marks)

Applying the right-hand rule, we see that to produce a centripetal force to keep the electron rotating in a circle in this direction will require a magnetic field directed upwards (out of the page). To calculate the magnetic field strength, we equate the magnetic force on charge to the centripetal force:

$$\frac{mv^2}{r} = qvB \text{ and hence } B = \frac{mv}{qr} = \frac{9.109 \times 10^{-31} \times 3.0 \times 10^6}{1.602 \times 10^{-19} \times 10} = 1.71 \times 10^{-6} \text{ Tesla} \quad \text{directed out of the page}$$

(3 marks)

Question 25 (Total 4 marks)

(a) X is a split-ring commutator. As the commutator rotates, the brushes which are connected to an external DC power supply make electrical contact with the turning commutator. This results in the direction of current in the armature coil being reversed each half cycle. This change in the current direction each half cycle ensures the current remains in one direction on the right-hand side of the coil and in the opposite direction on the left-hand side of the coil. This ensures the torque produced by the current in the coil in the magnetic field remains in one direction and turns the armature coil continually in that direction. *(2 marks)*

(b) As the rotational speed of the motor increases, the rate of change of magnetic flux linking the coil will also increase. By Faraday's law of electromagnetic induction, this change in magnetic flux will produce an electromotive force (emf) on the coil. Lenz's law states that, to conserve energy, the magnetic field produced by this induced emf must be in a direction that opposes the action that produced the changing magnetic flux. Hence, as the rotational speed of the armature increases, a back emf is induced that reduces the voltage across the armature coil ($V_{coil} = V_{applied} - emf_{back}$), which in turn reduces the current flowing in the coil. As the torque on the armature coil is proportional to the current in the coil ($\tau = nIA{\perp}B$), the torque on the coil will therefore decrease as the rotational speed increases. *(2 marks)*

Question 26 (Total 3 marks)

We must first calculate the mass change in the reaction.

Δm = *total mass of products – total mass of reactants*

$\Delta m = (9.013 + 4.003) - (12.064 + 1.008) = -0.056$ u

The negative sign indicates that the total mass of the products is less than the total mass of the reactants, and therefore some mass has been converted into energy.

As one atomic mass unit is equivalent to an energy of 931.5MeV/c^2, the energy released in the reaction is: $E = 0.056 \times 931.\ 5 = 52.164$ MeV

(3 marks)

Question 27 (Total 8 marks)

(a) The spectrum of a star consists of a continuous black-body spectrum superimposed by an absorption spectrum. The wavelength of peak emission can be identified from the star's continuous black-body spectrum. Wein's law ($\lambda_{max} = \frac{b}{T}$) can then be used to determine the surface temperature of the star. Alternatively, the star's black-body spectrum can be compared to computer-generated black-body spectra from bodies at different temperatures. The temperature of the star will be the same as the temperature of the body that has a matching black-body spectrum.

The absorption and emission spectrum of each type of element are unique to that element. The wavelengths of the absorption lines of the elements have been measured and catalogued by experiments in laboratories on Earth. The spectrum of a star consists of a continuum black-body spectrum superimposed by the absorption spectrum of the elements in the star. By comparing the absorption lines in the star's spectrum with the known absorption lines of elements, the elements present in the star can be determined. The intensity of the absorption lines can also be used to estimate the relative composition of elements in a star. *(4 marks)*

(b) Provided the axis of rotation of the star has some component perpendicular to the observer, the rotation of the star will broaden the spectral line to both longer and shorter wavelengths. This is because the rotation would cause one side of the star to move away from the observer and red-shift the light to longer wavelengths while the other side of the star would move towards the observer, blue-shifting the light to shorter wavelengths.

(4 marks)

Question 28 (Total 5 marks)

(a) Applying the transformer equation,

$\frac{V_P}{V_S} = \frac{N_P}{N_S}$ We can write $V_s = \frac{N_s V_p}{N_p} = \frac{50 \times 240}{300} = 40V$ *(2 marks)*

(b) When the switch is open, the secondary voltage (40 V) is applied across one globe. When the switch is closed, the same voltage will be applied across both globes and, as they are wired in parallel, the total resistance of the circuit will halve. As the secondary voltage remains constant, halving the resistance of the circuit will double the current (as $V = IR$) and will also double the power supplied to the circuit by the secondary coil (as $P = \frac{V^2}{R}$). Because the transformer is 'ideal', the input power must equal the output power ($I_p V_p = I_s V_s$) and so the input power must be doubled to double the output power. As the primary and secondary voltages are constant, doubling the input power and output power of the transformer by closing the switch will result in the current in the primary coil doubling. *(3 marks)*

Question 29 (Total 4 marks)

A polarising filter can be used to convert unpolarised light to polarised light. Light from an incandescent lamp is unpolarised and so the electromagnetic waves emitted are made up of changing electric and magnetic fields that are directed in all planes perpendicular to the wave velocity. In contrast, polarised electromagnetic waves have their changing electric field directed perpendicular to the wave velocity in the one plane (the plane of polarisation) only. The changing magnetic field in a polarised electromagnetic wave is perpendicular to the wave velocity in a plane that is also perpendicular to the plane of polarisation. The polarising filter only transmits that component of the incident unpolarised light that is in the direction of the plane of polarisation of the filter. The component of the incident waves that are perpendicular to the axis of polarisation of the filter are absorbed. If the electric field of the incident light is spread equally in all planes perpendicular to the wave velocity (i.e. is totally unpolarised), half the incident intensity will have a component in the direction of the axis of polarisation of the filter and half will have a component perpendicular to the axis. Thus half the light intensity will be transmitted by the polarising filter. (*4 marks*)

Question 30 (Total 8 marks)

(a) When the power is connected to the solenoid, the solenoid will produce (by the right-hand grip rule) a magnetic field in the coil directed towards the point *P* that increases at a constant rate from zero to 6×10^{-3} T after 0.03 seconds. This magnetic field will link with the ring to produce a changing magnetic flux ($\Phi = BA$) in the ring for the first 0.03 seconds. An emf will be induced in the ring by this changing magnetic flux (Faraday's law of induction) that will produce a current in the ring. Lenz's law tells us the current will be in a direction that will produce a magnetic field which will oppose the changing magnetic flux that produced the current. Hence, for the first 0.03 seconds, a clockwise current will be induced in the ring that will produce a magnetic field towards the solenoid. The magnetic field in the ring will interact with the magnetic field produced by the solenoid to produce a force on the ring directed away from the solenoid for the first 0.03 seconds. After this time there is no change in magnetic flux in the ring, as the magnetic field from the solenoid is constant. Therefore the ring will experience a force directed away from the solenoid for the first 0.03 seconds after the power supply is connected to the solenoid and then will experience zero force for the following 0.02 seconds. (Note we have neglected the force of gravity on the ring in this idealised experiment.)

(*4 marks*)

(b) (i) An emf will be induced in ring *Y* during the first 0.03 seconds after the power supply is connected to the solenoid because the magnetic flux in the ring will be changing during this period (by Faraday's law of electromagnetic induction). But the cut in the ring will prevent the current from flowing in the ring and therefore there will be no magnetic force on the ring. (Again we have neglected the gravitational force.) (*2 marks*)

(ii) The change in magnetic flux during the first 0.03 seconds in the ring will be given by:

$\Delta\Phi = \Delta BA = 6 \times 10^{-3} \times 4 \times 10^{-4} - 0 = 2.4 \times 10^{-6}$ Webers

and the induced emf in ring Y will be given by:

$$\varepsilon = N\frac{\Delta\Phi}{\Delta t} = \frac{2.4 \times 10^{-6}}{0.03} = 8.0 \times 10^{-5} V$$

(*2 marks*)

Question 31 (Total 5 marks)

Similarities

Both sets of data have the following similarities.

At time $t = 0$ s the roller coaster has zero acceleration. Between $t = 0$ s and $t = 0.75$ s the roller coaster experiences an increasing negative acceleration. The negative acceleration reaches a maximum at $t = 0.75$ s and then decreases at a slower rate than the rate it increased until it reaches zero at $t = 4$ s. Both curves appear to be identical between $t = 3$ s and $t = 4$ s.

We can explain each of these similarities as follows:

- The acceleration is zero at $t = 0$ s because the roller coaster magnets have not reached the braking fin and consequently there are no induced eddy currents in the fin and there is no retarding force on the roller coaster.
- After $t = 0$ s the roller coaster starts to pass over the braking fin, which will produce a changing magnetic flux on the aluminium breaking fin. This changing magnetic flux (by Faraday's law) will induce eddy currents in the breaking fin. These eddy currents will be in a direction that will produce a magnetic field which will oppose the change in magnetic flux that produced the eddy currents (by Lenz's law). Hence the magnetic field from the eddy currents will oppose the motion of the car, producing a negative acceleration on the car that will increase as more of the car's magnets interact with the braking fin. The car will therefore slow down as it passes the breaking fin and the negative acceleration will increase until at $t = 0.75$ s the magnets on the car start to leave the fin.
- Because the negative acceleration continually decreases the velocity of the car, it will take more time for the magnets to leave the fin than it took for them to move onto the fin. That is why the negative acceleration takes 3.25 s after its maximum value to return to zero at $t = 4$ s.
- Both curves have maximum negative acceleration at 0.75 s and return to zero acceleration after 4 s because, when the faster car passes the braking fin, the fin will experience a more rapidly changing magnetic field that will induce larger eddy currents and a greater retarding force on the faster moving car. Thus, even though it enters with a greater initial velocity, it slows faster, has a peak negative acceleration at the same time ($t = 0.75$ s) and leaves the braking fin at the same time ($t = 4$ s). Between 3 and 4 seconds both curves are identical, indicating the cars are travelling at the same speed. This is because the faster car has experienced a greater negative acceleration in the first 3 s.

Differences

The main difference between the curves is that the faster car has a greater negative acceleration than the slower car between $t = 3$ s and $t = 4$ s.

We can explain this difference as follows:

- As discussed above, this is because when the faster car passes the braking fin, the fin will experience a more rapidly changing magnetic field that will induce larger eddy currents and a greater retarding force on the faster moving car. This greater negative acceleration means both cars experience their maximum deceleration at the same time but the faster car is still moving faster than the slower car after 0.75 s. This is why its negative acceleration remains greater than that of the slower car until $t = 3$ s. *(5 marks)*

Question 32 (Total 7 marks)

When the ball is ejected upwards, it will undergo projectile motion.

The initial horizontal velocity of the projectile will be given by
$u_x = \omega r = 2\pi fr = 2\pi \times 3 \times 2 = 37.7$ ms^{-1} in a direction tangential to the disc from the position of release.

We can use the initial vertical velocity given in question ($u_y = 5.72$ ms^{-1}) to find the time of flight.

As the vertical velocity is zero at the top of the flight and calling upwards positive, the time to reach the maximum height can be calculated as follows:
$v = u + at$ and hence $t = \dfrac{v - u}{a} = \dfrac{0 - 5.72}{-9.8} = 0.584$ s

As the motion is symmetrical, the total time of flight will be twice the time to reach the top of the motion.

Total time of flight $2 \times 0.584 = 1.168$ s

The horizontal range of the projectile from the launch point will be given by:

Range $= u_x t = 1.168 \times 37.7 = 44$ m in a direction tangential to the release point.

When the ball reaches the ground, the disc will have rotated.

Number of rotations after ball is launched = time × frequency = $1.168 \times 3 = 3.5$ revolutions.

The launcher will therefore be on the opposite side of the disc when the projectile lands. We can apply Pythagoras' theorem to find the distance between the landing point and the launcher as follows:

Distance $= \sqrt{44^2 + 4^2} = 44.2$ m

Thus the landing point will be on a tangent to the disc that passes the launch point *X* to the right of position *X*. The landing point will also be 44.2 m from the new position of the launcher (which is on the side of the disc opposite the launch point *X*). *(7 marks)*

Question 33 (Total 9 marks)

Investigations by scientists involving the interaction of subatomic particles with fields and other particles have significantly improved our understanding of the properties of particles and our understanding of the physical world. Some of the investigations and observations made by scientists that support this statement include:

(a) Millikan's experiment to determine the charge on the electron. Millikan used the electric field between two parallel charged plates to balance the downwards force of gravity and hence suspend tiny, charged oil droplets between the plates. By equating the gravitational and electric forces on an oil droplet ($mg = \frac{qV}{d}$) he was able to determine the charge on the droplet and by finding the charge on many droplets he was able to determine the charge of a single electron. Knowing the charge of the electron was important in the development of modern theories of electromagnetism and atomic physics.

(b) JJ Thomson passed high-speed electrons through perpendicular electric and magnetic fields to determine their velocity and then, by measuring their radius of curvature in a magnetic field (using $\frac{mv^2}{r} = qvB$), was able to determine the charge to mass ratio of the electron. With Millikan's result this enabled the mass of the electron to be determined.

(c) Chadwick discovered the neutron by colliding what was at the time an unknown type of radiation with paraffin wax. By measuring the recoil velocity of hydrogen atoms ejected from the paraffin and applying the laws of conservation of energy and momentum he showed the unknown radiation consisted of particles that had a slightly larger mass than the proton. Finally, by showing the rays were not deflected by magnetic fields, he was able to deduce that the particles were not charged (neutral) and deduced the particles were the neutrons that had been predicted to exist by theory. This discovery led to a deeper understanding of the atomic nucleus and to investigations that enabled physical processes such as radioactivity in unstable nuclei to be understood.

(d) Davisson and Germer confirmed the wave nature of electrons, predicted by de Broglie, when they showed that a beam of high-speed electrons scattered by a nickel crystal formed an interference pattern. This confirmation led to a deeper understanding of atomic processes and paved the way for Schrodinger's theory of quantum mechanics.

(e) By examining the way alpha particles were scattered by atoms, Rutherford was able to predict that most of the mass of the atom was concentrated in a tiny positively charged nucleus. This discovery was instrumental in the development of modern atomic models of the atom.

(f) Inelastic scattering experiments using high-energy particle accelerators have been used to confirm that protons and neutrons are composite particles made of up and down quarks. Collisions in particle accelerators have produced a wide range of new particles that eventually led to the creation of the Standard Model of particle physics. Particle accelerators have since been used to confirm many of the predictions of the Standard Model including the existence and properties of many particles such as quarks, W and Z bosons and the Higgs boson.

(g) Experiments in which uranium was bombarded with neutrons led to the discovery of nuclear fission. This discovery led to a deeper understanding of the nucleus and its constituent nucleons. It also paved the way for understanding how nuclear energy could be released both explosively and in a controlled fashion.

All these observations and investigations illustrate how important the interaction of subatomic particles with fields, as well as other types of particles and matter, has been to the development of our current understanding of the properties of subatomic particles and the processes that occur in the physical world. *(9 marks)*

Question 34 (Total 9 marks)

(a) Because the satellite is in free flight in a gravitation field after the rocket burn, the total energy of the satellite ($K + U$) must remain constant. Increasing the velocity of the satellite when the engines are fired at P will result in the satellite having too much kinetic energy to remain in the original circular orbit. This additional kinetic energy will cause the satellite to move further from the Earth as it moves from P to Q. Because potential energy increases as the distance between the satellite and the centre of the earth increases, the potential energy of the satellite will increase as it moves from P to Q. Further, as the total energy of the satellite remains constant and energy is conserved, kinetic energy must be converted into an equal amount of potential energy as satellite moves from P to Q. Therefore, as the satellite moves from P to Q some of its kinetic energy is converted into potential energy, but the total energy of the satellite remains constant. *(2 marks)*

(b) After the rocket engines are fired, the total energy of the satellite will remain constant. Thus the kinetic energy of the satellite at Q can be calculated by deducting the potential energy of the satellite at Q from the total energy of the satellite after the rocket engines were fired at P.

Total energy of satellite at P after engine was fired:

$E_t = -1.195 \times 10^{10} + 5.232 \times 10^8 = -1.142\ 68 \times 10^{10}\ \text{J}$

Potential energy of the satellite at Q:

$$U_Q = \frac{-GMm}{r} = \frac{-6.67 \times 10^{-11} \times 6.0 \times 10^{24} \times 400}{6.850 \times 10^6} = 2.3369 \times 10^{10}\ \text{J}$$

Now we can find the kinetic energy of the satellite at Q as follows:

$E = U + K$ and hence $K = E - U = -1.142\ 68 \times 10^{10} - (-2.3369 \times 10^{10})$

$= 2.3369 \times 10^{10} - 1.142\ 68 \times 10^{10} = 1.194 \times 10^{10}\ \text{J}$ *(4 marks)*

(c) By equating the centripetal force on the satellite at Q to the gravitational force on the satellite, the kinetic energy for a satellite to travel in a uniform circular orbit at this radius can be determined:

$\frac{mv^2}{r} = \frac{GMm}{r^2}$ and therefore

$$K = \frac{mv^2}{2} = \frac{GMm}{2r} = \frac{6.67 \times 10^{-11} \times 6 \times 10^{24} \times 400}{2 \times 6.850 \times 10^6} = 1.1685 \times 10^{10}\ \text{J}$$

As the kinetic energy we calculated in part (b) is greater than this value, the satellite is going too fast at Q to move in a circular orbit with this radius. The satellite must therefore move further away from the Earth immediately after it passes point Q.

The kinetic energy required to escape from the Earth totally at Q would be equal to the absolute value of the gravitational potential energy at Q, which we calculated in part (b) to be $U = -2.3369 \times 10^{10}$ J

Therefore the kinetic energy required to escape from the Earth at Q is: $K = 2.3369 \times 10^{10}$ J.

As the kinetic energy of the satellite at Q is less than this value, the satellite will not escape from the Earth's gravitational pull but will remain in orbit around the Earth. (If the kinetic energy had been greater than the absolute value of the potential energy, the satellite would have assumed a hyperbolic path that would have taken it away from the Earth.)

Therefore, after passing Q, the satellite will initially move further from the Earth before moving into an elliptical orbit with the Earth at one of the foci of the ellipse. Note this is not reflected in the diagram that accompanies the question, which appears to show an elliptical orbit with its greatest distance from the Earth (apogee) at Q.* (*3 marks*)

OR

Students could refer to the diagram given and their knowledge of orbits to predict the physically correct orbit that would result from increasing the velocity instantaneously at P.

(c) Firing the rocket engines at P instantaneously in the direction of motion will change the original circular orbit to an elliptical orbit. The Earth will be at the focus of the ellipse closest to point P. The perigee (closest point to the central body) of the new elliptical orbit will be at P and the apogee (furthest point from the central body) will be at point Q. Half of the new elliptical orbit is shown by the dotted line on the diagram in the question. The other half of the ellipse, on the right-hand side of the diagram, would be the mirror image of the path shown by the dotted line on the left-hand side of the diagram. Thus, after passing point Q, the satellite will move closer to the Earth until it reaches point P and repeats its orbit.* (*3 marks*)

*Students should note that this is a problematic question, as the diagram accompanying it appears to show half of an elliptical orbit, with its closest approach (perigee) at point P and its furthest distance from the Earth at point Q (apogee). If a satellite was travelling in a uniform circular orbit at P and its velocity was instantaneously increased to some velocity below the escape velocity at P, the satellite would move into an elliptical orbit as shown. Every point in the orbit would be further from the Earth than point P but the satellite would continue to pass through the point P once each orbit. The satellite would reach the furthest point from the Earth halfway through its first elliptical orbit as the diagram in the question shows. The values in the questions, however, do not reflect this situation. In reality, the satellite would reach the opposite side of its orbit further from the Earth than indicated by the value given in the question. The kinetic energy of the satellite when it reached the opposite side of its orbit from P would then be less than the kinetic energy required for the satellite to assume a uniform circular orbit at that radius. Thus the satellite would move closer to the Earth until passing point P again and repeating the orbit. Because the diagram reflects physical reality but one of the values in the questions does not, HSC markers would allow either answer provided it was supported by calculations or references to the diagram in the question.

CHAPTER 10

Centre Number

Student Number

NSW GOVERNMENT

NSW Education Standards Authority

2024 HIGHER SCHOOL CERTIFICATE EXAMINATION

Physics

General Instructions

- Reading time – 5 minutes
- Working time – 3 hours
- Write using black pen
- Draw diagrams using pencil
- Calculators approved by NESA may be used
- A data sheet, formulae sheet and Periodic Table are provided at the back of this paper
- Write your Centre Number and Student Number at the top of this page

Total marks: 100

Section I – 20 marks

- Attempt Questions 1–20
- Allow about 35 minutes for this section

Section II – 80 marks

- Attempt Questions 21–33
- Allow about 2 hours and 25 minutes for this section

Section I

20 marks
Attempt Questions 1–20
Allow about 35 minutes for this section

Use the multiple-choice answer sheet for Questions 1–20.

---

1 The diagram shows an object, P, undergoing uniform circular motion.

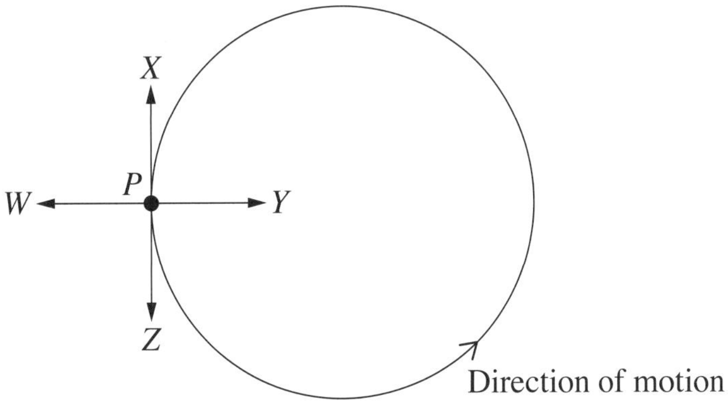

Which arrow shows the direction of the net force acting on P?

A. W

B. X

C. Y

D. Z

2 Which of the following provides evidence for the model of light proposed by Huygens?

A. Emission spectra

B. Diffraction of light

C. Black body radiation

D. The photoelectric effect

3 Which of the following is a fundamental particle in the Standard Model of matter?

A. Hadron

B. Neutron

C. Photon

D. Proton

4 A conducting coil is mounted on an axle and placed in a uniform magnetic field. The diagram shows different ways of connecting the coil to a power source.

Which setup allows the conducting coil to rotate continuously?

A.

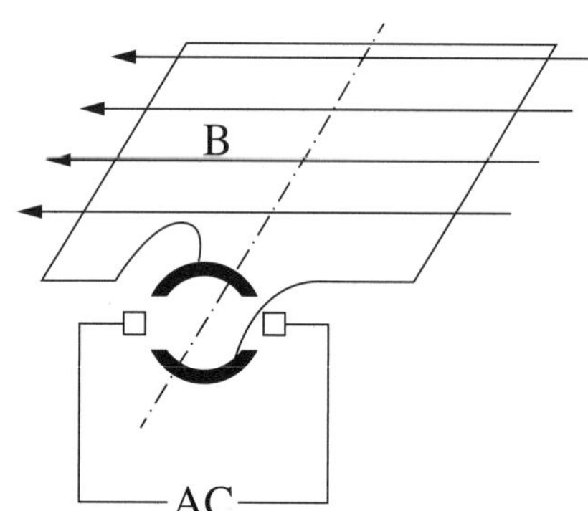

B.

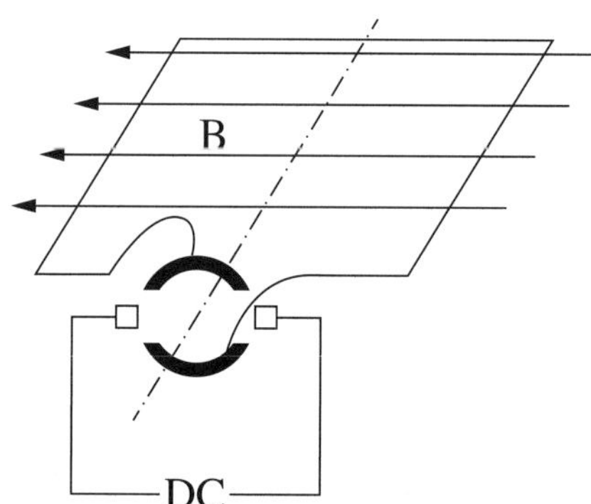

C.

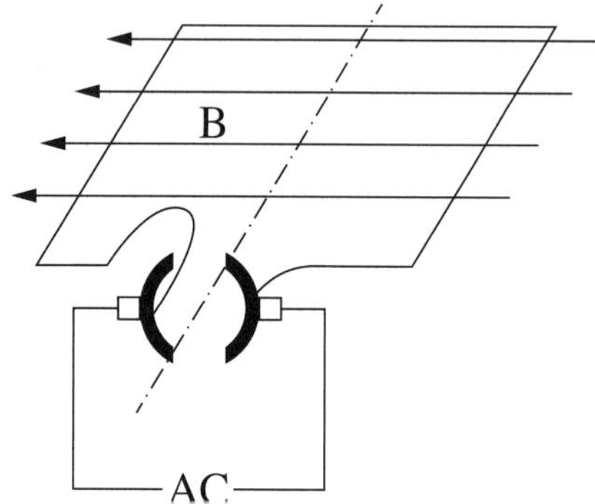

D.

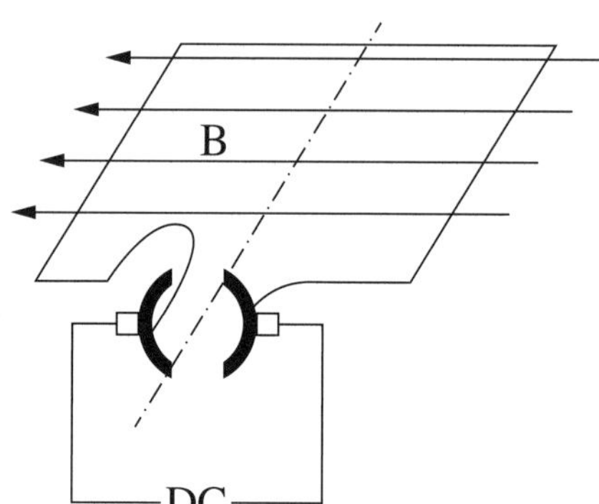

5 A star cluster is a group of stars that form at the same time. Hertzsprung–Russell diagrams for three star clusters, *X*, *Y* and *Z* are shown.

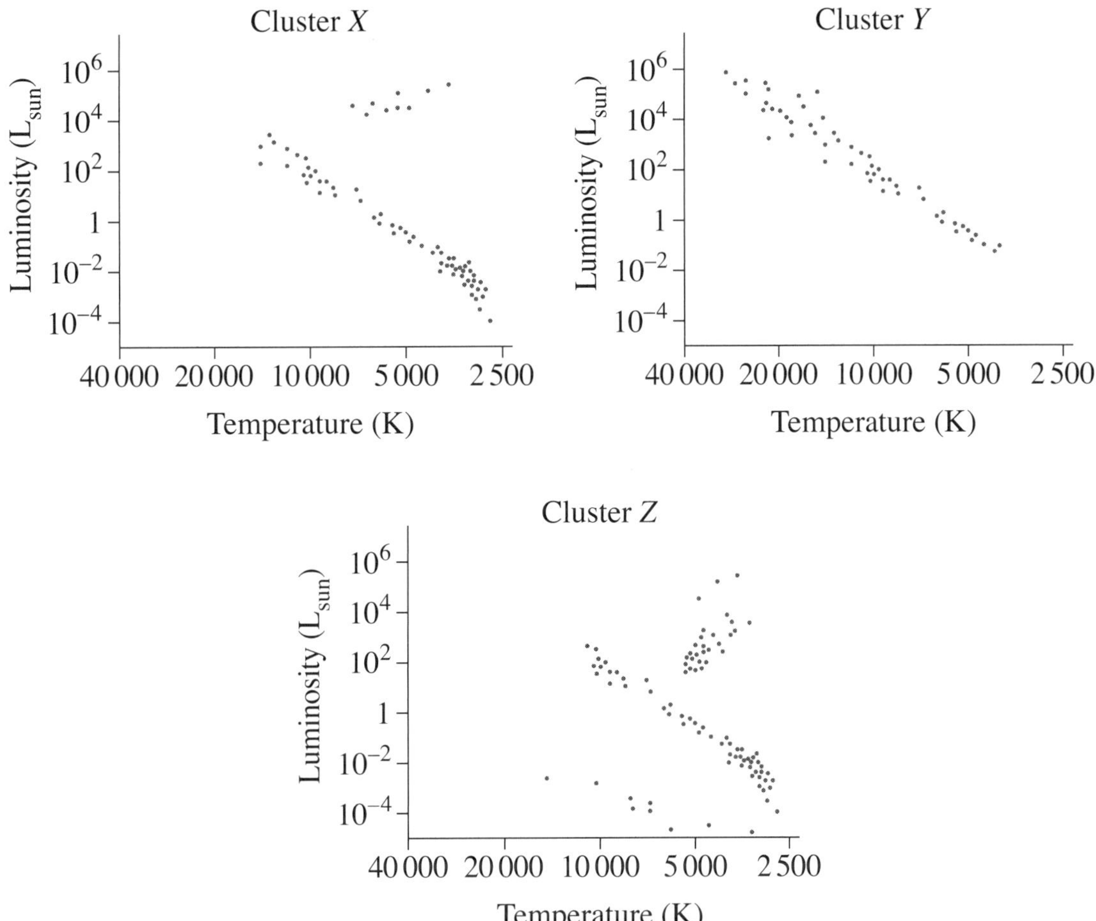

Which row of the table correctly shows the three star clusters from youngest to oldest?

	Youngest	⟶	*Oldest*
A.	*Y*	*X*	*Z*
B.	*Y*	*Z*	*X*
C.	*Z*	*X*	*Y*
D.	*Z*	*Y*	*X*

6 The photoelectric effect is mathematically modelled by the following relationship:

$$K_{max} = hf - \phi$$

In this model, the symbol ϕ represents the amount of energy

A. supplied by a photon to an electron.

B. retained by an electron after being hit.

C. required to release an electron from a material.

D. left over after a collision of a photon with an electron.

7 A pure sample of polonium-210 undergoes alpha emission to produce the stable isotope lead-206.

The half-life of polonium-210 is 138 days.

At the end of 276 days, what is the ratio of polonium-210 atoms to lead-206 atoms in the sample?

A. 1 : 4

B. 1 : 3

C. 1 : 2

D. 1 : 1

8 An ideal transformer produces an output of 6 volts when an input of 240 volts is applied.

What change would be needed to produce an output of 12 volts, using the same input voltage?

A. Increase the number of turns on the primary coil

B. Decrease the number of turns on the primary coil

C. Increase the resistance connected to the secondary coil

D. Decrease the resistance connected to the secondary coil

9 Object P is dropped from rest, and object Q is launched horizontally from the same height.

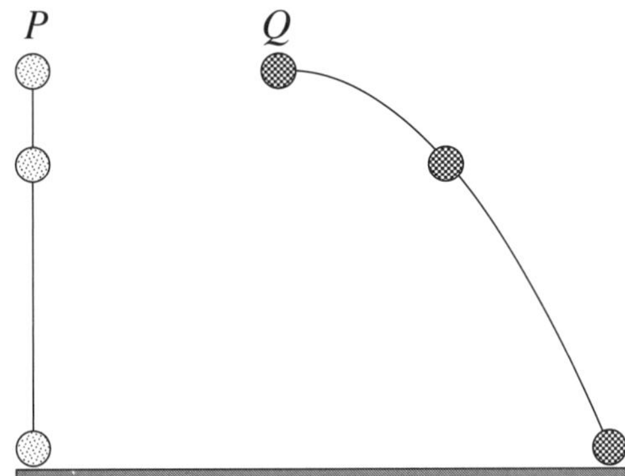

Which option correctly compares the projectile motion of P and Q?

A. The acceleration of P is less than the acceleration of Q.

B. The final velocity of Q is greater than the final velocity of P.

C. The time of flight of Q is greater than the time of flight of P.

D. The initial vertical velocity of P is less than the initial vertical velocity of Q.

10 A rod carrying a current, I, placed in a uniform magnetic field as shown, experiences a force F.

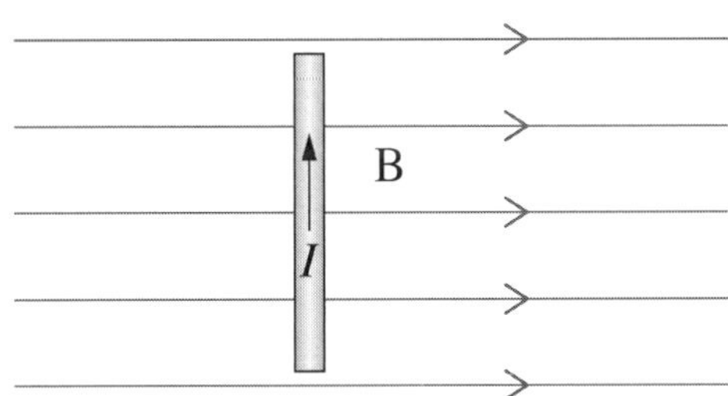

How many degrees must the rod be rotated clockwise so that it experiences a force $\frac{F}{2}$?

A. 30°

B. 45°

C. 60°

D. 90°

11 A satellite is in a circular orbit.

What is the relationship between its orbital velocity, v, and its orbital radius, r?

A. v is directly proportional to the square of r.

B. v is inversely proportional to the square of r.

C. v is directly proportional to the square root of r.

D. v is inversely proportional to the square root of r.

12 A rod has a length, L_0, when measured in its own frame of reference.

The rod travels past a stationary observer at speed, v, as shown in the diagram.

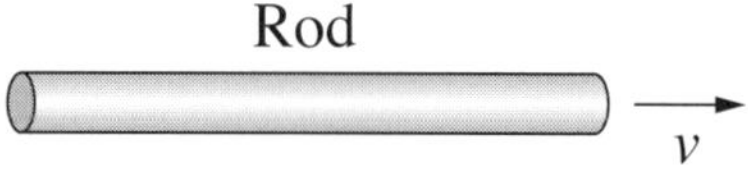

Which option represents the relationship between the speed of the rod, v, and the length of the rod as measured by the stationary observer?

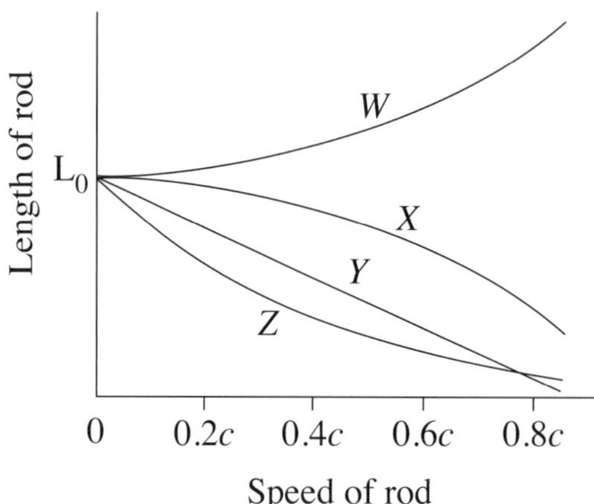

A. W

B. X

C. Y

D. Z

13 The diagram shows two identical satellites, *A* and *B*, orbiting a planet.

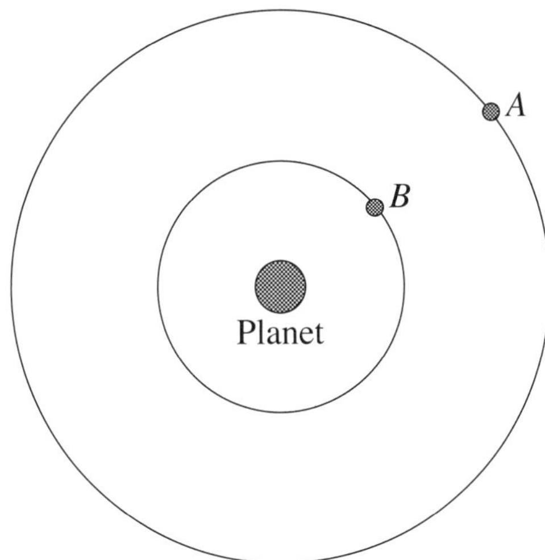

Which row in the table correctly compares the potential energy, *U*, and kinetic energy, *K*, of the satellites?

	Potential energy	*Kinetic energy*
A.	$U_A > U_B$	$K_A < K_B$
B.	$U_A < U_B$	$K_A > K_B$
C.	$U_A > U_B$	$K_A > K_B$
D.	$U_A < U_B$	$K_A < K_B$

14 The velocity of a proton $\left({}^1_1\text{H}\right)$ is twice the velocity of an alpha particle $\left({}^4_2\text{He}\right)$. The proton has a de Broglie wavelength of λ.

What is the de Broglie wavelength of the alpha particle?

A. $\dfrac{\lambda}{8}$

B. $\dfrac{\lambda}{2}$

C. 2λ

D. 8λ

15 A uniform magnetic field is directed into the page. A conductor PQ rotates about the end P at a constant rate.

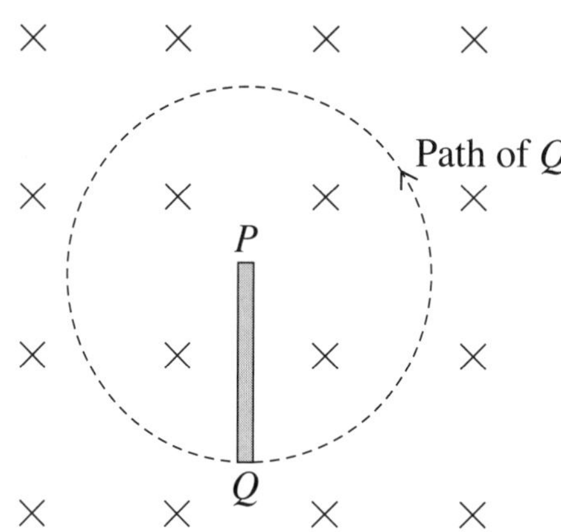

Which graph shows the emf induced between the ends of the conductor, P and Q, as it rotates one revolution from the position shown?

A.

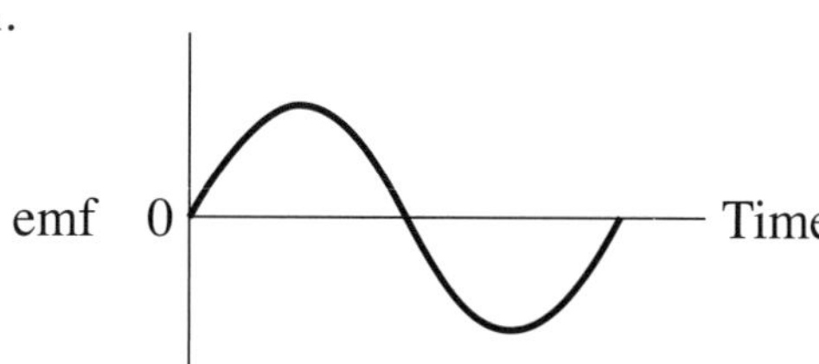

B.

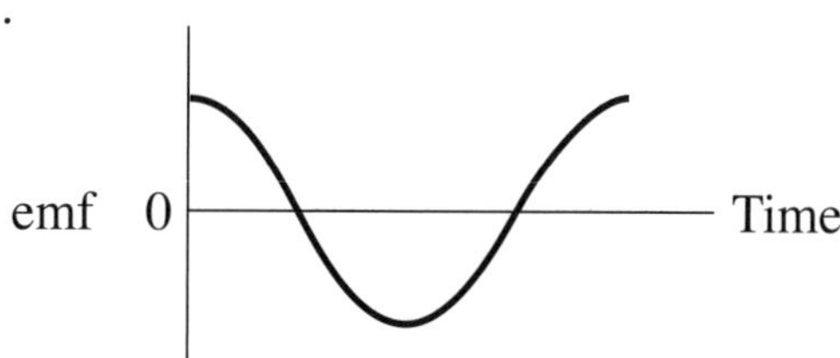

C.

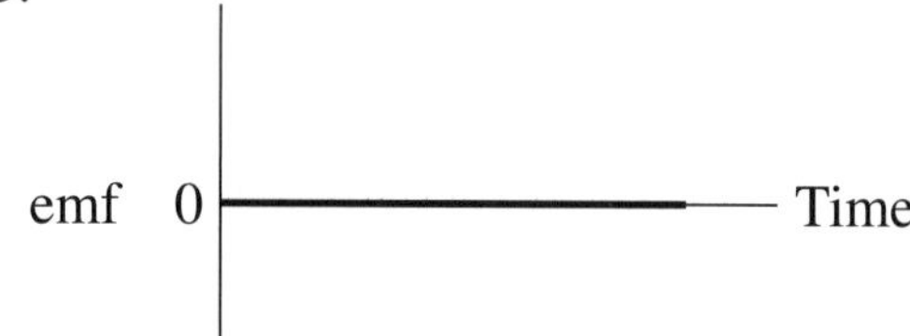

D.

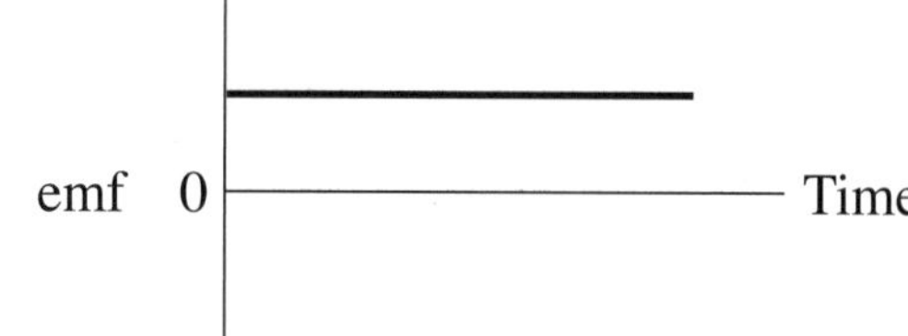

16 The graph shows the relationship between the maximum kinetic energy of emitted photoelectrons and the incident photon energy for four different metal surfaces.

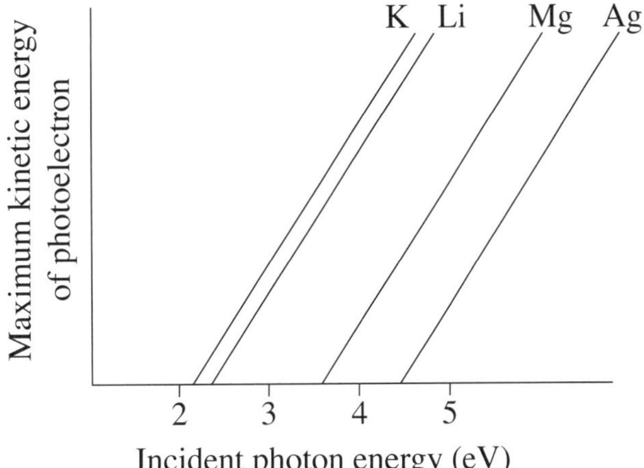

Light of frequency 7×10^{14} Hz is incident on the metals.

From which metals are photoelectrons emitted?

A. K, Li only

B. Mg, Ag only

C. All of the metals

D. None of the metals

17 The diagram shows a type of particle accelerator called a cyclotron.

Cyclotrons accelerate charged particles, following the path as shown.

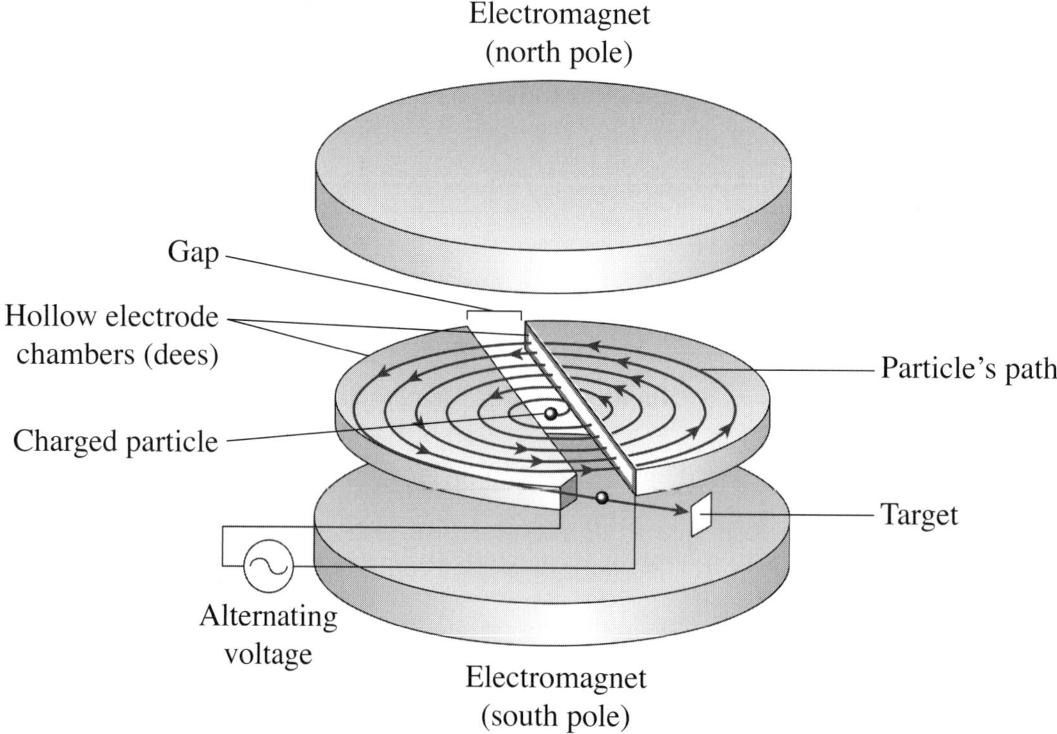

An electric field acts on a charged particle as it moves through the gap between the dees. A strong magnetic field is also in place.

Once a charged particle has the required velocity, it exits the accelerator towards a target.

Which of the following is true about a charged particle in a cyclotron?

A. It increases speed while inside the dees.

B. It only accelerates while between the dees.

C. It undergoes acceleration inside and between the dees.

D. It slows down inside the dees and speeds up between the dees.

18 The diagram shows a magnet moving towards a coil X.

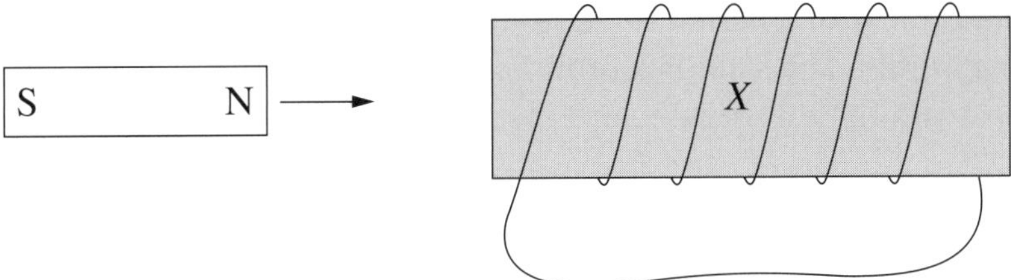

This action causes a current to be induced in the coil.

Which situation will induce a current in coil X that is in the same direction as the current induced by the movement of the magnet?

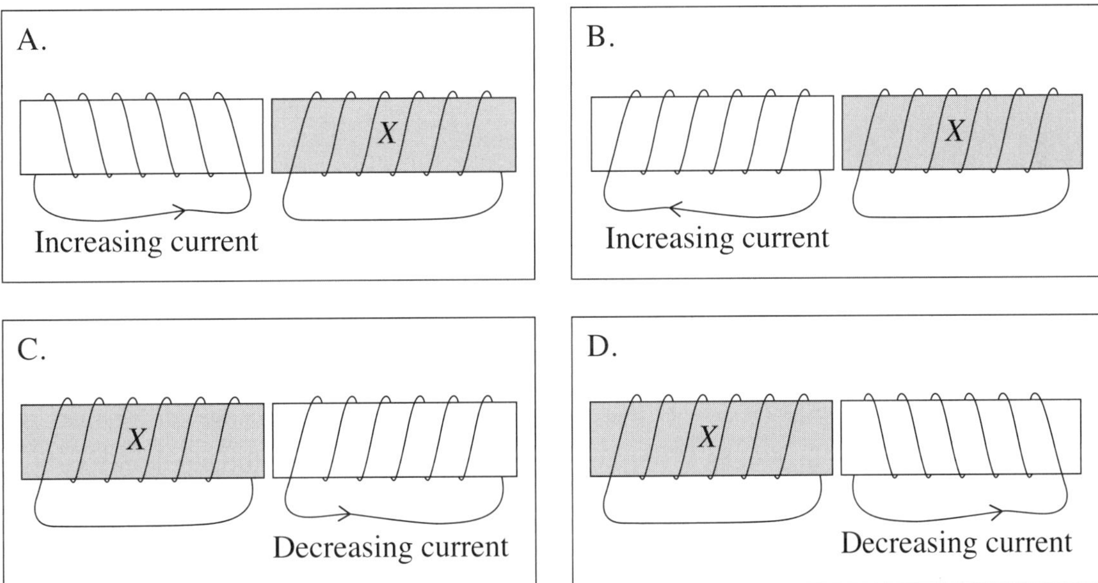

19 In a vacuum chamber there is a uniform electric field and a uniform magnetic field.

A proton having a velocity, v, enters the chamber. Its velocity remains unchanged as it travels through the chamber.

A second proton having a velocity, $2v$, in the same direction as the first proton, then enters the chamber at the same point as the first proton.

In the chamber, the acceleration of the second proton

A. is zero.

B. is constant in magnitude and direction.

C. changes in both magnitude and direction.

D. is constant in magnitude, but not direction.

20 Three identical atomic clocks are made so that they tick at precisely the same rate. One is kept in a laboratory, X, on Earth's equator. Another is placed on board a satellite, Y, in a circular orbit with a period of 12 hours. A third is placed in a satellite, Z, that is in a geostationary orbit. The satellites orbit Earth in the equatorial plane.

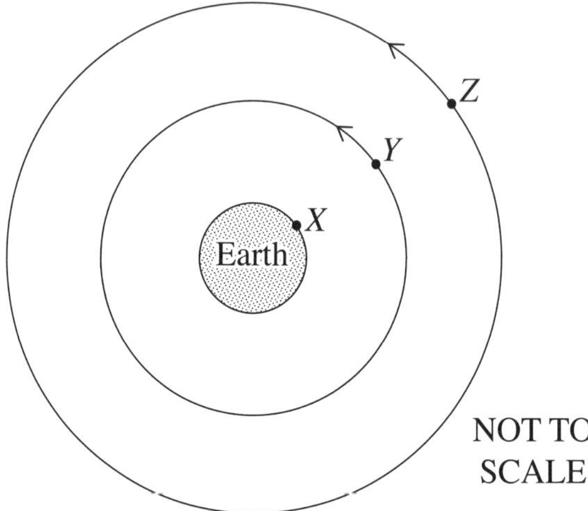

Assume that the satellites are inertial frames of reference and the clocks are affected ONLY by the predictions of special relativity.

Which statement correctly compares the rates at which the clocks tick, as determined by an observer at X, when the satellites are in the positions shown in the diagram?

A. The clock at Y ticks faster than either the clock at X or the clock at Z.

B. The clock at Y ticks slower than either the clock at X or the clock at Z.

C. The clocks tick at different rates, with X being the fastest and Y being the slowest.

D. The clocks tick at different rates, with Z being the slowest and X being the fastest.

2024 HIGHER SCHOOL CERTIFICATE EXAMINATION

Centre Numl

Student Numl

Physics

Section II Answer Booklet

80 marks
Attempt Questions 21–33
Allow about 2 hours and 25 minutes for this section

Instructions

- Write your Centre Number and Student Number at the top of th page.
- Answer the questions in the spaces provided. These spaces provide guidance for the expected length of response.
- Show all relevant working in questions involving calculations.
- Extra writing space is provided at the back of this booklet. If you use this space, clearly indicate which question you are answering.

Please turn over

Question 21 (6 marks)

To tighten a nut, a force of 75 N is applied to a spanner at an angle, as shown.

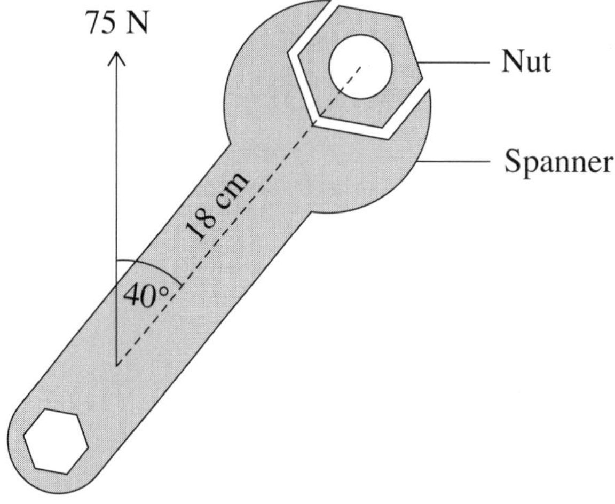

(a) Calculate the magnitude of the torque produced by the applied force. **2**

....

....

....

....

(b) Explain TWO ways in which torque can be increased in a simple DC motor. **4**

....

....

....

....

....

....

....

....

Question 22 (5 marks)

The following graph, based on the data gathered by Hubble, shows the relationship between the recessional velocity of galaxies and their distance from Earth.

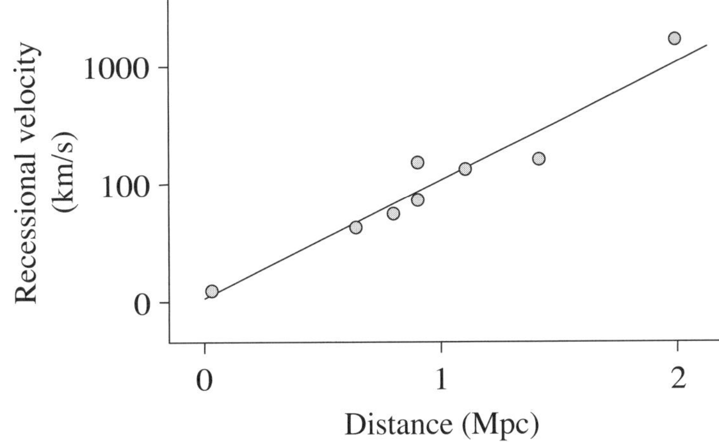

(a) Describe the significance of the graph to our understanding of the universe. **2**

....

....

....

....

(b) How were the recessional velocities of galaxies determined? **3**

....

....

....

....

....

....

Question 23 (9 marks)

Development of models of the atom has resulted from both experimental investigations and hypotheses based on theoretical considerations.

(a) A key piece of experimental evidence supporting the nuclear model of the atom was a discovery by Chadwick in 1932.

An aspect of the experimental design is shown.

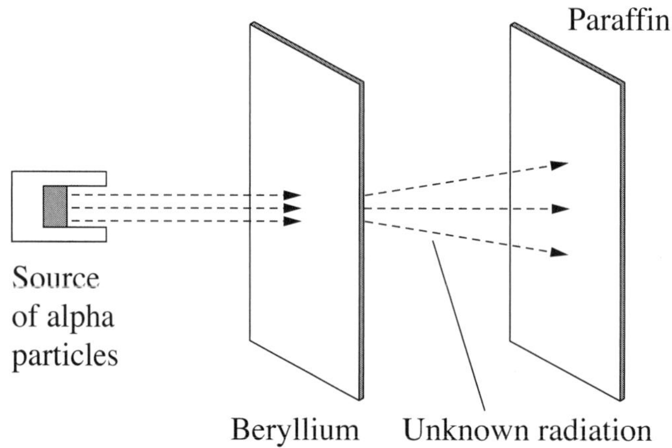

(i) What was the role of paraffin in Chadwick's experiment? **2**

....

....

....

....

(ii) How did Chadwick's experiment change the model of the atom? **3**

....

....

....

....

....

....

Question 23 continues on the following page

Question 23 (continued)

(b) Explain how de Broglie's hypothesis regarding the nature of electrons addressed limitations in the Bohr–Rutherford model of the atom. **4**

....

....

....

....

....

....

....

....

End of Question 23

Please turn over

Question 24 (8 marks)

An absorption spectrum resulting from the passage of visible light from a star's surface through its hydrogen atmosphere is shown. Absorption lines are labelled *W* to *Z* in the diagram.

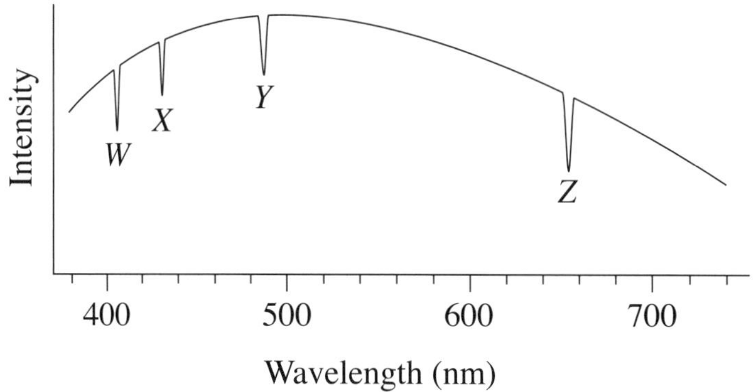

(a) Determine the surface temperature of the star. **2**

....

....

....

....

(b) Absorption line *W* originates from an electron transition between the second and sixth energy levels. Use $\frac{1}{\lambda} = R\left(\frac{1}{n_f^2} - \frac{1}{n_i^2}\right)$ to calculate the frequency of light absorbed to produce absorption line *W*. **3**

....

....

....

....

....

....

....

Question 24 continues on the following page

Question 24 (continued)

(c) Explain the physical processes that produce an absorption spectrum. **3**

....

....

....

....

....

....

....

End of Question 24

Please turn over

Question 25 (6 marks)

The mathematical model below shows the relationship between the orbital radius of a satellite and its period.

$$\frac{r^3}{T^2} = \frac{GM}{4\pi^2}$$

(a) By considering gravitational force, show how this model can be derived. **2**

....

....

....

....

....

....

Question 25 continues on the following page

Question 25 (continued)

(b) A planet with five moons is discovered. The following graph is produced from observations of the orbital radius of the moons and their orbital periods, measured in Earth days. **4**

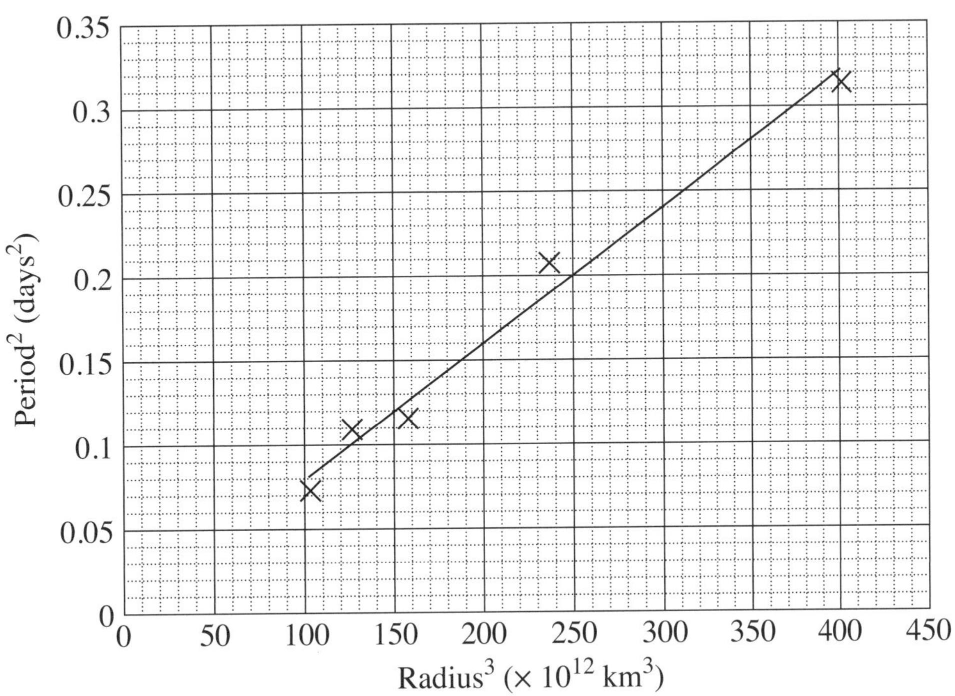

Use the graph to calculate the mass of the planet.

....

....

....

....

....

....

....

....

End of Question 25

Question 26 (3 marks)

Muons are unstable particles produced when cosmic rays strike atoms high in the atmosphere. The muons travel downward, perpendicular to Earth's surface, at almost the speed of light. **3**

Classical physics predicts that these muons will decay before they have time to reach Earth's surface.

Explain qualitatively why these muons can reach Earth's surface, regardless of whether their motion is considered from either the muon's frame of reference or the Earth's frame of reference.

....

....

....

....

....

....

....

....

Question 27 (7 marks)

The simplified model below shows the reactants and products of a proton–antiproton reaction which produces three particles called pions, each having a different charge.

$$\mathrm{p} + \bar{\mathrm{p}} \longrightarrow \pi^{+} + \pi^{0} + \pi^{-}$$

There are no other products in this process, which involves only the rearrangement of quarks. No electromagnetic radiation is produced. Assume that the initial kinetic energy of the proton and antiproton is negligible.

Protons consist of two up quarks (u) and a down quark (d). Antiprotons consist of two up antiquarks ($\bar{\mathrm{u}}$) and a down antiquark ($\bar{\mathrm{d}}$). Each of the pions consists of two quarks.

The following tables provide information about hadrons and quarks.

Table 1: Hadron information

Particle	*Rest mass* (MeV/c^2)	*Charge*
proton (p)	940	+1
antiproton ($\bar{\mathrm{p}}$)	940	−1
neutral pion (π^0)	140	zero
positive pion (π^+)	140	+1
negative pion (π^-)	140	−1

Table 2: Quark charges

Particle	*Charge*
down quark (d)	$-\frac{1}{3}$
up quark (u)	$+\frac{2}{3}$
down antiquark ($\bar{\mathrm{d}}$)	$+\frac{1}{3}$
up antiquark ($\bar{\mathrm{u}}$)	$-\frac{2}{3}$

Question 27 continues on the following page

Question 27 (continued)

(a) Identify the quarks present in the π^-, π^+ and the π^0 particles. **2**

....

....

....

....

(b) The energy released in the reaction is shared equally between the pions. **2**

Calculate the energy released per pion in this reaction.

....

....

....

....

(c) Calculation of the pions' velocities using classical physics predicts that each pion has a velocity, relative to the point at which the proton–antiproton reaction occurred, which exceeds 3×10^8 m s^{-1}. **3**

Explain the problem with this prediction and how it can be resolved.

....

....

....

....

....

....

End of Question 27

Question 28 (7 marks)

An electron gun fires a beam of electrons at 2.0×10^6 m s^{-1} through a pair of parallel charged plates towards a screen that is 30 mm from the end of the plates as shown.

There is a uniform electric field between the plates of 1.5×10^4 N C^{-1}. The plates are 5.0 mm wide and 20 mm apart. The electron beam enters mid-way between the plates. *X* marks the spot on the screen where an undeflected beam would strike.

Ignore gravitational effects on the electron beam.

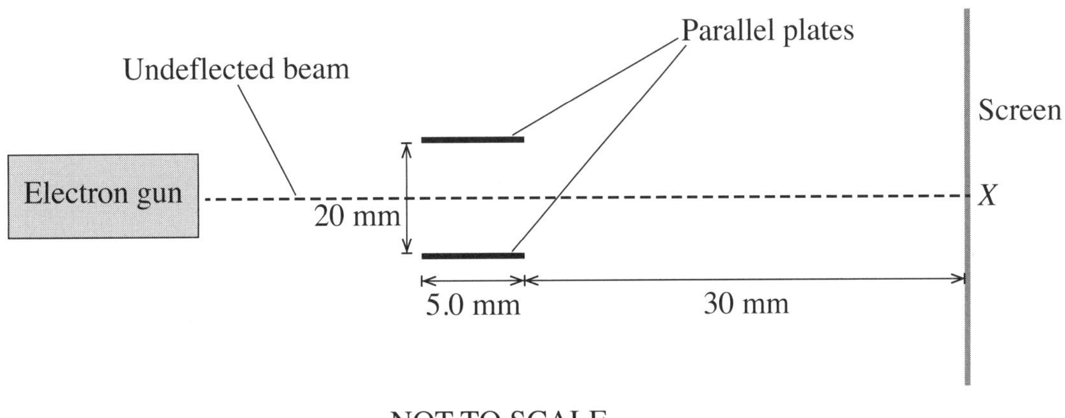

NOT TO SCALE

(a) Show that the acceleration of an electron between the parallel plates is 2.6×10^{15} m s^{-2}. **2**

....

....

....

....

(b) Show that the vertical displacement of the electron beam at the end of the parallel plates is approximately 8.1 mm. **2**

....

....

....

....

Question 28 continues on the following page

Question 28 (continued)

(c) How far from point *X* will the electron beam strike the screen? **3**

....

....

....

....

....

....

End of Question 28

Please turn over

Question 29 (6 marks)

Two horizontal metal rods, A and B, of different materials are resting on a frictionless table. Initially they are at rest in position 1.

Both rods are then connected to a battery using wires. After the switch is turned on, currents of different magnitude flow in each rod. The rods move to position 2 after time, t. In position 2, B has a larger displacement than A from position 1. The masses of the wires are negligible.

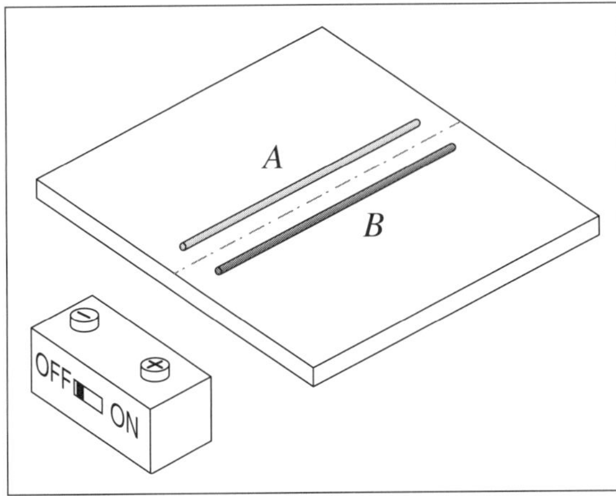

Position 1

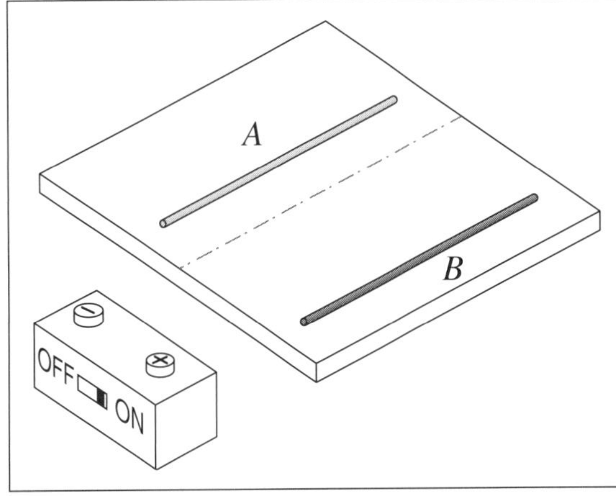

Position 2

(a) Position 1 is reproduced below. Draw wires to show how the battery must be connected to the ends of the two rods in order for the magnitude of the current in each rod to be different, and for position 2 to be reached. No components, other than the wires, are required. **2**

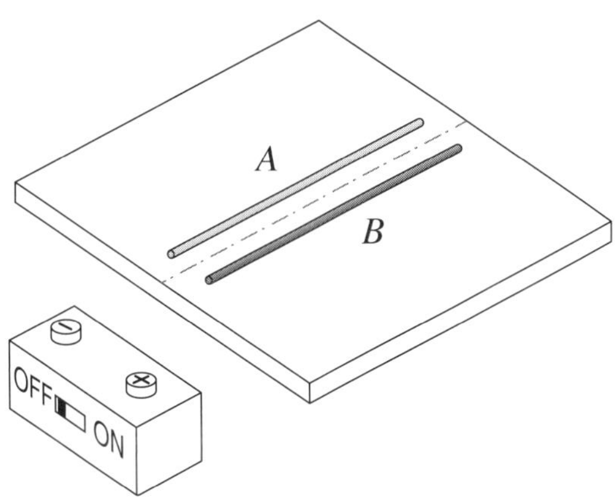

Question 29 continues on the following page

Question 29 (continued)

(b) When the switch is turned on, the current in rod *A* is greater than the current in rod *B*. **4**

Consider this statement.

> Position 2 results from the larger current in rod *A*, causing a larger force to act on rod *B*.

Evaluate this statement with reference to relevant physics principles.

....

....

....

....

....

....

....

....

End of Question 29

Please turn over

Question 30 (4 marks)

An object sits on the floor of a hollow cylinder rotating around an axis, as shown. The cylinder's rotation causes the object to undergo uniform circular motion. **4**

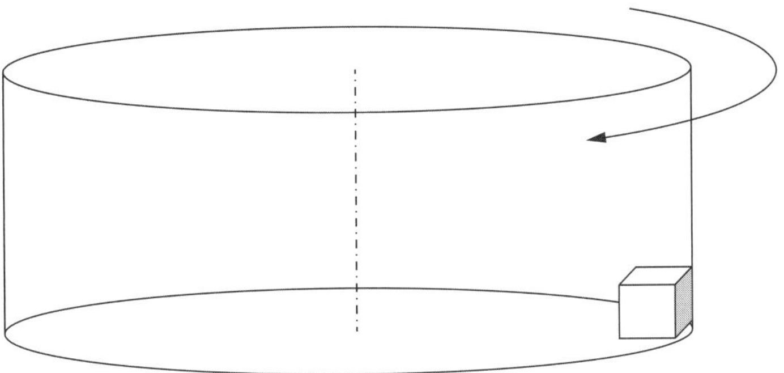

Explain the effect on all of the forces acting on the object if the period of the cylinder's rotation is halved. Ignore the effects of friction.

....

....

....

....

....

....

....

....

Question 31 (4 marks)

In a thought experiment, a projectile is launched vertically from Earth's surface. Its initial velocity is less than the escape velocity. 4

The behaviour of the projectile can be analysed by using two different models, Model *A* and Model *B* as shown.

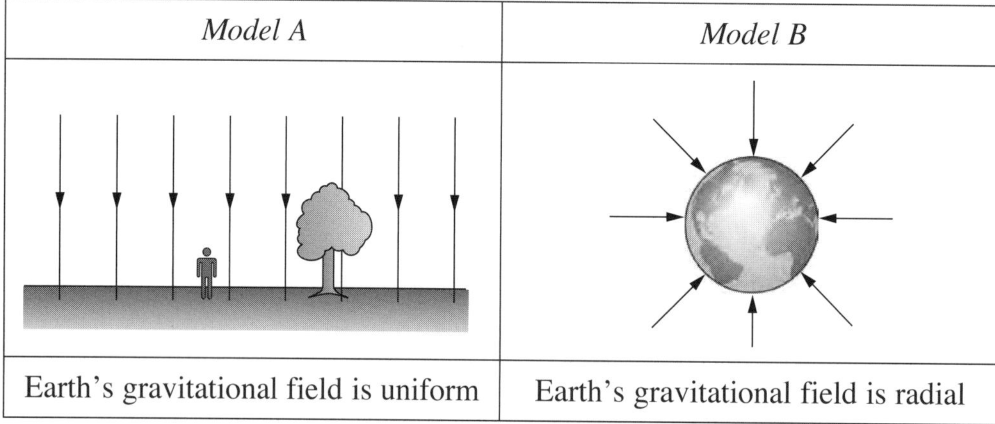

The effects of Earth's atmosphere and Earth's rotational and orbital motions can be ignored.

Compare the maximum height reached by the projectile, using each model. In your answer, describe the energy changes of the projectile.

Question 32 (8 marks)

Many scientists have performed experiments to explore the interaction of light and matter. **8**

Analyse how evidence from at least THREE such experiments has contributed to our understanding of physics.

Question 33 (7 marks)

A magnet is swinging as a pendulum. Close below it is an aluminium (non-ferromagnetic) can. The can is free to spin around a fixed axis as shown.

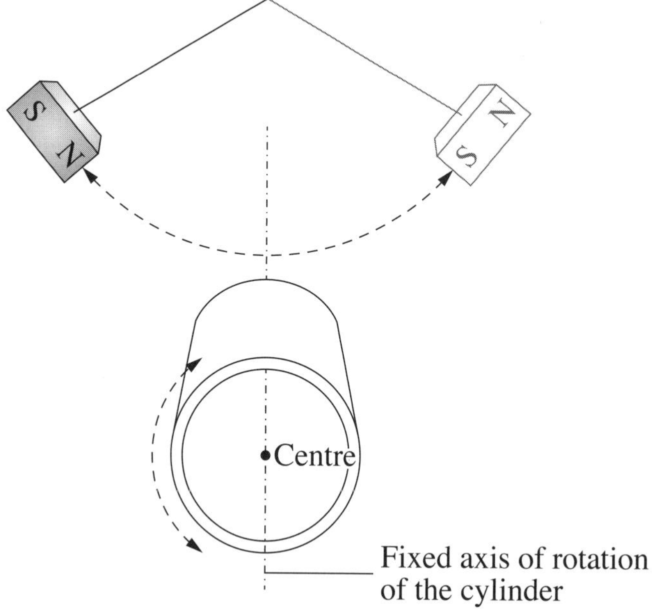

Analyse the motion and energy transformations of both the can and the magnet.

....

....

....

....

....

....

....

....

....

....

....

....

....

End of paper

2024 HSC Examination paper

Sample answers

Section I (Total 20 marks)

1 C Because the object is undergoing uniform circular motion, the net force on the object is the centripetal force. This force is directed towards the centre of the circle.

2 B Huygens proposed a wave model of light. Diffraction can be explained with a wave model of light, while the other options can only be satisfactorily explained using the particle (quantum) model of light.

3 C The photon is the only fundamental particle in this group. The other options are all composite particles, made up of two or more quarks.

4 D This is a DC motor, with a split ring commutator. Answers A and C are incorrect because the motor is connected to an AC power supply. Answer D is correct and B is incorrect because the commutator must reverse the current direction in the coil when the plane of the armature coil is perpendicular to the applied magnetic field. The motor in answer B would not continue to rotate as the commutator would change the direction of current in the armature coil when the plane of the coil was parallel to the applied magnetic field.

5 A *Y* is the youngest star cluster because all the stars are on the main sequence and none have evolved to become red giants or white dwarf stars. Star cluster *X* is older than *Y* because some of the stars have left the main sequence and evolved into red giant stars. (These are the very luminous, cooler stars at the top right of the HR diagram.) Star cluster *Z* is older than both *X* and *Y* because some of the stars in the cluster have left the main sequence and evolved to red giants and then to white dwarf stars. (These are the hot, less luminous stars in the lower left of the HR diagram.) We can also deduce the relative age of each cluster from the 'turn off point' from the main sequence because the turn off point is lower for older star clusters.

6 C The symbol ϕ represents the amount of energy required to release an electron from the surface of the metal. We can see this must be the case because if the incident photon had just enough energy to release an electron from the surface ($hf = \phi$), the equation shows that the photoelectron produced would have zero kinetic energy, as expected.

7 B Because 276 days is two half-lives, 75% of the initial sample of polonium-210 would have decayed to lead-206. This would leave 25% of the polonium-210. Hence the ratio of polonium-210 to lead-206 would be 1 : 3 after 276 days.

8 B By rearranging the transformer equation $\frac{V_P}{V_S} = \frac{N_P}{N_S}$, we see that $V_S = \frac{V_P N_S}{N_P}$. Hence reducing the number of turns on the primary coil will increase the voltage on the secondary coil.

9 B Both balls will have the same final vertical velocity. Ball P has no final horizontal velocity, but ball Q has its original horizontal velocity when it lands and hence its final velocity will be the vector sum of the initial horizontal velocity and the final vertical velocity. A is incorrect as both balls have the same acceleration (g downwards). C is incorrect as the balls will have the same time of flight (the time of flight is determined by the acceleration due to gravity, vertical displacement and initial vertical velocity only and these quantities are the same for both situations). D is incorrect as both balls have zero vertical velocity initially.

10 C We see from $F = lIBsin\theta$ that the force on the rod is proportional to the sine of the angle between the applied magnetic field and the rod. Therefore, if the force is F when the angle between the applied magnetic field and the rod is 90°, the force will be $\frac{F}{2}$ when the angle between the applied magnetic field and the rod is changed to 30°. Hence the wire must be rotated by 60° clockwise to halve the force on the wire.

11. D As gravity provides the centripetal force on a satellite we can write $\frac{GMm}{r^2} = \frac{mv^2}{r}$ and hence $v = \sqrt{\frac{GM}{r}}$. From this equation we can see the velocity of the satellite is inversely proportional to the square root of the radius of its orbit r.

12 B The length of the rod as measured by the stationary observer is given by $L = L_o\sqrt{1 - \frac{v^2}{c^2}}$. We see that as the rod approaches the speed of light its length approaches zero. Only curve X approaches zero as the speed of the rod approaches the speed of light (c). In addition, only X is concave down which is the shape of the graph of a function of the form $y = \sqrt{1 - x^2}$.

13 A The velocity and kinetic energy of a satellite decreases as the orbital radius increases (from $v = \sqrt{\frac{GM}{r}}$ and $k = \frac{GMm}{2r}$). Satellite B will therefore have the highest kinetic energy. According to the equation for potential energy $\left(U = -\frac{GMm}{r}\right)$, as r increases, U approaches zero (i.e. becomes less negative). Thus satellite B has a more negative potential energy value, i.e. $U_B < U_A$.

14 B The de Broglie wavelength is given by $\lambda = \frac{h}{mv}$. For an alpha particle with half the velocity and four times the mass of the proton, the new wavelength will be given by $\lambda(He) = \lambda\left(\frac{1}{\left(\frac{1}{2}\right)(4)}\right) = \frac{\lambda}{2}$.

15 D A test positive charge placed in the rod would experience a motor effect force, as a consequence of the rod's motion in a magnetic field ($F = qvBsin\theta$). This force will be constant in magnitude (as q, v, B and $sin\theta$ are unchanged) and the direction of the force will always be towards the end of the rod marked P. This force will result in the constant emf.

16 A The energy of the incident photon is given by $E = hf = (6.626 \times 10^{-34})(7 \times 10^{14}) = 4.64 \times 10^{-19}$ J. Converting this quantity to eV, $E = \frac{4.64 \times 10^{-19}}{1.602 \times 10^{-19}} = 2.9$ eV. Photoelectrons will only be released from potassium and lithium because the incident photons would not have enough energy to eject photoelectrons from magnesium or silver.

17 C The charged particles will accelerate (increase in speed is acceleration) between the dees as they are in an electric field ($F = qE$ and hence $a = \frac{qE}{m}$). The charged particles also accelerate when they are inside the dees (change in direction is acceleration) because they are moving perpendicularly to a magnetic field, which produces a centripetal force on the charged particles ($F = qvB$). This centripetal force produces a centripetal acceleration, causing the charged particles to move with uniform circular motion when they are inside the dees.

18 D By Lenz's law, moving the magnet towards the coil will induce a current in the coil in a direction that will oppose the change in magnetic flux that induced the current. Hence current will flow in a direction that will produce a north pole on the left end of the coil to oppose the approaching north pole produced by the moving magnet.

In A, an increasing current through the coil will produce a strengthening south pole at its right end. This is similar to an approaching south pole and will produce an opposing south pole at the left end of Coil *X*.

In B, even though the indicated increasing current seems to be going in the other direction, it will again produce a strengthening south pole at its right end. The outcome will therefore be the same as in A.

In C, the decreasing current through the coil will produce a diminishing south pole at its left end. This is similar to a south pole moving away and will produce an attracting north pole at the right end of Coil *X*.

Only in answer D will the induced current produce a magnetic field in coil *X* with its south pole at the right end of the coil. The applied magnetic field has a north pole at its left end that is decreasing and the induced current in coil *X* will produce a magnetic field with a south pole at its right end to oppose this change.

19 C In the vacuum chamber an electric field must be perpendicular to a magnetic field such that when a proton with velocity v enters the chamber, the magnetic and electric forces are equal in magnitude but operate in an opposite direction. This would ensure the net force on the proton was zero and hence its acceleration is also zero and it would move with constant velocity through the chamber. For this particle $qE = qvB$. If a proton entered the chamber in the same direction with a velocity of $2v$, the magnetic force on the proton would double and therefore be greater than the electric force and hence there would be a net force on the proton and it would begin to accelerate in the direction of the force.

If there was a magnetic field alone, the particle would move with uniform circular motion. This is because the magnetic force on the proton would have a constant magnitude and be directed perpendicularly to the velocity (i.e. it would be a centripetal force). But with a magnetic and electric field acting, the magnitude and direction of the net force on the proton would be given by the vector sum of the two forces. As the electric force remains in one direction but the magnetic force changes direction as the direction of the proton's velocity changes, the net force on the proton travelling at $2v$ would continually change in magnitude and direction and hence the acceleration of the proton would also change continuously in magnitude and direction.

20 B Time dilation occurs when a clock moves relative to an observer. The observer will see time passing more slowly in the frame moving with respect to the observer than time passing in the observer's frame. Now, as the velocity of a satellite is inversely related to the radius of the satellite's orbit $\left(v = \sqrt{\frac{GM}{r}}\right)$, satellite Y will be moving with the greatest velocity relative to X (note we are also told Y has a period of 12 hours and thus must move relative to X and Z which have periods of 24 hours). Hence an observer at X would see a clock on Y running slower than their own clock. If we assume the reference frames are inertial, X and Z will be in the same frame and the clock at Z will tick at the same rate as the clock at X. The only correct answer is therefore B.

Section II

Question 21 (Total 6 marks)

(a) The applied torque is given by $\tau = rF\sin\theta = 0.18 \times 75 \times \sin 40° = 8.7$Nm clockwise. *(2 marks)*

(b) The torque produced by a simple DC motor is proportional to the current in the armature, the number of turns on the armature coil, the cross-sectional area of the armature coil and to the applied magnetic field ($\tau = nIAB\sin\theta$). Hence the torque could be increased by (students could choose to answer using any **two** of the following):

(i) increasing the current in the armature coil by increasing the voltage applied across the armature coil. This would increase the motor force on each side of the armature coil and hence increase the torque of the motor

(ii) increasing the strength of the applied magnetic field which would also increase the motor force on each side of the armature coil and increase the torque of the motor

(iii) increasing the number of turns on the armature coil. This would increase the motor force on each side of the armature coil and hence increase the torque of the motor

(iv) introducing radial magnets in place of uniform field magnets. This would result in a torque closer to the maximum torque for a greater duration of the rotation cycle. *(4 marks)*

Question 22 (Total 5 marks)

(a) Hubble's results showed that the further away a galaxy, the faster it is moving away. In other words, he showed the universe was expanding. In addition, by graphing the results, he showed that the recessional velocity was proportional to the distance of the galaxy from the Earth. This is known as Hubble's law ($v = H_0D$) and the constant of proportionality (H_0), which is a measure of the rate of expansion of the universe, is called Hubble's constant. Hubble's discovery of an expanding universe also led to the Big Bang theory of the origin of the universe. (*2 marks*)

(b) Hubble used cosmological red shift to determine the recessional velocity of the galaxies he studied. Red shift refers to the increase in wavelength of light seen by an observer when the source of light is moving away from the observer. Hubble noted that the galactic spectrum of light from all the galaxies he observed was red-shifted, indicating that all the galaxies he observed were moving away from the Earth. Hubble then measured the red shift from each galaxy by comparing the wavelengths of atomic absorption lines in the galactic spectrum to the wavelength of the same atomic absorption lines measured in laboratories on Earth. Finally, he used the measured increase in wavelength (i.e. the red shift) from each galaxy and the Doppler shift equation for light to calculate the velocity with which the galaxy was moving away from the Earth (i.e. the recessional velocity). (*3 marks*)

Question 23 (Total 9 marks)

(a) The role of the paraffin (which is proton rich) was to provide protons that would be ejected after being hit by the energetic neutral radiation emitted from the alpha-irradiated beryllium. Chadwick hoped to show that the neutral radiation was made up of neutral particles with a similar mass to a proton, but he knew it would be difficult to detect or measure the properties of an uncharged particle directly. Because they were charged, the protons ejected by the neutral radiation from paraffin could be more easily detected and their energy and momentum more easily measured than the uncharged neutrons. Chadwick used the recoil velocity of the protons, with measurements of the recoil velocity of other atoms/molecules, to deduce (using conservation of energy and momentum) that the unknown radiation was made up of neutral particles with a mass slightly larger than a proton. Thus, experimentally, he confirmed the existence of the neutron. (*2 marks*)

(b) Rutherford's initial nuclear model of the atom was built using two fundamental particles: electrons and protons. His model could not explain the discrepancy between the atomic number and mass number of elements in the periodic table. Nor could the model explain the existence of isotopes—atoms of a specific element that had different masses. To improve his model, Rutherford proposed that a neutral particle, with a mass similar to a proton, must exist in the nucleus. He later called this particle a neutron. Chadwick's experiment proved the existence of neutrons, 12 years after they had been predicted by Rutherford. The discovery meant that atoms consisted of three rather than two fundamental particles: the proton, the neutron and the electron. This changed the model of the atom because, with protons and neutrons in the nucleus, the difference between the mass number and atomic number of the atoms in the periodic table could be explained by the presence of neutrons in the nucleus. It also could be used to explain isotopes, because a specific element could have a different number of neutrons in the nucleus and hence the same atomic number but a different mass number. *(3 marks)*

(c) A major limitation of the Bohr-Rutherford model of the atom was that there was no justification for the existence of quantised (stationary state) orbitals. Nor was there any justification for Bohr's postulate that these stationary states could only exist when the orbiting electron obeyed the condition that $mvr = \frac{nh}{2\pi}$.

De Broglie proposed that particles had wave properties and an associated wavelength related to the momentum of the particle by $\lambda = \frac{h}{mv}$. De Broglie explained the existence of stationary state orbitals by suggesting that electron orbitals could only exist when the circumference of the orbital was a whole number of electron wavelengths and that allowed orbitals corresponded to standing electron waves. Thus the first allowed orbital had a circumference that corresponded to one electron wavelength; the second allowed orbital had a circumference corresponding to two electron wavelengths, and so on. The allowed orbitals in de Broglie's standing wave model therefore occurred when,

$2\pi r = \lambda$ and as $\lambda = \frac{h}{mv}$

$$2\pi r = n\left(\frac{h}{mv}\right)$$

Rearranging this equation, we see that the condition for the existence of stationary orbit is $mvr = \frac{nh}{2\pi}$.

Thus de Broglie's hypothesis can be used to explain why the orbitals are quantised in the Bohr-Rutherford model and to derive the quantum condition that specifies what orbitals are allowed. *(4 marks)*

Question 24 (Total 8 marks)

(a) We can find the temperature of the star by reading the wavelength of maximum emission from the spectrum and applying Wien's Law.

The wavelength of maximum emission from the black body spectra is approximately 500 nm.

Therefore $\lambda_{max} = \frac{b}{T}$ and hence $T = \frac{b}{\lambda_{max}} = \frac{2.898 \times 10^{-3}}{500 \times 10^{-9}} = 5800\ K$ *(2 marks)*

(b) For a light wave, $c = f\lambda$ and hence $\frac{1}{\lambda} = \frac{f}{c}$.

We can use this relationship to rewrite the Rydberg equation to find the frequency.

$$\frac{1}{\lambda} = \frac{f}{c} = R\left(\frac{1}{n_f^2} - \frac{1}{n_i^2}\right)$$

Hence, $f = Rc\left(\frac{1}{n_f^2} - \frac{1}{n_i^2}\right) = (1.097 \times 10^7)(3 \times 10^8)\left(\frac{1}{6^2} - \frac{1}{2^2}\right) = 7.3 \times 10^{14}$ Hz

(We take the absolute value as the negative sign which indicates that radiation is absorbed rather than emitted.) *(3 marks)*

(c) An absorption spectrum is produced when a continuous spectrum is passed through a gas. A stellar absorption spectrum is produced when the continuous black body radiation emitted from the star's surface (photosphere) passes though the outer layers of the star's atmosphere. A continuous spectrum consists of a broad range of wavelengths (photon energies). Atoms and molecules in a gas can only absorb a photon of light if the energy of the photon is equal to the energy difference between the current stationary state of the electron and a higher energy stationary state. When an atom absorbs a photon, it excites the electron in the atom from one stationary state to a higher energy stationary state. If the photon does not have the exact energy required, it will not be absorbed and will pass through the gas. Hence an absorption spectrum is a continuous spectrum with certain wavelengths (photon energies) missing. As each atomic element absorbs a unique set of wavelengths, the missing wavelengths in the absorption spectra act as a signature for the types of atoms in the gas. *(3 marks)*

Question 25 (Total 6 marks)

(a) The centripetal force on a satellite is provided by its gravitational attraction to the planet it is orbiting, and hence we can write:

$$\frac{GMm}{r^2} = \frac{mv^2}{r} \text{ or } v^2 = \frac{GM}{r}$$

As $v = \frac{\textit{Circumference of the orbit}}{\textit{Period of the orbit}} = \frac{2\pi r}{T}$, we can substitute $\frac{2\pi r}{T}$ for v in the equation above to obtain: $\left(\frac{2\pi r}{T}\right)^2 = \frac{GM}{r}$ or $\frac{4\pi^2 r^2}{T^2} = \frac{GM}{r}$

And hence $\frac{r^3}{T^2} = \frac{GM}{4\pi^2}$ *(2 marks)*

(b) We can use the satellite equation to show the gradient of the graph is:

$$\textit{Gradient} = \frac{T^3}{r^3} = \frac{4\pi^2}{GM}$$

And hence $M = \frac{4\pi^2}{G(\textit{Gradient})}$

Reading the gradient of the line of best fit from the graph, as the line of best fit passes through the point (0, 0) the $\textit{Gradient} = \frac{(0.2 - 0)}{(250 \times 10^{12} - 0)}$ days^{2}km^{-3}.

Converting this gradient to units of seconds^{2}m^{-3}

$$\textit{Gradient} = \frac{0.2}{250 \times 10^{12}} \text{ days}^2\text{km}^{-3} = \frac{(0.2)(24 \times 60 \times 60)^2}{(250 \times 10^{12})(1000)^3} = 5.97 \times 10^{-15} \text{ s}^2\text{m}^{-3}$$

Finally, $M = \frac{4\pi^2}{(6.67 \times 10^{-11} \times 5.97 \times 10^{-15})} = 9.91 \times 10^{25}$ kg *(4 marks)*

Question 26 (Total 3 marks)

From the muon's frame of reference, the Earth's atmosphere is moving near the speed of light upwards. Because the length of a moving object contracts when viewed by a stationary observer, the thickness of the atmosphere will be reduced due to length contraction when viewed from the muon's frame of reference. Because the distance the muon must travel through the atmosphere is greatly reduced, it will reach the Earth's surface before it decays.

Because a moving clock appears to run slower to a stationary observer due to time dilation, an observer on the surface of the Earth would see time running much slower for the muon. Because time is running slower for the muon, much less time has passed when it moves through the atmosphere than for the stationary observer, so it can reach the surface before decaying.

Therefore length contraction enables the muon to reach the surface from the muon's frame of reference, or alternatively time dilation enables the muon to reach the surface when viewed from the frame of reference of an observer on the Earth's surface. *(3 marks)*

Question 27 (Total 7 marks)

(a) The proton contains two up quarks and a down quark and the antiproton contains two up antiquarks and one down antiquark.

The neutral pion has zero charge so must contain an up quark and an up antiquark.

The positive pion has a charge of +1 and hence must contain an up quark and a down antiquark.

The negative pion has a charge of –1 and hence must contain a down quark and an up antiquark. *(2 marks)*

(b) We can calculate the energy released in the reaction by finding the change in the rest mass of the particles.

Δ rest mass = Σrest mass products – Σrest mass of reactants

$$= (140 \times 3) - (940 \times 2) = -1460\ \frac{\text{MeV}}{c^2}$$

The negative sign tells us mass is lost (converted to energy) in this reaction.

The energy per pion is $\frac{1460}{3} = 487$ MeV *(2 marks)*

(c) Classical physics assumes that $p = mv$, kinetic energy is $K = \frac{1}{2}mv^2$ and bodies can travel at any velocity relative to an observer. These equations accurately predict the momentum and energy of a moving body at low velocities but become progressively more inaccurate as velocities increase towards the speed of light. Substituting the pion's energy into the kinetic energy equation above gives a value greater than the speed of light because these equations are only low-velocity approximations of the momentum and kinetic energy of a moving body.

Einstein's special relativity theory showed that a body could never reach the speed of light with respect to any observer. Because the speed of light could never be reached by a body, Einstein had to modify the classical equations for momentum and kinetic energy. For example, his relativistic equation for momentum is:

$$p_v = \frac{m_0 v}{\sqrt{\frac{1}{1 - \frac{v^2}{c^2}}}}$$

This equation shows us that the momentum and hence the kinetic energy of a moving body would be infinite if it reached the speed of light. This is why a body cannot accelerate to the speed of light, as it would require an infinite amount of energy to do so. Thus, while classical physics (incorrectly) would give a velocity greater than the speed of light for a 487 MeV pion, special relativity shows us that the velocity of a pion with this amount of kinetic energy will be less than the speed of light. *(3 marks)*

Question 28 (Total 7 marks)

(a) The force on the electron due to the electric field between the parallel plates will be given by $F = qE$ and, as $F = ma$, we can equate these equations to find the acceleration:

$$a = \frac{qE}{m} = \frac{(1.602 \times 10^{-19} \times 1.5 \times 10^{4})}{9.109 \times 10^{-31}} = 2.6 \times 10^{15}\ \text{ms}^{-2}$$

The acceleration will be perpendicular to the velocity, in the opposite direction to the electric field. *(2 marks)*

(b) Because the force on the electron remains in one direction and is perpendicular to the initial velocity, the electron will undergo projectile motion when it is between the plates.

The horizontal velocity (v_x) will therefore remain constant and the time for the electron to pass through the plates will be given by:

$t = \dfrac{L}{v_x}$, where L= 5.0 mm and is the length of the plates

As we know the time over which the electron is accelerated, we can calculate the electron's vertical displacement (s_y):

$$s_y = u_y t + \frac{1}{2}at^2$$

and as $u_y = 0$

$$s_y = \frac{1}{2}at^2 = \frac{1}{2}a\left(\frac{L}{v}\right)^2 = \frac{1}{2} \times 2.6 \times 10^{15} \times \left(\frac{0.005}{2.0 \times 10^6}\right)^2 = 8.1 \text{ mm, as expected.}$$ *(2 marks)*

(c) The vertical velocity when the electron leaves the plates will be given by: $v_y = u_y + at$

And as u_y=0

$$v_y = at = 2.6 \times 10^{15} \left(\frac{0.005}{2.0 \times 10^6}\right) = 6.5 \times 10^6\ \text{ms}^{-1}$$

After the electron leaves the plates, it will travel in a straight line at an angle (θ) from horizontal. We can find this angle by considering the vector addition of the horizontal and vertical velocities at the instant the electron leaves the plates.

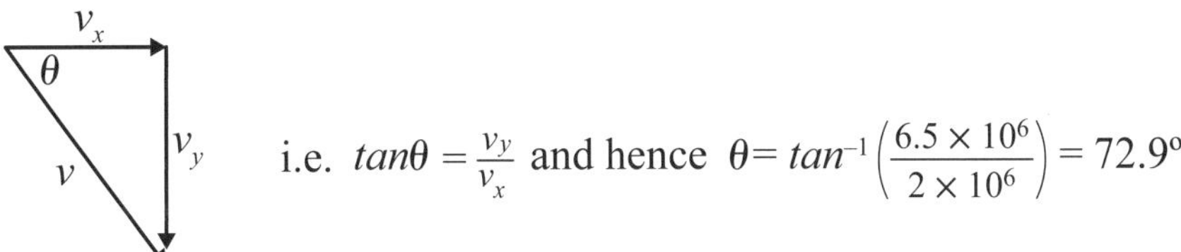

i.e. $tan\theta = \frac{v_y}{v_x}$ and hence $\theta = tan^{-1}\left(\frac{6.5 \times 10^6}{2 \times 10^6}\right) = 72.9°$

The additional vertical displacement of the electron (s_y) from when it leaves the plates until it reaches the screen will be given by:

$s_x = 30$ mm

72.9°

s_y

$tan\theta = \frac{s_y}{s_x}$, where s_x is the distance to the screen (30 mm)

Hence $s_y = s_x\, tan\theta = 0.03\tan(72.9°) = 0.097\ 51\text{m} = 97.5$ mm

The total distance from point X to the point where the electron will strike the screen will be given by its vertical displacement when it moved between the plates added to its vertical displacement after it left the plates.

Total vertical displacement from X is therefore: 8.1 mm + 97.5 mm = 106 mm *(3 marks)*

Question 29 (Total 6 marks)

(a) To repel one another, the currents in the rods must be in opposite directions and, to enable a different amount of current to flow in each rod, the rods must be wired parallel to one another. Both rods will have the same potential difference across them but different currents will flow in each if the resistance of each rod is different.

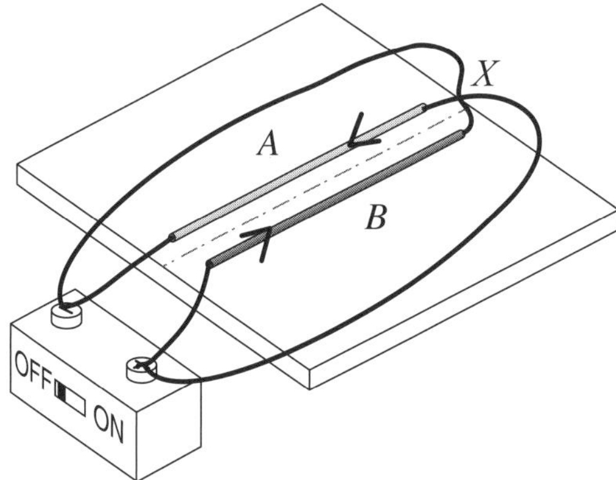

*Note the arrows show the direction of the current and the wires do not touch one another at point *X*. *(2 marks)*

(b) Because the displacement of rod *B* is greater than the displacement of rod *A* after the same time interval, either a different net force must have acted on each rod or they must have had different masses. The statement attempts to explain the different displacements by claiming the force (magnetic repulsion) on each rod would be different due to the different currents flowing in the rods. This statement is incorrect because, by Newton's third law, the electromagnetic repulsive force on each rod must be equal in magnitude and opposite in direction. The force on each rod would be equal in magnitude and opposite in direction, regardless of the magnitude of the respective currents.

We can also see this is the case by examining the electromagnetic interactions involved. The rods will repel one another if current flows in opposite directions in the rods. This is because, when current flows, each rod acts as a current-carrying wire in the magnetic field produced by the other wire. The right-hand grip rule shows that when currents travel in opposite directions in two parallel rods, each rod produces a magnetic field that will be perpendicular to the current flow in the adjacent rod. The right-hand palm shows that the direction of the magnetic field and current flow in each rod results in a force on each rod directed away from the adjacent rod. By combining the equation for the magnetic field and the force it produces on a parallel current-carrying rod, the theory of electromagnetism shows the force between two long straight rods of length l, separated by a distance r and carrying currents I_1 and I_2, is given by: $F = \frac{\mu_0 I_1 I_2 l}{2\pi r}$

Note that the same numbers will be substituted into this equation irrespective of which wire is being analysed and thus the magnitude of the force is the same on both wires.

The statement is therefore incorrect as a different current magnitude in each rod cannot explain the different displacement of each rod. The correct explanation for the different displacements must lie elsewhere. Perhaps, as $a = \frac{F}{m}$, the mass of *A* was greater than the mass of rod *B*, which would explain the smaller displacement of rod *A*. *(4 marks)*

Question 30 (Total 4 marks)

There are three forces operating on the object as it rotates in the cylinder.

Gravity acts downwards with a force $F_w = mg$ and a reaction force of the same magnitude acts upwards on the object from the floor ensuring the net vertical force on the object is zero.

In addition, the wall of the cylinder exerts a normal force on the block directed towards the centre of the cylinder. This normal force is the net horizontal force that is responsible for the block moving in a horizontal circle and thus $F_N = F_c = \frac{mv^2}{r}$.

If the period of the cylinder's rotation is halved, the gravitational force and the upwards reaction force on the object will not change but the centripetal force will increase. We can determine the amount the centripetal force will increase as follows:

The magnitude of the velocity of the block moving in a circle of radius r with period T is given by:

$$\text{Velocity} = \frac{\text{Circumference of circle}}{\text{Period}} = \frac{2\pi r}{T}$$

Substituting this value for the velocity into the centripetal force equation ($F_c = \frac{mv^2}{r}$) we see:

$$F_c = \frac{4\pi^2 mr}{T^2}$$

Note that the centripetal force is inversely proportional to the square of the period.

Therefore if the period of rotation is halved, the weight force and upwards reaction force remain unchanged, but the centripetal force exerted by the wall towards the centre of the cylinder will be four times greater. *(4 marks)*

Question 31 (Total 4 marks)

Because the only force operating in both models is the gravitational force, the total energy of the projectile will remain constant throughout the flight. As the projectile rises, its initial kinetic energy will be converted into potential energy. That is, the object does work against gravity as it rises and the gravitational force does an equal amount of work on the object when it falls. At the top of the flight all the kinetic energy will have converted into gravitational potential energy and the velocity will be zero. As the object falls, its gravitational potential energy will be converted to kinetic energy until it reaches the ground. As the total energy remains constant, when it reaches the ground, its final kinetic energy will be equal to its initial kinetic energy.

In model *A* the parallel, equally spaced field lines represent a uniform vertical gravitational field. The force of gravity on an object in this field ($F_w = mg$) and the acceleration due to gravity ($g = 9.8\ \text{ms}^{-1}$) would be constant at all points in the field.

In a uniform field, we can find the height reached by equating the initial kinetic energy and the final potential energy:

$K = U$ or $\frac{1}{2}mv^2 = mgh$ and hence, $h = \frac{v^2}{2g}$

In model *B* the field lines are radial and become further apart with height from the surface, indicating that the strength of the gravitational field decreases with height. In this model the weight force is given by $F_w = \frac{GMm}{(r_E + h)^2}$, where r_E is the radius of the Earth and h the height above the surface. Thus, in a radial gravitational field, the weight force $\left(F_w = \frac{GMm}{(r_E + h)^2}\right)$ and acceleration due to gravity $\left(g = \frac{GM}{(r_E + h)^2}\right)$ decrease with height above the surface. As in model *A* the total energy of the projectile will remain constant, its initial kinetic energy will be converted to potential energy as it rises and its potential energy will be converted into kinetic energy as it falls. The same amount of work will have to be done by the gravitational field to stop the projectile from moving upwards but, as the force decreases with height in model *B*, the projectile will have to travel further upwards before the field brings it to rest (i.e. $work = Fs$ and, if F is reduced, s must be increased for the field to do the same amount of work).

Therefore, because the force on the projectile decreases with height (rather than remaining constant), the projectile in model *B* will reach a greater height than the projectile in model *A*.

(4 marks)

Question 32 (Total 8 marks)

*Note that students could chose other light/matter experiments to discuss how evidence from these experiments contributed to our understanding of physics.

Many experiments have been conducted to explore the interaction of light and matter, which have resulted in evidence that has led to advances in our understanding of physics. The examples that follow illustrate how important these types of experiments have been to our understanding of physics. Others that could have been analysed include emission and absorption spectra, diffraction, polarisation and refraction.

(i) Newton's prism experiment

In the 17th century, Newton conducted an experiment in which a beam of white light was dispersed into a spectrum of colours by a glass prism. He then separated one colour at a time and showed that when it passed through a second prism no more colours were produced. That is, only the incident colour emerged from the second prism. He also used a second prism to show the spectrum in the first prism could be combined to produce white light again. The evidence from this experiment led to the realisation that the prism did not change white light into coloured light but that the colours were already in the white light and the prism simply separated them so they could be seen. White light was later shown to be made up of a series of wavelengths and each wavelength was refracted by a different amount when it entered a new medium. In addition, the experiment showed how a dispersive element, such as a glass prism, could be used to separate the components in a beam of light and allow them to be examined. This led to the development of the prism spectrometer, which itself led to many important discoveries, such as atomic emission and absorption spectra, black body radiation and stellar spectra. The spectral analysis of black body radiation and the emission and absorption spectra of elements were crucial to the development of quantum physics and the quantum model of the atom.

(ii) Black body radiation

In the 19th century several scientists conducted experiments to measure the spectral emission from hot bodies. They used prism spectrometers to measure the amount of radiation emitted at different wavelengths from bodies at different temperatures. They found that the wavelength of peak emission and the total radiation emitted increases with the temperature of the emitting body in a very specific way that was independent of the material of which the body was made. The 'black body' radiation spectra obtained in these experiments could not be explained by classical physics and led Planck to suggest that the only way to explain the results would be to assume energy was quantised. In addition, the experimental results enabled Planck to derive a fundamental constant of nature. This constant, which fixes the size of energy quanta, is now known as Plank's constant. Thus, evidence from black body spectra experiments led to the birth of quantum physics and a much deeper understanding of energy, light and matter.

(iii) The photoelectric effect

Experiments to examine the way electrons are ejected from a metal surface when light is incidental on the surface revealed that no electrons were emitted unless the incident light was above a specific threshold frequency, specific to the material. The experiments also found that the kinetic energy of the ejected electrons was proportional to the frequency of the incident light. The evidence obtained in these photoelectric effect experiments could not be explained by classical physics and, in conjunction with Planck's earlier work on quanta, led Einstein to develop a quantum (photon) model of light. The experiment supported the proposition that light interacted with matter as tiny energy packets (photons) rather than in continuous waves and that the energy of the photon is proportional to the frequency of the light. Quantisation of light has led to many advances in our understanding of physics, from quantised models of the atom to lasers. *(8 marks)*

Question 33 (Total 7 marks)

First consider the motion of the magnet. If the can was absent, the magnet would undergo a simple pendulum motion in which the total energy would remain constant (ignoring friction). The energy ($E = U + K$) would change from all potential energy ($U = mgh$) at the top of the motion to all kinetic energy ($K = \frac{1}{2}mv^2$) at the lowest point of the motion. Adding the aluminium can would add a dampening force to the motion. We can understand this by applying Lenz's law and Faraday's law of induction ($\varepsilon = N\frac{\Delta\Phi}{\Delta t}$). The moving magnetic field will induce an emf in the can as the magnet swings past the can. This potential difference in the can will create eddy currents that will produce a magnetic field in a direction that will oppose the change that produced the currents; that is, a force that opposes the motion of the magnet. This dampening force would cause some of the pendulum's energy to be transferred to the can as the pendulum swings. Thus work would be done by the moving magnet against the magnetic field that is produced by the induced eddy currents in the can. The work done by the moving magnet transforms some of the kinetic energy of the moving magnet into kinetic energy in the can and into heat produced by the induced currents in the can. This would reduce the height reached by the magnet in each oscillation. The magnet would still oscillate from all potential energy at the top to all kinetic energy at the bottom, but the total energy would decrease continually until the magnet came to rest.

From the can's perspective, a magnetic field is moving past and this moving field will induce an emf resulting in eddy currents that (by Lenz's law) will generate a magnetic field to oppose this change. This magnetic field will oppose the motion of the magnet. And by Newton's third law an equal and opposite force will act on the can that will move the top of the can in the same direction as the magnet when it passes. This force will cause the can to oscillate back and forth as the magnet passes in one direction and then the other. The top of the can will move in the direction of the magnet each time it passes.

The first swing will transfer some kinetic energy from the magnet to the can. Because the magnet/pendulum does work on the can in the first swing, it loses energy and will not reach the same height. It will therefore pass the can with a lower kinetic energy when it swings back in the opposite direction. But because the magnet and the top of the can are now moving in opposite directions, the relative velocity of the magnet from a frame of reference fixed to the top of the can is greater in the second swing than from the first swing. Thus larger eddy currents will be induced and hence more work will be done on the can by the magnet in the second swing than in the first swing. This work will result in the transfer of enough energy to stop the can's motion and to start it rotating in the opposite direction. That is, the top of the aluminium can will move in the direction of motion of the magnet. The maximum rate of rotation of the can will not be as great as in the first swing, as less than twice the work of the initial swing was done on the can because the pendulum had lost some energy to the can in the first swing. Subsequent swings will follow the same pattern and cause the can to rotate back and forth with the top of the can following the motion of the magnet (though slightly delayed).

Thus the pendulum's initial gravitational potential energy is converted to kinetic energy as it swings downwards. The moving magnet does work on the can by inducing currents in the can and therefore each time the pendulum swings, some of the pendulum's kinetic energy is converted into kinetic energy of the can and electrical energy in the can. The electrical energy in the can is converted to heat (due to resistance heating) and hence, even if we neglect friction, the pendulum's energy in time will be converted to heat and the motion will cease. If friction is included, the motion would cease even faster. Because, in addition to the resistance heating caused by the induced currents, some of the kinetic energy of the can and pendulum would be transformed via friction to heat each oscillation.

In summary, an observer will see the can oscillating back and forth around its axis as the magnet passes, and the amplitude of both the pendulum and can's oscillation will decrease until the motion ceases. The total energy of the system is conserved when the motion ceases because the initial gravitational energy of the pendulum will have been transformed into heat energy. (*7 marks*)

2024 HIGHER SCHOOL CERTIFICATE EXAMINATION

Physics

DATA SHEET

Charge on electron, q_e	-1.602×10^{-19} C
Mass of electron, m_e	9.109×10^{-31} kg
Mass of neutron, m_n	1.675×10^{-27} kg
Mass of proton, m_p	1.673×10^{-27} kg
Speed of sound in air	340 m s^{-1}
Earth's gravitational acceleration, g	9.8 m s^{-2}
Speed of light, c	3.00×10^{8} m s^{-1}
Electric permittivity constant, ε_0	8.854×10^{-12} $A^2 s^4 kg^{-1} m^{-3}$
Magnetic permeability constant, μ_0	$4\pi \times 10^{-7}$ N A^{-2}
Universal gravitational constant, G	6.67×10^{-11} N $m^2 kg^{-2}$
Mass of Earth, M_E	6.0×10^{24} kg
Radius of Earth, r_E	6.371×10^{6} m
Planck constant, h	6.626×10^{-34} J s
Rydberg constant, R (hydrogen)	1.097×10^{7} m^{-1}
Atomic mass unit, u	1.661×10^{-27} kg 931.5 MeV/c^2
1 eV	1.602×10^{-19} J
Density of water, ρ	1.00×10^{3} kg m^{-3}
Specific heat capacity of water	4.18×10^{3} J $kg^{-1} K^{-1}$
Wien's displacement constant, b	2.898×10^{-3} m K

FORMULAE SHEET

Motion, forces and gravity

$$s = ut + \frac{1}{2}at^2$$

$$v^2 = u^2 + 2as$$

$$\Delta U = mg\Delta h$$

$$P = \frac{\Delta E}{\Delta t}$$

$$\sum \frac{1}{2}mv^2_{\text{before}} = \sum \frac{1}{2}mv^2_{\text{after}}$$

$$\Delta\vec{p} = \vec{F}_{\text{net}}\Delta t$$

$$\omega = \frac{\Delta\theta}{t}$$

$$\tau = r_{\perp}F = rF\sin\theta$$

$$v = \frac{2\pi r}{T}$$

$$U = -\frac{GMm}{r}$$

$$v = u + at$$

$$\vec{F}_{\text{net}} = m\vec{a}$$

$$W = F_{||}s = Fs\cos\theta$$

$$K = \frac{1}{2}mv^2$$

$$P = F_{||}v = Fv\cos\theta$$

$$\sum m\vec{v}_{\text{before}} = \sum m\vec{v}_{\text{after}}$$

$$a_c = \frac{v^2}{r}$$

$$F_c = \frac{mv^2}{r}$$

$$F = \frac{GMm}{r^2}$$

$$\frac{r^3}{T^2} = \frac{GM}{4\pi^2}$$

Waves and thermodynamics

$$v = f\lambda$$

$$f = \frac{1}{T}$$

$$d\sin\theta = m\lambda$$

$$n_x = \frac{c}{v_x}$$

$$I = I_{\max}\cos^2\theta$$

$$Q = mc\Delta T$$

$$f_{\text{beat}} = |f_2 - f_1|$$

$$f' = f\frac{(v_{\text{wave}} + v_{\text{observer}})}{(v_{\text{wave}} - v_{\text{source}})}$$

$$n_1\sin\theta_1 = n_2\sin\theta_2$$

$$\sin\theta_c = \frac{n_2}{n_1}$$

$$I_1r_1^2 = I_2r_2^2$$

$$\frac{Q}{t} = \frac{kA\Delta T}{d}$$

FORMULAE SHEET (continued)

Electricity and magnetism

$$E = \frac{V}{d}$$

$$V = \frac{\Delta U}{q}$$

$$W = qV$$

$$W = qEd$$

$$B = \frac{\mu_0 I}{2\pi r}$$

$$B = \frac{\mu_0 N I}{L}$$

$$\Phi = B_{||}A = BA\cos\theta$$

$$\varepsilon = -N\frac{\Delta\Phi}{\Delta t}$$

$$\frac{V_p}{V_s} = \frac{N_p}{N_s}$$

$$\vec{F} = q\vec{E}$$

$$F = \frac{1}{4\pi\varepsilon_0}\frac{q_1 q_2}{r^2}$$

$$I = \frac{q}{t}$$

$$V = IR$$

$$P = VI$$

$$F = qv_{\perp}B = qvB\sin\theta$$

$$F = lI_{\perp}B = lIB\sin\theta$$

$$\frac{F}{l} = \frac{\mu_0}{2\pi}\frac{I_1 I_2}{r}$$

$$\tau = nIA_{\perp}B = nIAB\sin\theta$$

$$V_p I_p = V_s I_s$$

Quantum, special relativity and nuclear

$$\lambda = \frac{h}{mv}$$

$$K_{\max} = hf - \phi$$

$$\lambda_{\max} = \frac{b}{T}$$

$$E = mc^2$$

$$E = hf$$

$$\frac{1}{\lambda} = R\left(\frac{1}{n_f^2} - \frac{1}{n_i^2}\right)$$

$$t = \frac{t_0}{\sqrt{\left(1 - \frac{v^2}{c^2}\right)}}$$

$$l = l_0\sqrt{\left(1 - \frac{v^2}{c^2}\right)}$$

$$p_v = \frac{m_0 v}{\sqrt{\left(1 - \frac{v^2}{c^2}\right)}}$$

$$N_t = N_0 e^{-\lambda t}$$

$$\lambda = \frac{\ln 2}{t_{\frac{1}{2}}}$$

PERIODIC TABLE OF THE ELEMENTS

KEY

Atomic Number	79
Symbol	Au
Standard Atomic Weight	197.0
Name	Gold

1 H 1.008 Hydrogen																	2 He 4.003 Helium
3 Li 6.941 Lithium	4 Be 9.012 Beryllium											5 B 10.81 Boron	6 C 12.01 Carbon	7 N 14.01 Nitrogen	8 O 16.00 Oxygen	9 F 19.00 Fluorine	10 Ne 20.18 Neon
11 Na 22.99 Sodium	12 Mg 24.31 Magnesium											13 Al 26.98 Aluminium	14 Si 28.09 Silicon	15 P 30.97 Phosphorus	16 S 32.07 Sulfur	17 Cl 35.45 Chlorine	18 Ar 39.95 Argon
19 K 39.10 Potassium	20 Ca 40.08 Calcium	21 Sc 44.96 Scandium	22 Ti 47.87 Titanium	23 V 50.94 Vanadium	24 Cr 52.00 Chromium	25 Mn 54.94 Manganese	26 Fe 55.85 Iron	27 Co 58.93 Cobalt	28 Ni 58.69 Nickel	29 Cu 63.55 Copper	30 Zn 65.38 Zinc	31 Ga 69.72 Gallium	32 Ge 72.64 Germanium	33 As 74.92 Arsenic	34 Se 78.96 Selenium	35 Br 79.90 Bromine	36 Kr 83.80 Krypton
37 Rb 85.47 Rubidium	38 Sr 87.61 Strontium	39 Y 88.91 Yttrium	40 Zr 91.22 Zirconium	41 Nb 92.91 Niobium	42 Mo 95.96 Molybdenum	43 Tc Technetium	44 Ru 101.1 Ruthenium	45 Rh 102.9 Rhodium	46 Pd 106.4 Palladium	47 Ag 107.9 Silver	48 Cd 112.4 Cacmium	49 In 114.8 Indium	50 Sn 118.7 Tin	51 Sb 121.8 Antimony	52 Te 127.6 Tellurium	53 I 126.9 Iodine	54 Xe 131.3 Xenon
55 Cs 132.9 Caesium	56 Ba 137.3 Barium	57–71 Lanthanoids	72 Hf 178.5 Hafnium	73 Ta 180.9 Tantalum	74 W 183.9 Tungsten	75 Re 186.2 Rhenium	76 Os 190.2 Osmium	77 Ir 192.2 Iridium	78 Pt 195.1 Platinum	79 Au 197.0 Gold	80 Hg 200.6 Mercury	81 Tl 204.4 Thallium	82 Pb 207.2 Lead	83 Bi 209.0 Bismuth	84 Po Polonium	85 At Astatine	86 Rn Radon
87 Fr Francium	88 Ra Radium	89–103 Actinoids	104 Rf Rutherfordium	105 Db Dubnium	106 Sg Seaborgium	107 Bh Bohrium	108 Hs Hassium	109 Mt Meitnerium	110 Ds Darmstadtium	111 Rg Roentgenium	112 Cn Copernicium	113 Nh Nihonium	114 Fl Flerovium	115 Mc Moscovium	116 Lv Livermorium	117 Ts Tennessine	118 Og Oganesson

Lanthanoids

57 La 138.9 Lanthanum	58 Ce 140.1 Cerium	59 Pr 140.9 Praseodymium	60 Nd 144.2 Neodymium	61 Pm Promethium	62 Sm 150.4 Samarium	63 Eu 152.0 Europium	64 Gd 157.3 Gadolinium	65 Tb 158.9 Terbium	66 Dy 162.5 Dysprosium	67 Ho 164.9 Holmium	68 Er 167.3 Erbium	69 Tm 168.9 Thulium	70 Yb 173.1 Ytterbium	71 Lu 175.0 Lutetium

Actinoids

89 Ac Actinium	90 Th 232.0 Thorium	91 Pa 231.0 Protactinium	92 U 238.0 Uranium	93 Np Neptunium	94 Pu Plutonium	95 Am Americium	96 Cm Curium	97 Bk Berkelium	98 Cf Californium	99 Es Einsteinium	100 Fm Fermium	101 Md Mendelevium	102 No Nobelium	103 Lr Lawrencium

Standard atomic weights are abridged to four significant figures.
Elements with no reported values in the table have no stable nuclides.
Information on elements with atomic numbers 113 and above is sourced from the International Union of Pure and Applied Chemistry Periodic Table of the Elements (November 2016 version).
The International Union of Pure and Applied Chemistry Periodic Table of the Elements (February 2010 version) is the principal source of all other data. Some data may have been modified.